Blockchain and Smart-Contract Technologies for Innovative Applications

Nour El Madhoun • Ioanna Dionysiou •
Emmanuel Bertin
Editors

Blockchain and Smart-Contract Technologies for Innovative Applications

Springer

Editors
Nour El Madhoun
ISEP (Institut Supérieur d'Électronique de Paris)
Paris, France

Ioanna Dionysiou
Department of Computer Science
University of Nicosia
Nicosia, Cyprus

Emmanuel Bertin
Orange Innovation
Caen, France

ISBN 978-3-031-50027-5 ISBN 978-3-031-50028-2 (eBook)
https://doi.org/10.1007/978-3-031-50028-2

This Springer imprint is published by the registered company Springer Nature Switzerland AG
The registered company address is: Gewerbestrasse 11, 6330 Cham, Switzerland

Paper in this product is recyclable.

Preface

Blockchain and smart contracts represent a major technological advance, paving the way to innovative applications in a variety of fields. These technologies have not only redefined financial transaction methods, but also infused new dynamism into art, telecommunications, supply chain management, healthcare, and many other sectors. The book *Blockchain and Smart-Contract Technologies for Innovative Applications* offers a detailed exploration of blockchain and smart-contract technologies, highlighting the variety and richness of innovative applications that they enable.

Part I provides a general introduction. In chapter "Fundamentals of Blockchain and Smart Contracts", the fundamentals of blockchain and smart contracts are presented. Chapter "Smart Contract: The Contract Automation Climax. Back-End and Front-End Legal Implications" deals with contract automation focusing on the legal implications from a back-end and front-end perspective. Finally, chapter "Designing Future Sustainable Cryptocurrencies: Principles and Expectations" presents a forward-looking approach to the sustainable design of cryptocurrencies, anticipating the principles and expectations of the future.

Part II presents a series of in-depth analyses of blockchain and smart contract technologies. Chapter "A Taxonomy on Blockchain-Based Technology in the Financial Industry: Drivers, Applications, Benefits, and Threats" offers a detailed taxonomy of blockchain-based financial applications, identifying drivers, benefits, applications, and threats. Chapter "Scientific Workflows Management with Blockchain: A Survey" examines the potential of blockchain for scientific workflow management.

Part III explores the exciting world of NFTs. Chapter "Tokenization of Fine Arts: Revolutionizing the Fine Arts Industry with Blockchain" looks at the tokenization of fine arts, highlighting how blockchain is revolutionizing the art industry. Chapter "VeraciTIX: An NFT-Based Event Ticketing System" introduces VeraciTIX, an innovative proposal for event ticketing using NFTs.

Finally, Part IV focuses on ontologies and telecommunications applications using blockchain technology. Chapter "Auto-generation of Blockchain-Based Distributed Applications Using Ontologies" describes how ontologies enable the generation of distributed applications based on blockchain. Chapters "Blockchain and Smart Contracts for Telecommunications: The Whys and Wherefores" and "Distributed Ledger Technology for Next-Generation Cellular Networks: A SWOT Analysis" offer a comprehensive perspective on the growing importance of blockchain in the telecommunications sector, including a SWOT analysis of next-generation cellular networks.

The book *Blockchain and Smart-Contract Technologies for Innovative Applications* is an essential resource for all those aiming to deepen their understanding of blockchain and smart-contract technologies. In addition to demystifying these technologies, the book offers a detailed exploration of their innovative applications. It is aimed not only for those who want to understand the technological fundamentals, but also for those who are curious to discover how blockchain and smart contract technologies are revolutionizing many sectors.

Paris, France — Nour El Madhoun
Nicosia, Cyprus — Ioanna Dionysiou
Caen, France — Emmanuel Bertin

Acknowledgments

We express our sincere gratitude to all the authors for their significant contributions to each chapter. We also thank the editors and reviewers for their proofreading work, ensuring that each chapter met high standards:

Badis Hammi, EPITA Engineering School, France
Baptiste Hemery, Orange Labs, France
Daniel Alejandro Maldonado Ruiz, Escuela Politécnica Nacional (EPN), Ecuador
Davide Frey, INRIA, France
Damien Graux, Trinity College Dublin, Ireland
Dimitri Saingre, Davidson consulting, France
El-Hacen Diallo, Université Claude Bernard Lyon 1, France
Emmanuel Bertin, Orange Labs, France
Fariba Ghaffari, IRT b-com, France
Guy Pujolle, Sorbonne Université, France
Ioanna Dionysiou, University of Nicosia, Cyprus
Jenny Torres, Escuela Politécnica Nacional (EPN), Ecuador
Kevin Atighehchi, Université Clermont Auvergne, France
Khaldoun Al Agha, Université Paris-Saclay, France
Line Guffond Le Goanvic, 360 Vision & Perspectives, France
Lina Mroueh, ISEP Engineering School, France
Lucas Sguerra, Eniblock, France
Meroua Moussaoui, Orange Labs, France
Mohamed Ali Kandi, IRIT, France
Mustapha-Kamal Benramdane, Conservatoire National des Arts et Métiers (CNAM), France
Nischal Aryal, Orange Labs, France
Nour El Madhoun, ISEP Engineering School, France
Nouredine Tamani, ISEP Engineering School, France
Rachid Guerraoui, EPFL, Switzerland
Rashid Boudour, Université Badji Mokhtar-Annaba, Algeria
Saad El Jaouhari, ISEP Engineering School, France

Samia Bouzefrane, Conservatoire National des Arts et Métiers (CNAM), France
Sara Tucci-Piergiovanni, CEA-List, France
Shyam Mohan, Sri Chandrasekarendra Saraswathi Viswa Maha Vidyalaya, India
Souvik Sengupt, IONOS, Germany
Tiphaine Henry, CEA-List, France
Tristan Bilot, Université Paris-Saclay, France
Wafa Njima, ISEP Engineering School, France
Yackolley Amoussou-Guenou, Université Paris Panthéon-Assas, France

Finally, we are deeply grateful to Mary James, Olivia Ramya Chitranjan, John Barthelemy, and Brian Halm at Springer for their constant help and support throughout this project.

Contents

Part I
Introduction to Blockchain and Smart-Contract Technologies

Fundamentals of Blockchain and Smart Contracts

Fahad Rahman, Chafiq Titouna, and Farid Nait-Abdesselam

Abstract In this chapter, we explore the core principles of Blockchain technology and smart contracts, offering insights into their functionality and implementation across different sectors. Blockchain technology stands distinguished for its superior performance compared to its counterparts. It is rooted in a trustless framework, guaranteeing immutable records, transparency, unwavering availability, and pseudonymity in the recording of transactions. Its various forms—Public, Private, and Consortium—have been effectively applied in real-world scenarios including Cryptocurrencies (programmable money), FinTech, and Healthcare. One of the most impactful features of blockchain is the smart contract. These are autonomous codes programmed to execute automatically when specific predetermined conditions are met. Smart contracts contribute to the reduction of transaction expenses, performance enhancement, and augmentation of trust and transparency in business dealings. The combined innovations of blockchain and smart contracts are poised to transform the modes of business and interpersonal interactions. Even though Blockchain technology confronts certain challenges, ongoing research is dedicated to resolve these issues. The potential for incorporating Blockchain and Smart Contracts in Web 3.0 and Industry 4.0 is immense and is anticipated to expand progressively.

Keywords Blockchain technology · Smart contracts · Decentralization · Cryptography · Transaction validation · Blockchain challenges · Blockchain use cases

---

F. Rahman (✉)
Université Paris Cité, Paris, France
e-mail: fahad.rahman@etu.u-paris.fr

C. Titouna
ESIEE, University Gustave Eiffel, Paris, France
e-mail: chafiq.titouna@esiee.fr

F. Nait-Abdesselam
University of Missouri Kansas City, Kansas City, MO, USA
e-mail: naf@umkc.edu

N. El Madhoun et al. (eds.), *Blockchain and Smart-Contract Technologies for Innovative Applications*, https://doi.org/10.1007/978-3-031-50028-2_1

1 Introduction

Over time, various instruments of trust have emerged to facilitate exchange and protect buyers and sellers, including minted coins, paper money, and banking systems. Innovations like telephones, credit cards, Internet, and mobile technologies have improved the efficiency of transactions and facilitated global economic growth. However, limitations in current payment systems such as long settlement times, intermediaries, fraud, and cyberattacks create the need for more efficient and secure payment systems. Ownership of assets like money, land, and vehicles must be proven and updated during transfers or inheritance, and digital ledgers have been increasingly used for structured record maintenance over the past few decades.

Assets such as money, land, vehicles, and agricultural land require ownership records to be maintained for transfer purposes, which is done through a ledger system. This record-keeping has traditionally been done manually using spreadsheets but they have limitations in handling large volumes of records and are not tamper-resistant. But with the advancement of database technologies, digital ledgers were created providing improved record management and security. Distributed Ledger Technology (DLT) is the evolution of digital ledgers, the decentralization of ledgers, and the elimination of the need for a clearing house [1].

Blockchain, in its essence, represents a disruptive innovation, bringing a paradigm shift in our digital world. It is a distributed and decentralized ledger technology that securely records and verifies transactions across multiple systems. Its inception, attributed to an entity known as Satoshi Nakamoto, was initially aimed at supporting cryptocurrencies like Bitcoin. However, its potential extends far beyond digital currency. From healthcare to supply chain, blockchain is revolutionizing various sectors by promoting transparency, security, and eliminating intermediaries. Smart contracts, built atop blockchain, further augment its capabilities, automating contractual agreements. Yet, this technology is not devoid of challenges, ranging from scalability issues to regulatory uncertainties. This introduction seeks to unfold the intricacies and potentials of blockchain in the modern digital landscape.

In 2008, a pseudonymous person, Satoshi Nakamoto published a paper about Bitcoin. He introduced a P2P system for transactions used for electronic cash without depending on centralized approving authority, i.e., (banks). Satoshi defined Blockchain as a distributed storing information. Bitcoin is just the first prominent example of how it can be used. It suggests that Blockchain can be applied to other areas beyond cryptocurrency and has immense potential to revolutionize various industries. In this chapter, the authors have two words Block and chain without space, so it became Blockchain. It is a combination of two simple terms [2]: (1) Block: A batch of transactions, (2) Chain: Encrypted connections between records.

The technical definition of Blockchain is the "Use of distributed ledger in a decentralized way that records transactions in chronological order." The main difference between distributed ledger and Blockchain is consensus algorithms, which are used to develop trust among the nodes of the network. The new block

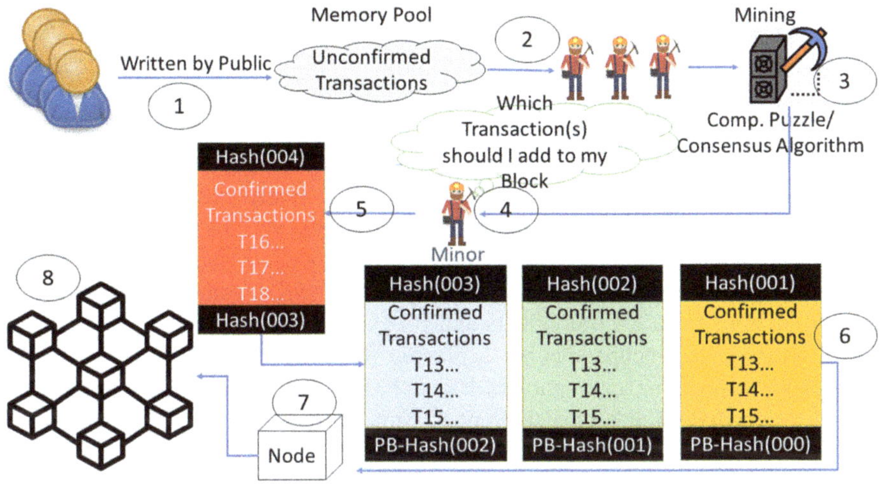

Fig. 1 A new block in blockchain

Table 1 Blockchain evolution steps

Year	Event
1990	Concept of distributed computing
1991	Early concepts and cryptography
2008	Satoshi Nakamoto (Pseudonym) published the idea
2009	Satoshi Nakomoto created and started of the Bitcoin network
2010	Fist cryptocurrency stock exchange is launched
2011	One Bitcoin equals one US Dollar
2012	The deployment of cryptocurrency in applications related to cash
2013	Early adoption, currency transfer, and digital payments systems
2014	Financial markets including Microsoft using blockchain
2015	Introduction and launching of Ethereum and smart contracts
2017	1 Bitcoin equals 10,000 US Dollars
2022	1 Bitcoin equals 42,272 US Dollars
Present	Expansion and diversification

creation time directly depends on the consensus algorithm. The new block creation is depicted in Fig. 1.

The concept of a blockchain, though most commonly associated with cryptocurrencies like Bitcoin, has a history that extends back well before the inception of digital currency. Understanding this history allows for a broader view of this influential technology and its potential applications. Brief history of Blockchain is mentioned in Table 1:

Blockchain is a unique form of DLT. It allows decentralized data storage across various computers, negating the need for a central authority. Each data block contains transaction records and is securely linked to the preceding block via a unique code called a hash. Blockchain's security relies on cryptography, ensuring changes to block data are easily detectable. For the addition of a new block, network consensus is required, typically achieved through protocols like Proof of Work or Proof of Stake. Blockchain's hallmark feature is its immutable nature, meaning that once a block is added, altering it is extraordinarily difficult. This technology promotes transparency with every transaction verifiable and permanently stored. It can also facilitate smart contracts, which automatically execute when specific conditions are met. Blockchain's application spans various sectors, including Banking and Finance, Energy, Healthcare, and Supply chain management.

This chapter is comprised of the following sections: Key concepts of Blockchain are described in Sect. 2. Components and Blockchain architecture are discussed in Sect. 3. Consensus mechanisms are explained in Sect. 4. Types of Blockchain are apprised in Sect. 5. The application of Blockchain in Real life is explored in Sect. 6. Section 7 described Blockchain challenges. Smart contracts are described in Sect. 8. Blockchain Oracle is described in Sects. 9 and 10 presents the conclusion.

2 Key Concepts of Blockchain

In this section, we present some useful concepts of Blockchain. We summarize them into five elements: Decentralized Control, Integrity/ Transparency, Confidentiality, Enhanced Security, and Faster Processing.

2.1 Decentralized Control

Blockchain enables multiple parties to share information among themselves eliminating the risks of centralized control and providing cost savings on security. The network provides a shared system of record for all participants, and trust is established through the use of cryptographic protocols. Changes require agreement from all parties, making it highly secure [3].

2.2 Integrity and Transparency

The Blockchain is growing like an ever-expanding archive of history, while the Merkle tree ensures data integrity by hashing transactions to a single root. Blockchain technology offers public verifiability through integrity and transparency, distinguishing it from traditional database technology. Users can trust the data to be uncorrupted and unaltered since its recording and verification are appended to the Blockchain.

2.3 *Confidentiality*

Although Blockchain technology is typically an openly distributed ledger, it is also possible to establish a private system that prioritizes confidentiality. This is achieved by ensuring that only authorized individuals or organizations have access to the data. Permissioned Blockchains are being emerged as an alternative to Public Blockchains, especially for enterprise needs. Confidentiality can be maintained in this system using solutions, for example, Hyperledger Fabric Blockchain and Blockstream.

2.4 *Enhanced Security*

Blockchain transactions are secured through encryption and interconnected with the preceding transaction since the data is distributed and stored among a network of computers. The utilization of cryptographic protocols helps in preventing fraudulent and unauthorized activities and safeguards against DOS attacks. Each block has a unique cryptographic fingerprint (hash), ensuring the data's security.

2.5 *Faster Processing*

The use of Blockchain technology has significantly reduced the settlement time required from days to mere minutes or seconds. Information is accessible to all participants so it becomes simpler to establish trust without the need for intermediaries. Uploading product data to the Blockchain can also streamline the tracking process. The trustless system and digital assets ensure secure and efficient transactions.

3 Components and Blockchain Architecture

Each block of Blockchain consists of two main parts: header and body part. The header part has block hash, previous block hash, nonce, timestamp, and Merkle root, whereas the body part keeps the transaction data. The following terms are common and frequently used in Blockchain [4]. Table 2 summarizes them.

3.1 *Private Keys*

A private key generates a signature for each transaction on the Blockchain, confirming that it has come from a specific user and preventing alteration by

Table 2 Blockchain key terms

Term	Description
Node	The end devices connected with blockchain like PCs, laptops, mobile phones
Ledger	A decentralized, distributed database is called a ledger
Block	A batch of transactions
Mining	The procedure of adding a new block to the system
Miner	The person or machine doing the mining
Transaction	Every record in the system
Nonce	Target difficulty level set by the network

malicious entities. Private keys are used to sign the cryptocurrencies being sent to others, and if someone obtains the private key, they can send the cryptocurrencies to themselves resulting in a security breach. This has been involved in many hacks worldwide.

3.2 Public Keys

The public keys can be generated from the private one mathematically by using Rivest, Shamir, Adleman (RSA)/ Elliptic Curve Cryptography (ECC), etc. algorithms, which can be distributed to everyone. While it is easy to derive a public key from a private key, it would take millions of years to do the opposite. Public keys are practically irreversible, and a unique public key is generated for each private key. An example of a public key is 0327F47F4SSF760BF5BA223515E63B7B2A1621AB71AE7631386738B3BC4D1 B73SE.

3.3 Addresses

The public key is derived from cryptographic hash functions which represent cryptocurrency address. In Bitcoin, addresses from public keys are generated using "Secure Hash Algorithm 256" (SHA-256) and the "RACE Integrity Primitives Evaluation Message Digest 160" (RIPEMD-160). The address typically appears in a transaction, signifying the recipient of the funds. An example of a cryptocurrency address is 1JPgMJuAvYJU6mxxbJdmf1XBd7bBPdPV3a.

3.4 *Basics of Cryptographic Techniques*

The secure and transparent nature of blockchain technology relies heavily on cryptographic techniques. The following are the fundamental cryptographic techniques commonly used in blockchain:

3.4.1 Hash Functions

A hash function takes an input (or "message") and produces a fixed-length string of characters, which is a digest that represents the data. The same input will always produce the same output, but even a tiny change in the input will produce a significantly different output. The use cases are data integrity verification, forming blocks by hashing transaction data and metadata, creating transaction IDs. The common algorithms are SHA-256 (Secure Hash Algorithm 256-bit).

3.4.2 Public-Key Cryptography (Asymmetric Cryptography)

In public-key cryptography, each participant has a pair of cryptographic keys: a public key, which is openly distributed, and a private key, which is kept secret. The use cases are transaction verification (Digital Signatures) and secure data transmission. The common algorithms are RSA and ECC.

3.4.3 Digital Signatures

Digital signatures provide a way to verify the authenticity of data or transactions. A sender uses their private key to create a digital signature on a message, and the receiver uses the sender's public key to verify the signature. The use cases are Transaction authorization and data integrity.

3.4.4 Merkle Trees

Merkle trees are a type of binary tree composed of a set of nodes with the leaf nodes representing data blocks. Each non-leaf node is the cryptographic hash of its children. This structure allows for efficient and secure verification of large sets of data. The use cases are efficient data verification and forming the block structure in some blockchain architectures.

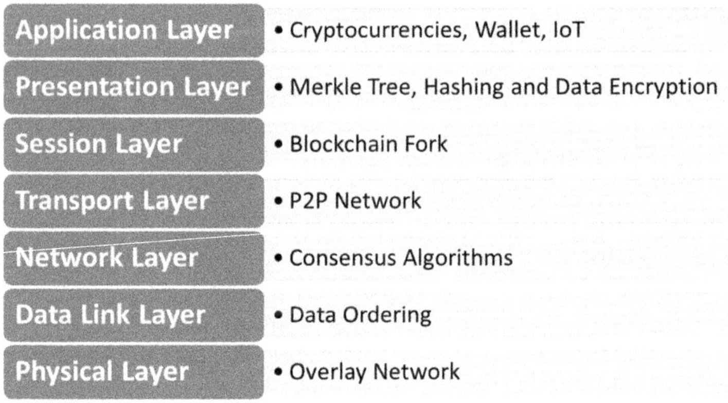

Fig. 2 Blockchain and OSI model

3.4.5 Symmetric Cryptography

Although less commonly used in blockchain than asymmetric cryptography, symmetric cryptography uses the same key for both encryption and decryption. The use case is secure data transmission within closed networks. The common algorithm is AES (Advanced Encryption Standard).

3.5 Blockchain Data Flow in OSI Model

The Open Systems Interconnection (OSI) is a famous conceptual model that provides a basic understanding of data flow. Figure 2 depicts the Blockchain theoretical working given the OSI model.

4 Consensus Mechanisms

A Blockchain consensus mechanism is a protocol that allows participants in a decentralized network to agree on the contents of a shared digital ledger. It ascertains the integrity and security of the Blockchain by preventing fraudulent transactions and double-spending. Proof of Work (PoW), Proof of Stake (PoS), and Delegated Proof of Stake (DPoS) are frequently used Consensus algorithms. These mechanisms contribute to the trustworthiness of a Blockchain by requiring validators to perform specific tasks or hold a certain amount of cryptocurrency. Overall, consensus mechanisms promote a stable, secure, and decentralized ecosystem for digital transactions [5]. Some commonly used consensus algorithms are mentioned in Table 3.

Table 3 Consensus mechanisms

Sr. no.	Algorithm name	Mechanism
1	Proof of Work (PoW)	It is a pioneer consensus algorithm in Blockchain. PoW creates a competitive environment among the miners to validate transactions and generate new blocks. In this way, miners can earn rewards for their efforts
2	Proof of Stake (PoS)	The proof of stake (PoS) concept is based on the idea that a person's ability to validate block transactions or mine new blocks is directly proportional to the number of coins they hold. In other words, a miner who owns a greater amount of Bitcoin or altcoins will have more mining power than someone who holds fewer coins
3	Practical Byzantine Fault Tolerance (PBFT)	One of the most popular byzantine fault tolerance (BFT) consensus algorithms is practical byzantine fault tolerance (PBFT), which achieves consensus through a series of message exchanges between nodes, ensuring that particular number of nodes agree on the same value before finalizing it
4	Proof of Burn (PoB)	Proof of burn (PoB) is a different type of consensus algorithm that aims to solve the energy consumption problem associated with proof of work (PoW) system. PoB is sometimes referred to as a PoW system that does not waste energy. Its underlying concept is to allow miners to "burn" tokens
5	Raft consensus	Raft is an algorithm for achieving distributed consensus. It operates by selecting a leader within the cluster who is responsible for processing client requests and replicating log entries to the other servers. The data in Raft is unidirectional, flowing solely from the leader to other servers
6	Federated consensus	The conventional approach for achieving consensus in Blockchain technology is through a fundamental concept, the Byzantine agreement. In this method, nodes on the Blockchain validate blocks of data by collectively agreeing on the solution to a given problem. A quorum, which is a minimum number of nodes, must agree that the presented solution is correct to reach a Byzantine agreement, validate a block, and append in the blockchain
7	Proof of Activity	Proof of Activity (PoA) verifies the authenticity of transactions as well as facilitates consensus among miners. It has features of both PoW and PoS. Decred (DCR) is an autonomous cryptocurrency that utilizes PoA
8	Proof of elapsed time	It is a consensus algorithm that aims to prevent excessive resource utilization and energy consumption while maintaining process efficiency through a fair lottery system. In this system, players may either choose their own set of six numbers, including five regular and one Powerball or allow computer terminals to select them at random. If the numbers on a ticket match the winning numbers in the exact order they are drawn and the player wins the jackpot prize

(continued)

Table 3 (continued)

Sr. no.	Algorithm name	Mechanism
9	Proof of Capacity	Proof of capacity (PoC) is a consensus algorithm employed in Blockchain networks, which permits mining devices to determine mining rights using their hard drive space instead of the computational power of mining device or miner's stake like PoW and PoS, respectively
10	Proof of importance	Proof of importance is a mechanism utilized to ascertain the eligibility of network participants (nodes) to append a new block referred to as "harvesting" by NEM. Upon successfully harvesting a block, nodes are rewarded with transaction fees within that block

5 Types of Blockchain

5.1 *Public Blockchain*

A Public Blockchain is a type of Blockchain that is accessible to everyone, allowing anyone to read, write, and audit it. It operates without a central authority, and decisions are based on complicated mechanisms. Public Blockchains have safeguards in place to protect against malicious actors. They are resource hungry to mine and add a block over the network. Bitcoin and Ethereum [6] are famous examples of public Blockchains.

5.2 *Private Blockchain*

A Private Blockchain is owned and operated by an individual or organization. It has an administrator/anchor who manages permissions and identities, and the consensus is achieved at the discretion of the central in-charge who can grant mining rights to anyone or not at all. Private Blockchains are faster and cheaper than public Blockchains because they do not require much energy, time, and money to reach a consensus. However, they are less secure than Public Blockchains. Examples of Private Blockchains include Bankchain and Medichain.

5.3 *Consortium Blockchain*

A Consortium Blockchain is a hybrid of Private and Public Blockchains, where a group of companies or representatives come together to make decisions for the network's benefit, rather than relying on one central authority. This approach speeds up decision-making and mitigates single points of failure. Consortium Blockchains offer options for rights and access management while leveraging the benefits of Blockchain technology. Examples of consortium Blockchains include R3 and EWF.

6 Application of Blockchain in Real life

Blockchain is ideal for use cases where security, transparency, and decentralization are critical, such as financial services, supply chain management, and identity verification [7].

6.1 Finance

The loan process involves numerous intermediaries and bank-related obstacles, adding to the overall complexity of the process. The proposal suggests the deployment of a peer-to-peer mortgage lending platform using Blockchain, which involves creating a smart contract between borrowers and lenders through token-based loan issuance. The platform of Blockchain technology would enable real-time reporting of transactions, making the process cost-effective and eliminating the need for intermediaries making the system transparent for both borrowers and lenders. This would increase transaction performance and reduce the time taken for transactions, resulting in streamlined and efficient processing of mortgages [8].

6.2 Healthcare

Healthcare billing encounters various issues such as billing inaccuracies, fraud, and slow processing time. A smart contract can be used in Blockchain-based solutions for automatic payment processing, audit, and transparency. Smart contracts can also be initiated to provide insurance to the patients. Storing billing information in a Blockchain would provide a trustworthy source of data for claims adjudication, reduce costs, and save time for funds transfer. It would also minimize billing fraud because the information stored in the Blockchain is immutable [9].

6.3 Digital Privacy

Digital media created for human consumption can be easily copied and distributed, which leads to monetary loss for content creators. A Blockchain-based media tracking technology assigns a unique ID to video content and stores it on the Blockchain. Content can be consumed using tokens, and a smart contract can be triggered to monitor copyright infringement. The smart contract can automatically trigger copyright actions if duplicate content is found. The use of Blockchain technology enables tracking the entire lifecycle of any content, while also allowing instant tracking of piracy. Additionally, micro-consumption of content can be achieved using this technology [10].

6.4 Royalty

In a digital world where it is easy to share and distribute copies, it is challenging to pay creators fairly for their work. A transparent Blockchain-based ledger can be used to store media assets and their rights holders. By utilizing smart contracts, media owners can distribute rights and automate royalty payments based on consumption, including streaming. The use of Blockchain technology provides transparency to content developers on the consumption and payments of their media, enabling instant payments based on usage [11].

6.5 Crowdfunding

Independent filmmakers often struggle to raise funds for their niche projects. A Blockchain-based crowdfunding solution enables the public to invest in projects using smart contracts and tokens, with potential returns offered in the form of percentages. The solution also allows for the tracking of investments through KYC verification and enables the public to earn from content market capture. Blockchain-powered crowdfunding offers benefits such as privacy and transparency, eliminating the need for external payment or verification solutions.

6.6 Real Estate

The issue at hand is that multiple listing services are providing incomplete lists, and there is a lack of clear definitions when it comes to releasing apartment ownership or documents. The proposal is to use a public Blockchain for listing and storing rental documents and service payments. This would provide a centralized location for all rental records and enable independent verification and automatic reconciliation of transactions and listings. The rental process would be facilitated by self-executing smart contracts, which would automate transactions between tenants and landlords, eliminating the need for a real estate agent. The proposal suggests that using a Blockchain can mitigate the risks of double-spending, fraud, abuse, and manipulation of transactions, thereby increasing the security of data sharing in property rental agreements. This would reduce barriers for both property owners and renters, ultimately making the rental process simpler and more efficient.

7 Blockchain Challenges

The challenges that are being faced by Blockchain technology include scalability issues, energy consumption concerns, and regulatory uncertainty. Additionally, the lack of interoperability between different Blockchain networks remains a significant challenge for widespread adoption [12]. Blockchain trilemma is the biggest problem of Blockchain mentioned in Fig. 3.

7.1 Decentralization

The decentralization of Blockchain networks means that all participants have equal access and influence, but maintaining it can be challenging as the network grows. The concentration of mining power, governance issues, and user adoption are some of the challenges faced by the decentralization of Blockchain. The possession of mining power by a few large miners can potentially compromise the network's security. Governance issues can lead to disagreements and forks in the Blockchain, reducing its effectiveness. Finally, user adoption can be limited due to the complexity and technical expertise required to use Blockchain effectively.

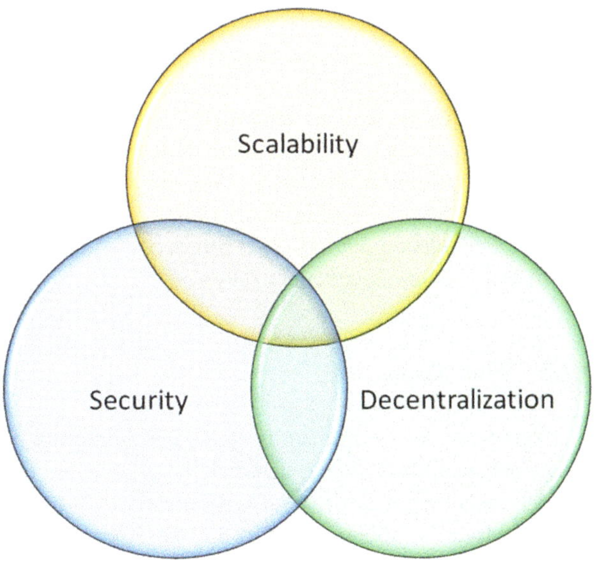

Fig. 3 Blockchain trilemma

Achieving and maintaining decentralization requires careful consideration of technical, economic, and social factors, and there are trade-offs between characteristics such as scalability, decentralization, and security. Balancing these trade-offs is crucial to create a Blockchain suitable for the specific use case [13].

7.2 Security

Blockchain security refers to the measures and techniques related to integrity, confidentiality, and availability of data stored on a Blockchain network, including the use of cryptographic algorithms and consensus mechanisms to ensure trust and immutability [14].

7.2.1 51% Attack

A 51% attack occurs when a certain number of miners take over more than fifty percent of a Blockchain network's computing power so they can effectively control the network. An example of this type of attack was carried out on Bitcoin Gold, which resulted in a loss of $18 million. This type of attack enables the attackers to block new transactions and reverse confirmed transactions, resulting in double-spending [15].

7.2.2 Eclipse Attack

The Distributed Application Architecture is the basis for the described attack, which partitions tasks among peers without a central server. The attack involves compromising a node's ability to communicate with others on the network by exploiting limitations in its identification and selection. Bitcoin and Ethereum have a limited number of connections per node, which means that a single node relies on the other connected nodes for a complete network view, presenting an opportunity for hackers. The potential damage resulting from this type of attack includes double-spending, attacks on second-layer protocols, and potential attacks on smart contracts due to inconsistent Blockchain views.

7.2.3 Sybil Attack

In a Sybil attack, a single entity generates numerous false identities to exert undue influence or control over a network or system, increasing the likelihood of other nodes connecting with the attacker's nodes. In Bitcoin, there is no record of nodes, and if a node is isolated from the honest network, it can increase the chance of execution of other attacks. The potential damage from a Sybil attack includes the attacker's ability to refuse the relay blocks and only relayed blocks that are created by them [16].

7.2.4 Time Jacking Attack

A type of Sybil attack in which all nodes keep a network time clock that depends on the median time of their peers which should not be more than 70 minutes. An attacker can alter the network time clock by connecting to several peers and providing false timestamps, causing it to slow down or speed up. This type of attack can result in a time value distortion of up to 70 minutes, causing a difference of up to 140 minutes between nodes, which can lead to potential damage [17].

7.3 Blockchain Scalability

Blockchain scalability is critical for mainstream adoption but increasing transactions per second (TPS) may compromise security and decentralization. Cryptographic algorithms ensure security but can affect scalability and decentralization. The property of Blockchain to manage a large number of transactions without compromising performance is called scalability. Limited block size and network congestion cause scalability issues, leading to high fees and slow transactions [18]. Blockchain scalability solutions can be categorized into three layers, as shown in Fig. 4.

7.3.1 Blockchain Scalability Layer-2 Solutions

Layer 2 scaling solutions are famous and necessary to increase transaction speed on major Blockchains, which are limited to around 7–15 TPS, compared to Visa's 100,000 TPS. There are two ways to scale: scaling the base layer or assigning some work to be completed by the new layer. However, researchers have not found a way to optimize decentralization, security, and scalability all simultaneously, known as the Blockchain trilemma. Layer 2 solutions like roll-ups, side chains, plasma, and channels each have their own place in scaling a Blockchain and work outside the actual Blockchain code. Roll-ups combine transactions into a single transaction, while side chains run parallel to the main chain and use resources to offload work. Plasma employs child chains to broadcast vital operations to the main chain, and channels are a mechanism to lock up money and conduct virtual currency trading. Each solution has its benefits and drawbacks and is best suited for specific situations [19].

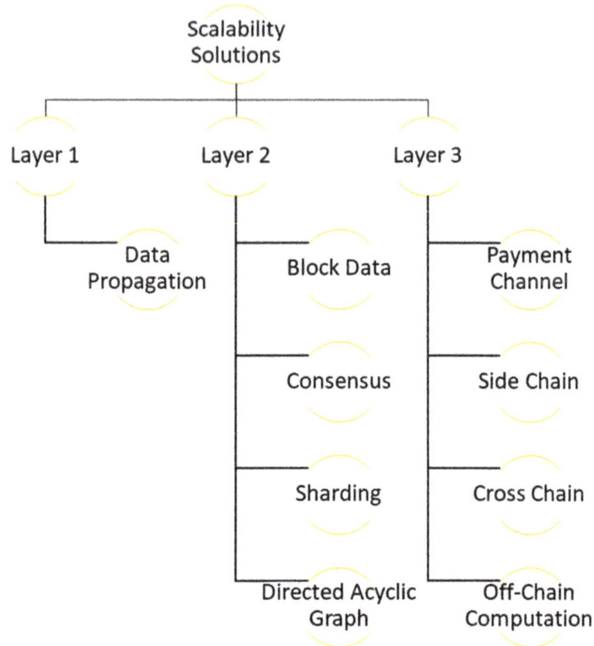

Fig. 4 Blockchain scalability solutions

7.3.2 Sharding

This process divides the Blockchain into petty and easily administrative parts known as shards. Each shard independently processes a particular subset of transactions.

7.3.3 Off-Chain Solutions

Off-chain protocols are used to process transactions outside the Blockchain, thereby reducing the network's load. Payment channels, such as the Lightning Network, and state channels, such as Plasma, are examples of off-chain protocols.

7.3.4 Protocol Upgrades

Blockchains can be upgraded to improve their scalability, such as increasing the block size or changing the consensus algorithm to reduce the time needed to validate transactions.

7.3.5 Hybrid Solutions

Combining multiple scaling solutions to achieve greater scalability, such as using sharding and off-chain solutions together, is called Hybrid solutions. In summary, Blockchain technology is facing an ongoing challenge of scalability, but several solutions are currently being developed and implemented to enhance the performance of Blockchain networks.

8 Smart Contracts

Smart contracts are Blockchain-based digital agreements between parties that are carried out automatically by code without discretion, because of Blockchain technology's autonomy, self-sufficiency, and decentralization. They differ from traditional contracts by being algorithmically followed and not needing trust. However, smart contracts raise several considerations, including the need for a new body of law and regulation and their impact on contract law and the social contract. Smart contracts are applicable in various scenarios, i.e., automatic payments for betting, establishing inheritance gifts, and pledge systems like Kickstarter. The widespread use needs a lot of adjustment, but they can offer significant advantages over traditional contracts [20]. The conceptual information flow of a smart contract is mentioned in Fig. 5.

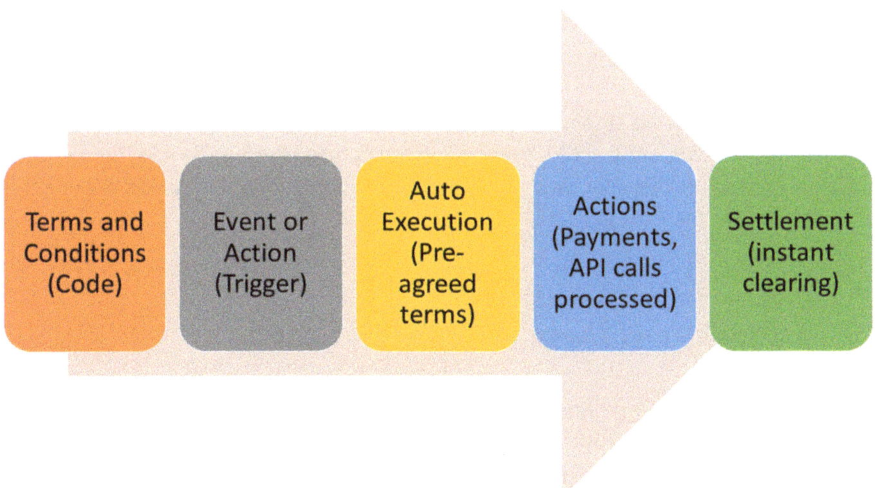

Fig. 5 Smart contract information flow

8.1 Architecture of Smart Contracts

The architecture of smart contracts is based on several key components, including:

- Code: Smart contracts are written as code, typically using programming languages such as Solidity or Vyper. The code defines the rules and conditions of the contract, such as the parties involved, the actions to be taken, and the triggers that will initiate those actions.
- Blockchain: Smart contracts are saved in a deactivated form on a blockchain, which is a decentralized and distributed ledger that stores all transactions in a safe, immutable, and transparent manner. The blockchain ensures that the contract is immutable and tamper-proof, as all parties can see and verify the contract's execution.
- Nodes: The blockchain is maintained by a network of nodes, which are computers that validate transactions and execute smart contracts. Each node has a replica of the blockchain, and any modification to the contract must be verified by the majority of the nodes before it is accepted.
- Cryptography: Smart contracts use cryptography to ensure security and privacy. For example, public-key cryptography can be used to authenticate the parties involved in the contract, while hash functions can be used to ensure that the contract's data has not been tampered with.
- Oracles: Smart contracts can also interact with external data sources through oracles. Oracles are trusted third-party services that provide data to the smart contract, allowing it to execute actions based on real-world events.

8.1.1 Smart Contract Layers

Smart contracts can be implemented on different layers including the application layer, security layer, contract layer, and infrastructure layer, each serving a specific function to support the operation and execution of the contracts on the blockchain, as shown in Fig. 6.

8.2 Activation and Execution of Smart Contract

By default, the smart contracts are in deactivation form on the Blockchain network. For activation of smart contract, some conditions should be fulfilled, i.e., time or the situation, for example, it is written in the smart contract that this contract will execute on 1st June 2023 so that contract will be executed on this date (condition). The second way to activate the smart contract is situation which means that user action is required, for example, if a user presses a certain button (condition) then the smart contract will execute. After activation, the execution of the smart contract will be done in the following steps, as shown in Fig. 7.

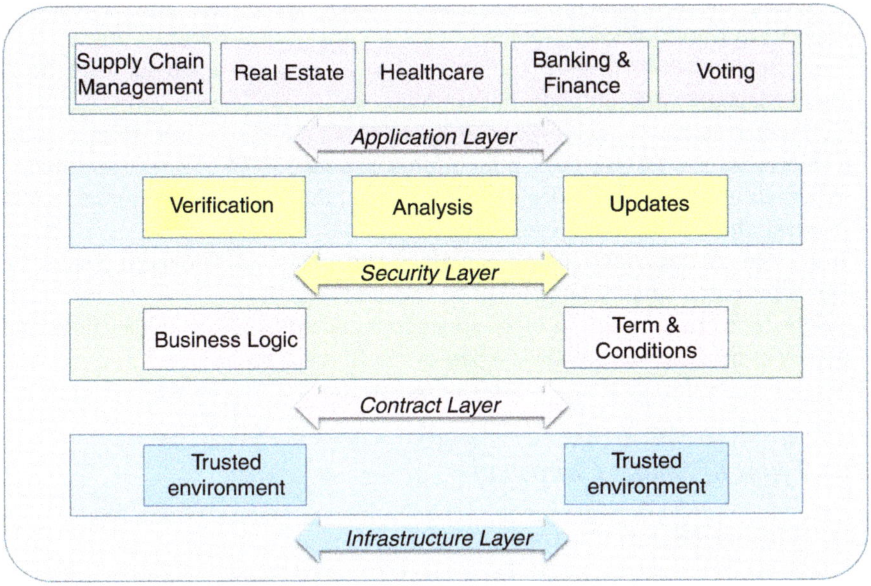

Fig. 6 Basic research framework of smart contracts

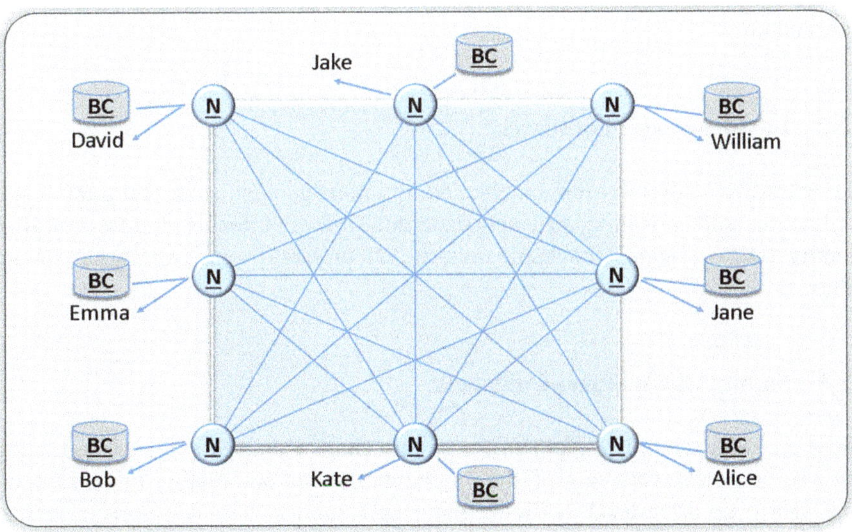

Fig. 7 Smart contracts architecture

1. If Bob wants to execute the smart contract, he will broadcast the Public key to the Blockchain network.

2. In return, the Blockchain network will provide the Private key of the smart contract to Bob.
3. Now Bob has both (Public and Private) keys of smart contract whereas the Blockchain network (all nodes) has a Public key.
4. Bob will generate an encrypted message by using the Private key which contains a smart contract. For the decryption of this message, a Public key is required.
5. When all nodes receive an encrypted message, these are capable to decrypt the message through their received Public key.
6. If the majority of nodes (more than 50%) verify and validate the message then Bob's proposed smart contract will be executed.
7. All the terms and conditions of the smart contract will apply on all the Blockchain networks because every node is a part of the Blockchain.

8.3 *Types of Smart Contracts*

8.3.1 Financial Smart Contracts

These are smart contracts that involve financial transactions, such as payments, loans, insurance, and investment contracts. They are often used in decentralized finance (DeFi) applications to automate financial operations and reduce the need for intermediaries.

8.3.2 Identity Smart Contracts

These are smart contracts that involve identity management, such as digital identity verification, authentication, and authorization. They are often used in Decentralized Identity (DID) systems to enable secure and privacy-preserving digital identity solutions.

8.3.3 Supply Chain Smart Contracts

These are smart contracts that involve supply chain management, such as tracking and tracing goods, verifying product authenticity, and automating logistics operations. They are often used in Blockchain-based supply chain solutions to increase transparency, efficiency, and trust.

8.3.4 Real Estate Smart Contracts

These are smart contracts that involve real estate transactions, such as property sales, leases, and rental agreements. They are often used in Blockchain-based real estate solutions to reduce costs, increase transparency, and streamline processes.

8.3.5 Governance Smart Contracts

These are smart contracts that involve governance and decision-making, such as voting, consensus, and dispute resolution. They are often used in decentralized autonomous organizations (DAOs) to enable decentralized decision-making and community governance.

8.4 Challenges in Smart Contracts

The main challenge in smart contracts is ensuring that the code is free from errors and vulnerabilities, as even small mistakes can have significant consequences. Another challenge is the legal and regulatory framework surrounding smart contracts, as they are still a relatively new and untested technology in many jurisdictions [21].

8.4.1 Security Vulnerabilities

Smart contracts are vulnerable to security attacks, which can result in the loss of funds or data. These attacks can occur due to coding errors, software bugs, or malicious attacks by hackers.

8.4.2 Legal Uncertainty

There is a lack of legal clarity surrounding smart contracts, as they are not yet widely recognized by regulatory authorities. This can result in disputes over the enforceability of smart contracts and the legal remedies available in case of breaches.

8.4.3 Interoperability Issues

Smart contracts are often built on different Blockchain platforms, which can result in interoperability issues. This can make it difficult to integrate smart contracts with other systems or migrate them from one Blockchain platform to another.

8.4.4 Complexity

Designing and implementing smart contracts can be a complex task, leading to potential errors and delays. This can make it challenging for developers to create and maintain smart contracts that are secure, dependable, and efficient.

8.4.5 Lack of Standardization

Smart contract development lacks standardization, which can cause interoperability problems and inconsistencies in their implementation. As a result, it may be challenging for users to validate the authenticity and dependability of smart contracts.

8.4.6 Limited Scalability

The current Blockchain infrastructure poses a scalability limitation to smart contracts in that there exists a restriction on the number of transactions that can be handled during a specific timeframe. As a result, delays and higher transaction fees may occur, which can make using smart contracts less cost-effective for certain applications.

8.5 Security and Privacy of Smart Contracts

Smart contract security involves ensuring that the code is free from vulnerabilities and that it functions as intended to prevent loss of funds or unintended outcomes. Best practices include code audits, testing, and following standardized templates.

1. One of the main security risks associated with smart contracts is the possibility of bugs or vulnerabilities in the code. Once a smart contract is deployed, it becomes immutable, meaning that it cannot be changed. If there is a bug or vulnerability in the code, it can be exploited by malicious actors, potentially resulting in loss of funds or other negative consequences.
2. To reduce the risk factors related to smart contracts, it is important to follow best practices for secure software development. This includes conducting code reviews, testing the contract thoroughly before deployment, and implementing security measures such as access control and encryption.
3. Another important factor in measuring the security and privacy of smart contracts is the choice of a blockchain platform. Different platforms have different security features and protocols, and it is important to choose one that is appropriate for the specific use case.
4. In addition to technical measures, it is also important to consider the legal and regulatory aspects of smart contract security. Smart contracts may need to conform to various rules and regulations, such as data protection laws and anti-money laundering regulations.

Overall, the security of smart contracts is a complex and evolving issue and requires a multi-faceted approach to mitigate the risks.

8.6 Use Cases of Smart Contracts

Smart contracts can be used in the finance industry for the automatic execution of agreements, payments, and transactions based on predetermined conditions. They can also be used in supply chain management to ensure transparency and automate the tracking of goods and services, mitigating the chances of fraud and errors. There are various practical use cases for smart contracts, including:

8.6.1 Real Estate

The automation of buying, selling, and renting property using smart contracts can reduce the reliance on intermediaries and improve the efficiency of the process.

8.6.2 Healthcare

Smart contracts can be used to securely and efficiently manage patient health records, automate insurance claims processing, and guarantee the safe exchange of medical information between healthcare professionals.

8.6.3 Banking and Finance

Financial transactions, such as loans and insurance contracts, can be automated using smart contracts and make sure that payments are made automatically when certain conditions are met.

8.6.4 Voting

By using smart contracts, it is possible to create voting systems that are secure and transparent, making them less susceptible to fraud and manipulation.

8.7 Unified Modeling Language (UML) Class Diagram and Coding Examples of Smart Contracts

Smart contracts are often used in blockchain-based systems such as Ethereum to automate the process of verifying and enforcing contractual agreements. Some examples of smart contracts include:

8.7.1 Insurance Contracts

Smart contracts can be used to automatize the way of paying out insurance claims. When an insured event occurs, the contract can automatically trigger the payout to the policyholder without the need for manual intervention. Unified Modeling Language (UML) class diagram of an insurance contract in smart contract is mentioned in Fig. 8.

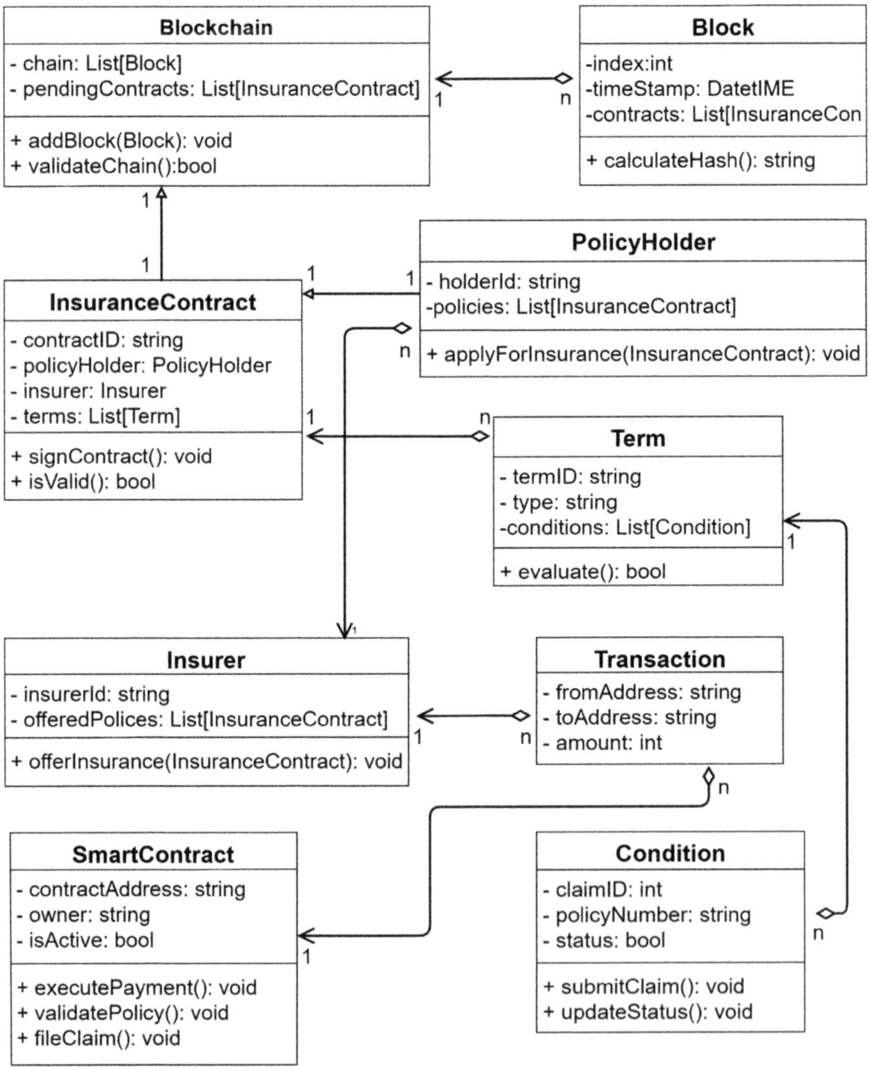

Fig. 8 UML class diagram for insurance smart contract

Table 4 Description of classes

Class	Description
Blockchain	Manages the entire chain and pending insurance contracts. Responsible for adding new blocks and validating the blockchain
Block	Represents a single block in the blockchain. It contains multiple insurance contracts and has methods for calculating its own hash
InsuranceContract	Represents a single insurance contract in the blockchain, which includes details like the policyholder, insurer, and terms
PolicyHolder	Represents the individual or entity holding the policy. They can apply for insurance
Insurer	Represents the insurance company offering the policy. They can offer insurance contracts
Term	Represents the terms and conditions of the insurance contract
Smart contract	Contains the logic for executing transactions based on predefined conditions
Condition	Represents a condition in a term that can be evaluated
Transaction	Represents a transaction in the system, containing details like the sender and receiver addresses, and the amount transferred

Table 5 Description of methods

Methods	Description
addBlock, validateChain, calculateHash	These are related to blockchain operations
signContract, isValid, applyForInsurance, offerInsurance, evaluate, execute, signTransaction	These methods are for managing insurance contracts and transactions

As shown in Fig. 8, in this contract, the **insurer** and **employee** addresses are defined as the parties involved in the insurance agreement. The **premium** value is the amount of ether that the employee needs to pay to initiate the insurance coverage. The **payPremium**() function allows the employee to pay the premium amount to the contractor. Once the premium is paid, the contract is considered active. If an accident occurs during fieldwork, the **employee** can call the **fileClaim**() function to initiate a claim. If the claim is valid and has not been previously processed, the contract pays out 1000 ether to the employee and updates the **isClaimed** flag. Finally, the **insurer** can withdraw the remaining funds from the contract by calling the **withdrawFunds**() function, but only after a claim has been processed. Tables 4 and 5 summarize the descriptions of classes and methods, respectively. In addition, the following code represents a sample insurance smart contract.

```
pragma solidity ^0.8.0;

contract fieldworkInsurance {
  address public insurer;
  address public employee;
  uint256 public premium;
  bool public isClaimed;
```

```
constructor(address _insurer, address _employee,
          uint256 _premium) {
  insurer   = _insurer;
  employee  = _employee;
  premium   = _premium;
  isClaimed = false;

}

function payPremuim() external payable {
  require(msg.sender == employee,
         "Only employee can pay the premium.");
  require(msg.value  == premium,
         "Incorrect premium amount.");
}

function fileClaim() external {
  require(msg.sender == employee,
         "Only employee can file a claim.");
  require(!isClaimed, "Claim already processed.";

  isClaimed = true;
  payable(employee).transfer(1000 ether);
}

function withdrawFunds() external {
  require(msg.sender == insurer,
         "Only insurer can withdraw funds.");
  require(isClaimed, "No claim has been processed
             yet.");

  payable(insurer).transfer(address(this).balance);
}
}
```

8.7.2 Supply Chain Management

UML class diagram of a supply chain management system in smart contract is mentioned in Fig. 9. As shown in this contract, the **buyer** and **seller** addresses are defined as the parties involved in the supply chain transaction. The **penalty** value is the amount of ether that the buyer needs to pay if the equipment is not delivered, while the **payment** value is the amount of ether that the buyer needs to pay if the

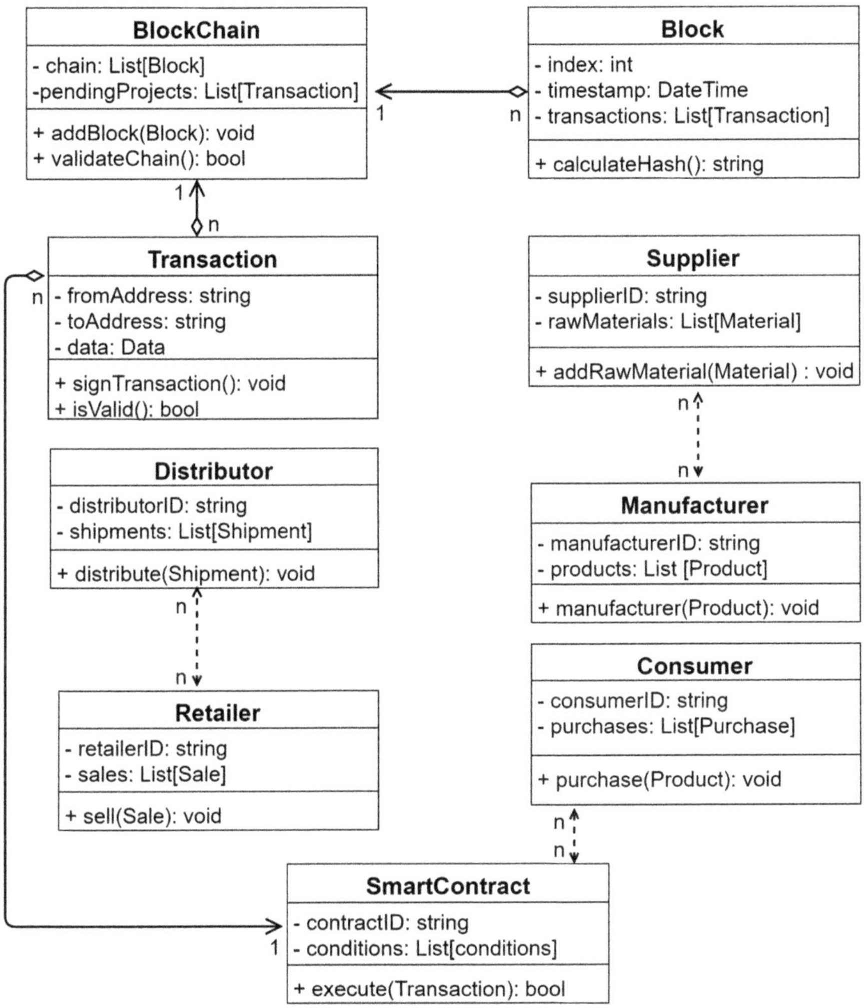

Fig. 9 UML class diagram for supply chain management smart contract

equipment is delivered. The **confirmDelivery()** function allows the buyer to confirm that the equipment has been delivered. Once the equipment is delivered, the contract updates the **isDelivered** flag. If the equipment is delivered, the **releasePayment()** function can be called by the buyer to release the payment to the seller. If the equipment is not delivered, the **imposePenalty()** function can be called by the seller to impose a penalty on the buyer. Note that this is a simplified example and in a real-world scenario, the contract would need to include additional functionality and error handling to ensure the security and reliability of the transaction. Tables 6 and 7 summarize the descriptions of classes and methods, respectively. In addition,

Table 6 Description of classes

Class	Description
Blockchain	Manages the entire chain and pending transactions. Responsible for adding new blocks and validating the blockchain
Block	Represents a single block in the blockchain. It contains multiple transactions and has methods for calculating its own hash
Transaction	Represents a single transaction in the blockchain, which could be anything from adding a raw material to selling a finished product
Supplier	Represents the supplier in the supply chain. They add raw materials
Manufacturer	Represents the manufacturer in the supply chain. They manufacture products from raw materials
Distributor	Represents the distributor in the supply chain. They distribute products to retailer
Retailer	Represents the retailer in the supply chain. They sell products to consumers
Consumer	Represents the end consumer in the supply chain. They purchase products from retailers
Smart contract	Contains the logic for executing transactions based on predefined condition

Table 7 Description of methods

Methods	Description
addBlock, validateChain, calculateHash	These are related to blockchain operation
addRawMaterial, manufacture, distribute, sell, purchase	These methods are for managing the supply chain
signTransaction, isValid	These methods are for managing transactions
Execute	This is a method in the smart contract class to execute transactions based on predefined conditions

the following code represents a sample code for a supply chain management system smart contract.

```
pragma solidity ^0.8.0;

contract EquipmentDelivery {
  address public buyer;
  address public seller;
  uint256 public penalty;
  uint256 public payment;
  bool public isDelivered;

constructor(address _buyer, address _seller,
          uint256 _penalty, uint256 _payment) {
  buyer       = _buyer;
  seller      = _seller;
  penalty     = _penalty;
  payment     = _payment;
```

```
    isDelivered = false;
  }

  function confirmDelivery() external {
  require(msg.sender == buyer,
          :Only the buyer can confirm delivery).");
  required(!isDelivered, "Equipment has already
          been delivered.");

  isDelivered = true;
  }

  function releasePayment() external {
    require(msg.sender == buyer,
          "Only the buyer can release payment.");
    require(isDelivered, "Equipment has not been
          delivered yet.");

    payable(seller).transfer(payment);
  }

  function imposePenalty() external {
       require(msg.sender == seller,
              "Only the seller can impose a penalty.");
       require(!isDelivered, "Equipment has
              been delivered.:);

    payable(buyer).transfer(penalty);
  }
  }
```

8.7.3 Crowdfunding Smart Contract

UML class diagram of a Crowdfunding system in a smart contract is mentioned in Fig. 10. As shown in this smart contract, the **projectCreator** address is defined as the creator of the project, while the **fundingGoal** value is the amount of ether that needs to be raised to consider the funding goal reached. The **totalFunding** variable keeps track of the total amount of ether raised so far, while the **isFundingGoalReached** flag indicates whether the funding goal has been reached. The **fundProject()** function allows anyone to fund the project by sending ether to the contract. The function first checks that the funding amount is greater than 0 and that the funding goal has not already been reached. If the conditions are met, the contract adds the funding amount to the **fundingAmount** mapping and updates

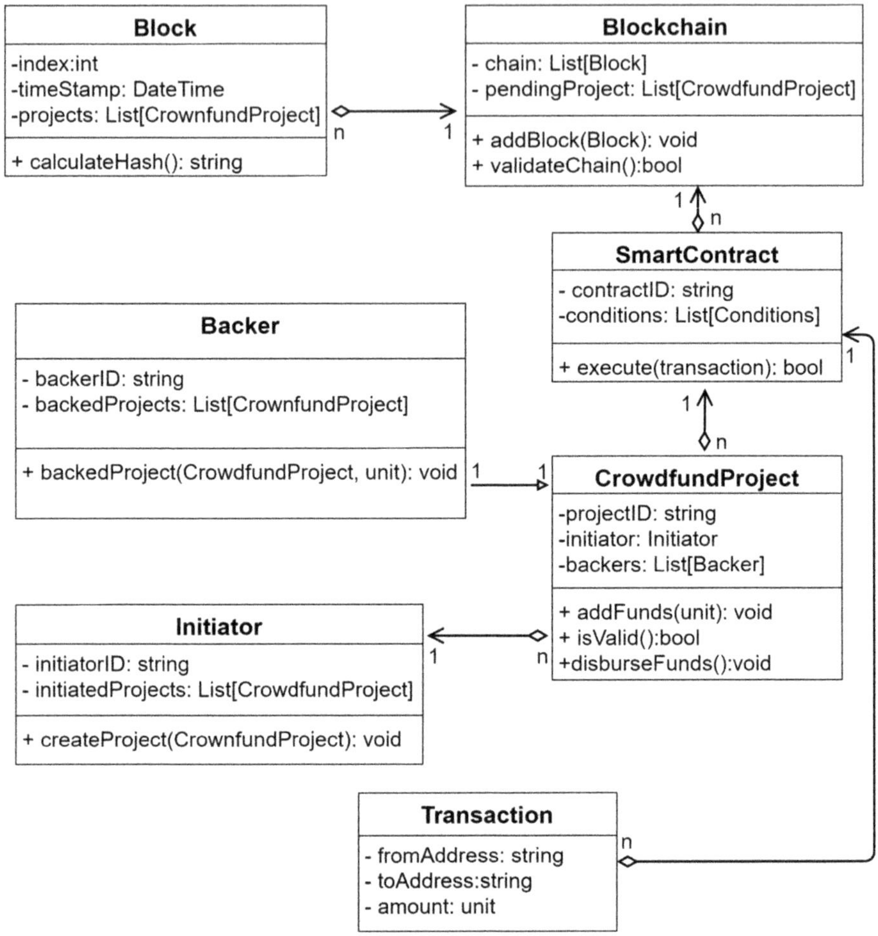

Fig. 10 UML class diagram for crowdfunding smart contract

the **totalFunding** variable. If the **totalFunding** variable is greater than or equal to the **fundingGoal**, the **isFundingGoalReached** flag is set to true and the contract automatically transfers the total balance of the contract to the **projectCreator**. Note that this is a simplified example and in a real-world scenario; the contract would need to include additional functionality and error handling to ensure the security and reliability of the fundraising process. Tables 8 and 9 summarize the descriptions of classes and methods, respectively. In addition, the following code represents a sample code for a crowdfunding smart contract.

```
pragma solidity ^0.8.0;

contract ProjectFunding {
```

Table 8 Description of classes

Class	Description
Blockchain	Manages the entire chain and pending projects. Responsible for adding new blocks and validating the blockchain
Block	Represents a single block in the blockchain. It contains multiple crowdfunding projects and has methods for calculating its own hash
CrowdfundProject	Represents a crowdfunding project with details like the project ID, initiator, backers, goal amount, and current amount. It contains methods for adding funds, validating the project, and disbursing funds
Initiator	Represents an individual or entity initiating a crowdfunding project. They can create new projects
Backer	Represents an individual or entity backing a crowdfunding project. They can back projects with a specified amount
Smart contract	Contains the logic for executing transactions based on predefined conditions
Transaction	Represents a transaction in the system, containing details like the sender and receiver addresses, and the amount transferred

Table 9 Description of methods

Methods	Description
addBlock, validateChain, calculateHash	These are related to blockchain operations
addFunds, isValid, disburseFunds	These methods are for managing crowdfunding projects
createProject	This is a method for initiators to create new crowdfunding projects
backProject	This is a method for backers to back projects with funds
Execute	This is a method in the smart contract class to execute transactions based on predefined conditions

```
    address public projectCreator;
    uint256 public fundingGoal;
    uint256 public totalFunding;
    bool public isFundingGoalReached;

    mapping(address => uint256) public fundingAmount;

  constructor(address _projectCreator,
      uint256 _fundingGoal) {
    projectCreator        = _projectCreator;
    fundingFoal           = _fundingGoal;
    totalFunding          = 0;
    isFundingGoalReached = false;

  }
```

```
function fundProject() external payable {
  require(msg.value > 0,
          "Funding amount must be greater than 0.");
  require(!isFundingGoalReached,
          "Funding goal has already been reached.");

  fundingAmiount[msg.sender] += msg.value;
  totalFunding += msg.value;

if (totalFunding >+ fundingGoal) {
    isFundingGoalReached = true;
    payable(projectCreator).transfer
           (address(this).balance);
}
}
}
```

8.8 Programming Languages for Writing a Smart Contract

Smart contracts can be developed in various programming languages, depending on the specific blockchain technology and the requirements of the application. Some of the most common programming languages for smart contract development are:

1. Solidity: This is the most popular programming language for developing smart contracts on the Ethereum blockchain. Solidity is a contract-oriented, high-level language with syntax similar to that of JavaScript.
2. Vyper: This is a high-level programming language for Ethereum smart contract development. Vyper is designed to be more secure and simple than Solidity, with less room for error and vulnerabilities.
3. Rust: This is a systems programming language that is gaining popularity for smart contract development on the Polkadot and Kusama blockchains. Rust is known for its performance, safety, and memory management capabilities.
4. Go: This is a popular programming language for smart contract development on the Hyperledger Fabric blockchain. Go is known for its simplicity, scalability, and performance.
5. Java: This is a widely used programming language for developing enterprise-level applications, including smart contracts on the Corda blockchain. Java is known for its reliability, security, and cross-platform compatibility.
6. JavaScript: This is a popular programming language for developing web applications and is also used for smart contract development on the EOS and TRON blockchains. JavaScript is known for its versatility, ease of use, and compatibility with web technologies.

Table 10 Smart contracts and decentralized applications (DApps) comparison

Smart contracts	Decentralized applications
A smart contract is a self-executing contract with the terms of the agreement directly written into code	A DApp is a broader software application that operates on a decentralized network and is often composed of a frontend and one or more smart contracts acting as the backend logic
It acts as a set of rules and automated actions that are executed when predefined conditions are met	It provides a user interface and additional functionalities that interact with smart contracts
Generally focused on specific tasks or agreements, like transferring tokens if certain conditions are met	Can be a full-fledged application offering a range of features and services, like a decentralized exchange or a voting system
Smart contracts are immutable once deployed, ensuring that the agreed-upon logic will be executed	Operates in a trustless manner thanks to the underlying smart contracts and blockchain technology
Written in programming languages specific to blockchain platforms, like solidity for Ethereum	The frontend is often written in common web languages like HyperText markup language (HTML), Cascading style sheets (CSS), and JavaScript, while the backend logic is in smart contracts

8.9 Smart Contracts and Decentralized Applications (DApps)

Smart contracts and decentralized applications (DApps) are closely related concepts in the blockchain ecosystem. Smart contracts provide the rule-based logic that DApps utilize to offer broader functionalities in a decentralized manner. Smart contracts and DApps serve different roles, as mentioned in Table 10. DApps interact with smart contracts to facilitate decentralized functions. Smart contracts handle the business logic, while DApps manage user interactions. Think of smart contracts as the backend and DApps as the full application stack. DApps often utilize multiple smart contracts to provide complex functionalities. Actions initiated via a DApp's frontend interface trigger smart contract functions, which then execute automatically on the blockchain.

9 Blockchain Oracle

Blockchain oracles provide a secure and decentralized way for Blockchain applications to interact with external data sources, allowing smart contracts and dApps to access and utilize real-world data. They are crucial components of many Blockchain applications, such as DeFi, supply chain management, and insurance. Oracles can be centralized or decentralized and can provide a wide range of data types, including price and weather data. Decentralized oracles use consensus algorithms to ensure data accuracy and integrity, while centralized oracles rely on a single entity or source, creating a potential central single point of failure. As the Blockchain continues to proliferate, the demand for reliable and trustworthy oracles is also increasing [22–24]

Chainlink, Band Protocol, Tellor, Oraclize, API3, DIA, and Augur are the commonly used Blockchain Oracles depending on the specific blockchain technology and use case.

10 Conclusion

The journey through this chapter has served as a profound initiation into the essentials of blockchain technology and smart contracts. These innovative and disruptive mechanisms have laid the groundwork for a new wave of digital operations. In their capacity to engender trust in an otherwise trustless environment, they have the potential to redefine business operations, governance structures, and personal data handling. At its core, blockchain's decentralization, immutability, and transparency foster a unique form of digital interaction where intermediaries are no longer a necessity. Its potential reach extends far beyond the financial sphere, touching upon diverse sectors such as supply chain logistics, healthcare, and public services.

Complementing blockchain, smart contracts automate contractual obligations, the encoded terms of which self-execute upon fulfillment. Their impact in mitigating fraud, curtailing expenses, and streamlining operations is noteworthy, despite challenges pertaining to coding flaws, scalability, and legal validation. Closing this chapter, it is apparent that the transformative power of blockchain technology and smart contracts is still in its infancy, promising a bright horizon for these digital solutions. As we continue to refine and enhance these technologies, we can anticipate a more secure, decentralized, and efficient digital ecosystem.

In the subsequent chapters, we will delve further into the intricate dynamics of these digital marvels, examine their real-world implementations, and scrutinize the evolving regulatory frameworks that govern them. The foundational understanding gleaned from this chapter will equip us to navigate the intricacies and possibilities of blockchain and smart contracts in our increasingly digital age.

References

1. N.U. Hassan, C. Yuen, D. Niyato, Blockchain technologies for smart energy systems: fundamentals, challenges, and solutions. IEEE Ind. Electron. Mag. **13**(4), 106–118 (2019)
2. P. Mukherjee, C. Pradhan, Blockchain 1.0 to blockchain 4.0-the evolutionary transformation of blockchain technology, in *Blockchain Technology: Applications and Challenges* (Springer, Berlin, 2021), pp. 29–49
3. B. Cao, Y. Li, L. Zhang, L. Zhang, S. Mumtaz, Z. Zhou, M. Peng, When internet of things meets blockchain: challenges in distributed consensus. IEEE Netw. **33**(6), 133–139 (2019)

4. M.H. Rehmani, *Blockchain Systems and Communication Networks: From Concepts to Implementation* (Springer, Boston, 2021)
5. S. Wan, M. Li, G. Liu, C. Wang, Recent advances in consensus protocols for blockchain: a survey. Wirel. Netw. **26**(2), 5579–5593 (2020)
6. B.K. Mohanta, D. Jena, S.S. Panda, S. Sobhanayak, Blockchain technology: a survey on applications and security privacy challenges. Int. Things **8**, 100107 (2019)
7. T. Choudhury, A. Khanna, T.T. Toe, M. Khurana, N.G. Nhu, *Blockchain Applications in IoT Ecosystem* (Springer, Berlin, 2021)
8. M. Xu, X. Chen, G. Kou, A systematic review of blockchain. Financial Innovat. **5**(1), 1–14 (2019)
9. I. Yaqoob, K. Salah, R. Jayaraman, Y. Al-Hammadi, Blockchain for healthcare data management: opportunities, challenges, and future recommendations. Neural Comput. Appl. **34**, 1–16 (2021)
10. J. Guo, C. Li, G. Zhang, Y. Sun, R. Bie, Blockchain-enabled digital rights management for multimedia resources of online education. Multimedia Tools Appl. **79**, 9735–9755 (2020)
11. T. Alladi, V. Chamola, R.M. Parizi, K.-K.R. Choo, Blockchain applications for industry 4.0 and industrial IoT: a review. IEEE Access **7**, 176935–176951 (2019)
12. S. Namasudra, G.C. Deka, P. Johri, M. Hosseinpour, A.H. Gandomi, The revolution of blockchain: state-of-the-art and research challenges. Arch. Comput. Methods Eng. **28**, 1497–1515 (2021)
13. U. Khalid, M. Asim, T. Baker, P.C. Hung, M.A. Tariq, L. Rafferty, A decentralized lightweight blockchain-based authentication mechanism for iot systems. Cluster Comput. **23**(3), 2067–2087 (2020)
14. N. Anita, M. Vijayalakshmi, Blockchain security attack: A brief survey. In: 2019 10th International Conference on Computing, Communication and Networking Technologies (ICCCNT) (IEEE, Piscataway, 2019), pp. 1–6
15. I. Yaqoob, K. Salah, M. Uddin, R. Jayaraman, M. Omar, M. Imran, Blockchain for digital twins: recent advances and future research challenges. IEEE Netw. **34**(5), 290–298 (2020)
16. S. Zhang, J.-H. Lee, Double-spending with a sybil attack in the bitcoin decentralized network. IEEE Trans. Ind. Informat. **15**(10), 5715–5722 (2019)
17. M. Saad, J. Spaulding, L. Njilla, C. Kamhoua, S. Shetty, D. Nyang, D. Mohaisen, Exploring the attack surface of blockchain: a comprehensive survey. IEEE Commun. Surv. Tutor. **22**(3), 1977–2008 (2020)
18. A.A. Mazlan, S.M. Daud, S.M. Sam, H. Abas, S.Z.A. Rasid, M.F. Yusof, Scalability challenges in healthcare blockchain system-a systematic review. IEEE Access **8**, 23663–23673 (2020)
19. A. Singh, R.M. Parizi, M. Han, A. Dehghantanha, H. Karimipour, K.-K.R. Choo, Public blockchains scalability: An examination of sharding and segregated witness, in *Blockchain Cybersecurity, Trust and Privacy* (Springer, Cham, 2020), pp. 203–232
20. S.N. Khan, F. Loukil, C. Ghedira-Guegan, E. Benkhelifa, A. Bani-Hani, Blockchain smart contracts: applications, challenges, and future trends. Peer-to-peer Netw. Appl. **14**, 2901–2925 (2021)
21. A. Bhardwaj, S.B.H. Shah, A. Shankar, M. Alazab, M. Kumar, T.R. Gadekallu, Penetration testing framework for smart contract blockchain. Peer-to-Peer Netw. Appl. **14**, 2635–2650 (2021)
22. K. Mammadzada, M. Iqbal, F. Milani, L. García-Bañuelos, R. Matulevičius, Blockchain oracles: A framework for blockchain-based applications, in *Business Process Management: Blockchain and Robotic Process Automation Forum: BPM 2020 Blockchain and RPA Forum, Seville, Spain, September 13–18, 2020, Proceedings 18* (Springer, Berlin, 2020), pp. 19–34
23. S.K. Ezzat, Y.N. Saleh, A.A. Abdel-Hamid, Blockchain oracles: state-of-the-art and research directions. IEEE Access **10**, 67551–67572 (2022)
24. M. Taghavi, J. Bentahar, H. Otrok, K. Bakhtiyari, A reinforcement learning model for the reliability of blockchain oracles. Expert Syst. Appl. **214**, 119160 (2023)

Smart Contract: The Contract Automation Climax. Back-End and Front-End Legal Implications

Roberto Alma and Lorenzo Piatti

Abstract Smart Contracts are a core part of blockchain technology use cases and usability. Even if they are essentially simply pieces of code, thanks to the raise of contract—and transaction—automation they became a concrete everyday tool in the exchange of value in the decentralized world. This chapter covers what smart contracts are and why they are relevant in the blockchain context: the approach starts with some legal observations and then evolves to a technical description. Smart contracts have brought legal-contract automation to its climax: starting with a division into three phases—Instrumental, Telematica, and Autómata—the present work leads the reader toward an awareness of this legally relevant evolution. This chapter closes with a description of a key component of smart contracts: front-end. The graphical interface for the user, in fact, is fundamental aspect of the usability—and legality—of the software itself. This contribution is intended to be useful both for a legal reader, who wants to dive into the legal aspects related to smart contracts, and for a general reader, who wants to take the first steps in this blockchain-based automation context.

Keywords Decentralized ledger · DLT · Smart contract · Contract automation · Digital signature · eIDAS

1 Introduction

Smart contracts are pieces of software. They are written in code—the most notorious is Solidity, but different languages are available for developers—and deployed on a blockchain. Once published, they can be executed by simply calling one or more

R. Alma
Attorney, Rome, Italy

L. Piatti (✉)
InfoCert S.p.A., Milano, Italy
e-mail: lorenzo.piatti@infocert.it

N. El Madhoun et al. (eds.), *Blockchain and Smart-Contract Technologies for Innovative Applications*, https://doi.org/10.1007/978-3-031-50028-2_2

functions: predefined rules and conditions that are encoded in their text. When these conditions are met, the smart contract automatically executes the agreed-upon actions, such as transferring funds or releasing assets [1].

One of the key benefits of smart contracts is that they are immutable and tamper-proof. Once a smart contract is deployed on a (proper) blockchain, it cannot be altered or deleted, even if it can be halted or paused. This makes smart contracts interesting both from a commercial and legal point of view: highly secure—if properly written—resistant to fraud and corruption [2].

Smart contracts can incorporate—or execute—legal binding agreements and they have a wide range of applications, from automating financial transactions to streamlining supply chain management. They are particularly well-suited for use cases that involve multiple parties and complex transactions, where traditional legal contracts can be slow, costly, and prone to errors. As blockchain technology continues to evolve, smart contracts are expected to become an increasingly important tool for businesses and organizations looking to streamline their operations and improve the efficiency of their transactions [3].

This chapter aims to explore smart contract from a legal point of view, trying in the process to reveal the complexity behind lines of code and their concrete implications. Smart contract, by itself, is not easily usable by the end user: to ensure true mass adoption it needs a graphical interface. This chapter, therefore, seeks to cover—in its first part—the aspects related to the back-end code, the actual smart contract, and then describe the front-end concept and its legal implications.

It is important to note that this chapter has been written with European regulations in mind. While contracts are a universal concept, and what follow can be easily applied worldwide, the legal framework surrounding them can vary significantly depending on the jurisdiction. As such, this work focuses on the legal principles and practices that are relevant to contracts within the European context, and mostly with reference to the Italian one. It is important to always check with a legal advisor specific to each jurisdiction to ensure that the legal requirements and implications of any contracts are fully understood.

The chapter is organized as follows. Section 2 presents the necessary background on contract automation and some legal aspects. Section 3 analyzes the changes we have witnessed in past years in the contract automation evolution, while Sect. 4 introduces some contracts standard. Section 5 analyzes the importance of the graphic interface both from the usability and legal point of view. Section 6 concludes the work.

2 The Evolution of the Computer Contract: The Form and Expression of Will in the Digital Age

This section introduces the background that is necessary to understand the subsequent analysis. We start with the relevant background contract law Sect. 2.1, followed by a brief description of European Union Regulation and finishing with the technical implication of Decentralized Ledger Technology (DLT) and Smart Contract Sect. 2.2. This approach is useful to fully understand the concrete implication and usage of Smart Contract in a regulated environment. Those confident with EU contract regulation can skip this section and go directly on Sect. 3.

2.1 Legal Basic Knowledge

Generally speaking, a contract is a legally binding agreement between two or more parties that describes terms and conditions of their—typically economical—relationship. Its purpose is to create legal obligations between the involved parties: this means that if one party fails to meet their obligations under the contract itself, the other party can take legal action against a single Court to enforce the terms of the agreement and grant obligations to be fulfilled [4].

A contract can have verbal or written form, depending on the individual state's rules: usually, written form is mandatory for real estate or risky relationship, but a written is generally recommended to avoid disputes. In fact, contracts serve as a tool for risk management, as by defining the terms of an agreement in advance, parties can minimize the risks associated with a particular transaction: the choice about the contract form usually depends on the risk appetite of each party, and industry. Without contracts it would be difficult for businesses and individuals to engage in transactions with third parties with a good degree of confidence: economic activity would be severely limited [4].

Contracts are so important that they are used in a lot of different levels of our society: both in the physical and digital frameworks. For example, a construction company can enter into a contract with a supplier that specifies the price and delivery date of materials granting a correct project planning and avoidance of delays or cost overruns. Also, the digital world—a part from the code itself—is based on contracts: for example, a software company may include a warranty in their contract with a customer that specifies the level of performance that the software must meet. This approach helps protect the customer if the software fails to meet its stated performance standards. Typically, also end-users, during the purchase of a service or an application, bind themself to that specific service signing—or accepting—specific Terms and Conditions.

From the regulatory point of view, the European legislator has intervened in several aspects of contracting, electronic and analog:

1. Regulation (EU) No. 910/2014 (eIDAS): This Regulation defines a legal framework for electronic identification/authentication means and trusted services for electronic transactions in the European Union. eIDAS sets conditions and requirements for the issuance of electronic signatures, seals, time stamps, delivery service, and web browser authentication. It poses the ground for secure electronic contracts in the digital single market.

2. Directive 2000/31/EC (E-Commerce Directive): This Directive gives a legal basis for electronic commerce in the European Union. It sets rules regarding the provision of information society services, including the legal validity of contracts concluded through electronic means.
3. Regulation (EU) 2016/679, the General Data Protection Regulation (GDPR): This Regulation governs the protection of personal data in the European Union. The GDPR has a significant impact on digital contracts as it establishes the basis for the collection, processing, and protection of personal data of EU citizens which are usually needed for the creation and conclusion of contracts.
4. Directive 2011/83/EU (Consumer Rights Directive): This Directive harmonizes consumer rights in the European Union and sets rules on pre-contractual information, withdrawal rights, and consumer protection in distance or off-premises contracts.
5. Regulation (EU) 2019/1151 (Platform-to-Business Regulation): This Regulation provides rules for the provision of online intermediation services in the EU, including contracts concluded through digital platforms. It aims to promote consumer trust in online transactions and ensure fair competition conditions in the digital market.

2.2 *Contract Evolution*

There are three conditions that have made possible the slow evolution from the analog contract to the electronic contract: in fact, European and national legislator did not immediately accept the moving from papers to bits as contract support. It has been a long and winding journey. In fact, the definition of contract can be found in different legal sources, starting with The Principles of European Contract Law and going through specific and each Member State legislation: different levels of legislation, doctrine and jurisprudence have stratified over the years. Nowadays, we can find a specific definition of electronic document in art. 3(1)(35) of the eIDAS Regulation, as well as the techno-legal aspects of the electronic signature, references to which are scattered among technical standards, national legislation, and European legislation. The analysis that follows will be all the more interesting when the reader's purpose is to study the validity of a contract for which the legal system—or common sense—requires a certain degree of formality: increasingly essential user interfaces and the use of simple devices for complex transactions often make us forget how much technical and legal complexity lies behind a single "click" [5].

As is well known, the general principle of freedom of contractual forms applies in most of the European legal system: unless otherwise provided, an agreement can be concluded (and proven) verbally. This approach is, of course, discouraged in all those cases where the obligatory relationship is particularly risky or if one wants to take certain precautions in protecting one's interests: concretely, producing a piece of paper (or a file) in Court is easier—and cheaper—than conducting testimonial evidence for each executed agreement or transaction. For this reason, a large part of contracts—even when there is no rule that imposes certain formal constraints on the parties—are concluded in writing. In the physical world, the written form mostly consists of drawing up on a physical medium the terms of the agreement and having the parties sign them: simplifying, a private writing consists of a sheet of paper containing the mutual obligations and terms of the agreement with the handwritten signatures of the contracting parties at the bottom. In addition, in our legal system there are special "enhanced" formalities-for example, the public deed or notarized private writing-necessary for acts to which the legislature has recognized certain criticality, such as donations or real estate transfers. In the face of this formal complexity, the legislature also provided for certain instruments that could reproduce in the electronic world the guarantees of security and legal certainty present in the physical world: with the electronic document and electronic signatures, in fact, almost all acts that must-or can-be in writing in the physical world find their correspondent in the digital world. The exceptions to this principle are very limited: for example, the "holographic will" cannot be drafted by computer, given the regulatory constraint that it must also be entirely drafted by the testator in his or her own hand. It is not of interest here to go over the evidentiary legal value of the different electronic signatures provided for by the national legislation and by the eIDAS Regulation. It seems important, however, for the reader to be reminded of how far the evolution of the digital contract has come: the legal system is going through a phase in which it is permitted by the legislature and accepted by jurisprudence that an electronically signed computer document—in compliance with the relevant technical and legal regulations—is valid for all legal purposes.

Having resolved the formal issue, it is now necessary to address the one relating to the expression of the will of the parties. As is well known, the understanding between the parties is the first and foremost among the elements that make up the very definition of a contract: to grasp its true value, this element must be traced back to its capacity to constitute, regulate, or extinguish a legal relationship. Indeed, the propensity of the contractual parties is to express a specific will: the understanding of a particular agreement is the common will to affect their—legal and patrimonial—relations. In the physical context, the "expression of will" can be made expressly—for example, through signature—or by concluding facts, allowed in some national legislation. In the digital landscape, the modes of expression of will do not change: the parties may expend their electronic signatures or conclude contracts—where permitted by law—with implied acceptances. For years jurisprudence and doctrine have debated the validity of the expression of consent in this context, but today market and—most importantly—regulatory developments recognize legal validity to the most varied ways of accepting contractual terms [4]. The evolution of the form of the electronic contract, together with that of the expression of will in the digital context, has been going on for almost thirty years and can be divided into three phases: Instrumental, Telematica, and Autómata.

In the mid-1990s, we witness a first phase of computerization of legal relations in which the main actor is the evolution of the medium on which the terms of the contract are crystallized. We begin, in fact, to draft contracts with electronic tools and then print them out and sign them with a handwritten signature. The epiphany of contract computerization sees technology reduced to a mere writing and printing tool: at this stage—what we can call "Instrumental"—the autonomy and self-sufficiency of the contract are reduced to a minimum. The contractual relationship-including its execution-depends entirely on human presence and interaction. A contract understood in these terms, in fact, can be drafted, modified, implemented, or extinguished only with the active and timely intervention of a subsequent interaction with the contractual object: contractual terms can only be integrated through the repetition of the procedure that generated them. At this stage, the form of the contract does not change, and the will of the parties is expressed by the affixing of a handwritten signature on a paper medium. Toward the end of the millennium, with the issuance of Directive 1999/93/EC on the subject of electronic signatures, we witness the beginning of a new phase: although the practical consequences of this rule will delay at least a decade, in this period European and Member States legislators write the rules so that the analog document is entirely abandoned in favor of the electronic document. Thanks to the rapid evolution of e-commerce, we are not limited to a mere transposition of physical content to a digital medium, but we try to replace the from file produced by a text editor with an HTML text (which stands for Hyper Text Markup Language, one of the longest-lived languages of the web), and then in a real web page. This second phase witnesses the slow emancipation from a paper-based conception of documents and the creation of new contractual arrangements that guarantee the parties the proper expression of their will. According to eIDAS Regulation, contractors can affix an electronic signature—simple, advanced, qualified/digital—to electronic documents: the introduction of a formally strong instrument such as the digital signature, the use of which is justified in those contracts that require written form (think, for example, of the opening of a bank account), while in other areas—closely related to e-commerce—the checking of a box with a tick is sufficient (think, for example, of online purchases). At this stage, the qualities of a handwritten signature begin to be reproduced digitally, allowing even qualified individuals such as notaries, magistrates, and lawyers to eliminate printing from their document production process. Here we can observe an important paradigm shift in electronic contracting: while it is true that the contract in the strict sense is still formalized in human readable characters (regardless of the medium, physical, or electronic), much of the conclusion and execution—in the broadest sense—of the agreement is entrusted to software, i.e., machine readable instructions. When a user buys online, in fact, he agrees to contractual conditions that are—potentially—clearly visible (either directly on the web page or on a PDF file that is made available to the party), but the contract is formalized with a click: once the consent to proceed is given, back-end processes for payment and order finalization are activated that are not controllable by the buyer and, in some cases, not even (directly) by the seller. These processes are mostly entrusted to scripts: simplifying, a script is a computer program written in a computer code that is indirectly machine readable and allows impressive automation. E-commerce platforms communicate with each other and with those responsible for logistics and dispatch almost in unison with the newly concluded contract between service provider and buyer. In this sense—

compared to the Instrumental phase—the automation coefficient of this evolutionary phase, which we can call "Telematica," increases considerably. However, the level of autonomy of the contract still remains limited: withdrawal, modification, and return of an order still depend entirely on the parties, through e-mails or request forms where the user interfaces with operators. Moreover, although the conclusion of the contract and its own execution are closely related, they are still inherently distinct: the expression of will can be simplified—reduced to the enhancement of a check box or the pressing of a button—and the medium has become almost irrelevant. The content of the contract is mostly crystallized on the web page or a PDF document intended for human reading only, and the execution of the contract itself depends on manual—even if in some cases highly automated—intervention. In this second phase, the contract form is fully electronic; the signed document consists of electronic elements, and the expression of will is embodied in the affixing of an electronic signature on a document that is fully intelligible to the parties: the signing instrument changes, but the contract is only human readable.

The natural evolution of this approach is the merging of the two contractual moments that remain separate in the Telematica phase: contract formation and its execution. In fact, the innovation brought by Decentralized Ledger Technology (DLT) [3], and the entry of the binary code into the Civil Code, are paving the way for an extreme contract automation, given an intrinsic connection between the terms of the agreement and their execution. In this sense, we can speak of a third phase—calling it the "Autómata" phase—where we are witnessing a twofold paradigm shift: on the one hand, a content shift, where the language used in the contract is no longer (only) human readable but also becomes machine readable, and on the other hand—perhaps the greatest innovation—this language shift makes the execution of the contract immediately enforceable, opening up a number of scenarios that need in-depth legal analysis. These innovations are particularly relevant: the first paves the way for interesting scenarios of human-machine or even machine-only contractual interaction, while the second makes it possible to imagine greater reliability of contractual transactions, provided that the underlying computer code is technologically and legally sound. It becomes, then, clear the paradigm shift from the previous two phases: although the contractual tools remain unchanged—the document is digitized and the signature is electronic—the parties find themselves reading a new and mostly unfamiliar language. This approach may raise some critical issues: the pursuit of contractual automation risks weakening the actual fairness of the expression of will by contracting parties who do not understand the computer code outlining their contract, leading to pathological consequences of voidability or even nullity of the contract. The following section briefly analyzes the enabling object of this third phase: the Smart Contract.

3 Smart Contract and Smart Legal Contract: A New Contract Paradigm

This section describes on a high level how we moved from "simple" software to Smart Contract (Sect. 3.2), and then why this becomes relevant from a legal point of view (Sect. 2.2). Last section gives a glance of how a Smart Contract comes to life: this is relevant to fully understand each step and relevant touch point in the Smart Contract life-cycle (Sect. 3.3).

3.1 From Software to Smart Contract

The diffusion of blockchain and Decentralized Ledgers Technology has introduced new technological concepts [3] that in some cases have attracted the attention of jurists: in the context of IT contracting, the term that is certainly most interesting is "smart contract." The word itself, however, runs the risk of being misleading: therefore, it seems useful to make a brief historical, lexical, and content analysis of it with the aim of circumscribing how much this instrument can be considered "smart" and how much it can be considered a "contract." From an historical perspective, smart contracts appeared in the literature in 1996, when Nick Szabo-a computer scientist and cryptographer-introduced [6]. the concept of expressing promises between parties in electronic form. Szabo put forth a number of technological ideas to enable what we have called above the Autómata phase of contract computerization. The word "contract" is used with Szabo's knowledge, to effectively denote a computerized tool capable of managing obligatory relationships between parties. However, in recent years the term has been used generically by market participants and the literature in the field: care should, then, be taken to keep of smart contract and smart legal contract distinct [7]. From a formal point of view, in fact, in its contemporary meaning the smart contract is nothing more than a computer program (or "script"): that is, a set of strings of code that, given a certain input, process it following certain instructions to return an output. This aspect has some non-trivial substantive implications, including the fact that a smart contract cannot always be said to be a contract understood in the purely legal sense of the term. On the one hand, in fact, the concept of contract is exhaustively defined by each Member State Civil Code (or equivalent), accompanied by a series of legal implications that elude the simple computer code. On the other hand, a script as such has its own dignity and autonomy as code, without any need to become—also—a contract: a set of instructions that—as in the Telematica phase explained above—can accompany a legal transaction without, however, being part of the very contractual conditions that characterize it. To enrich the code can, in addition, be invoked trusted external elements—so called Oracles—that communicate the coming true of a certain event [7].

The definition component is fundamental and reflects the paradigm shift that is taking place: if smart contract were not intended to flank or replace the current conception of contracts, there would have been no need to coin a dedicated semantics for an object—the script—that technologically already permeates our society, including legal society. In order to give the words the importance they deserve, it seems necessary today to distinguish the subset of scripts that can be considered real contracts from those that are merely software tools, bringing back in words dear to jurists the concepts expressed in the past about such a complex object: in the first case we will have smart legal contracts, while in the second we can continue to speak of smart contracts, while being aware that in the everyday use of the term to the difference can be really subtle. As for its assumed intelligence, a smart contract merely receives instructions and executes them: the code contains—or should contain, even through interaction with other scripts -all the possible solutions of a given sequence of events, not being able to invent new content. In this sense, then, one cannot speak of an object evolving or even thinking: there is no intelligence guiding the choices or consequences of an input, other than that of the programmer/jurist who wrote the instruction sequence. The reader, then, should not be misled by the words: the smart aspect of the contract is simply semantic, and while it may be useful in giving the idea of an important evolution in computer contracting, it should not lead one to think that the contract writes and executes itself. This type of tool has been called "smart" by Szabo simply because of its digital and ameliorative nature, compared to analog documents: for the author, the automation of certain clauses and—consequently—transactions can only increase the ease of relations between the parties [8].

3.2 From Smart Contract to Smart Legal Contract

The above brief analysis shows that, although the name is commercially appealing, the scripts in question rather than being "smart" seem to be "dumb," and-most importantly-only in some cases can they be considered "contract" [7]. With this remark, one certainly does not want to diminish the role of contract automation, which makes this tool extremely interesting: the reader simply wants to take note that the code we are dealing with does not think autonomously and is absolutely unable—for now—to transform a piece of information into knowledge. In this sense it can be said that there is no "thinking" of the contract (or code) but only a mere execution of given and predetermined instructions. These considerations make clear to the reader the extreme delicacy of the role of the person—hopefully a jurist—who drafts a smart legal contract: the contractual clauses represented by the code are not validated by some superior artificial intelligence but are self-regulating according to the instructions imposed by the person who wrote them. Until the rise Telematica phase, the pathological stage of the contract is managed by the parties, or by the Judge, according to well-known principles of contractual interpretation, non-performance, and sanction. In the Autómata phase, in contrast, contractual evolution permits that the clauses become an integral part of the instrument—the computer code—that is used to execute them, making the entire agreement directly enforceable: there is no longer any room for either interpretation or non-performance. These features, however, may not be entirely consistent with a complex legal system such as the one in which we find ourselves operating: while eliminating non-performance from the contractual cycle may have positive effects for the successful completion of transactions, it requires scrupulous regulatory intervention to revise a system that rests—also—on the opportunity of one of the parties not to perform its obligation.

To give a degree of concreteness to what has been written so far, Fig. 1 is an example of a smart contract, written in the Solidity language, a syntax used by the Ethereum platform [1] Ethereum—unlike Bitcoin, which is primarily a cryptocurrency exchange network—is designed to be a decentralized infrastructure on which to run complex programs such as smart contracts. Except for the parts following the "//" symbols, the text is read and compiled by the machine:

The above contract allows a pizza and unicorn purchase. It states that the price of the pizza is 0.1 Ether (the currency on the Ethereum blockchain) and that the seller is the creator of the contract, which means only him can perform some action in according to the contract itself (such as withdraw of the Ether stored in it). The buyer can purchase the pizza by calling the *buyPizza()* function and transferring the set price. Once the purchase has been confirmed, the funds are held in the contract until the seller calls the *withdraw()* function, thus transferring the funds to his address. If the buyer wishes to request a refund, he can do so by calling the *refund()* function, which will transfer the pizza price back to the buyer. When the buyer buys the pizza and unicorn by calling the *buyPizza()* function, the buyer's name is concatenated with "'s Unicorn" to create the new unicorn name: this grants some kind of customization of the item. If the buyer requests a refund by calling the *refund()* function, the unicorn's name is restored to its original and general value "Unicorn."

Given the extreme complexity of platform operation and contract programming, development, and deployment, it is not possible here to dwell on the technological details. However, it is important to note that while not traditionally defined as a contract, this computer program contains the instructions and executes what would traditionally be a complex contract: think of the disbursement and receipt of funds, in an apparent total absence of cause. To make this code relevant for contractual purposes, certain clauses need to be rephrased or added: such as indications of the parties, time of execution, eventual restitution of value with interest. This is not an oversight, as it is simply an embryonic form of contract. The legal practitioner's choices, when faced with such an object, are twofold: use it as is, to accompany a broader contractual set, written perhaps in traditional language; or enrich the code with all those details that may allow a contract to be configured under the related legal framework. For now, it seems that the first of these two approaches prevails.

From a strictly legal point of view, the inherent autonomy of this type of instrument deserves special analysis. Assuming that the degree of autonomy of the smart contract is directly proportional to the degree of decentralization of the

```
pragma solidity ^0.8.0;

// Contract definition
contract PizzaUnicorn {

    address payable public seller; // Seller address
    address payable public buyer; // Buyer address
    uint public price; // Prezzo della pizza in wei
    string public unicornName; // Unicorn name
    string public unicornColor; // Unicorn color

    // Log events
    event PurchaseConfirmed(address buyer, uint amount);
    event Refund(address buyer, uint amount);

    // Contract constructor
    constructor(string memory _unicornName, string memory
    _unicornColor) {
        seller = payable(msg.sender);
        price = 100000000000000000; // 0,1 ether
        unicornName = _unicornName;
        unicornColor = _unicornColor;
    }

    // Function for buying pizza and unicorn
    function buyPizza(string memory _buyerName) public payable {
        require(msg.value == price, "Pizza and unicorn price
        is 0,1 ether.");
        buyer = payable(msg.sender);
        emit PurchaseConfirmed(buyer, price);
        // Add buyer's name to the unicorn
        unicornName = string(abi.encodePacked(_buyerName,
        "'s Unicorn"));
    }

    // Refund function
    function refund() public {
        require(msg.sender == buyer, "Only yhe buyer
        can ask for a refund.");
        payable(msg.sender).transfer(price);
        emit Refund(buyer, price);
        buyer = payable(address(0));
        // Reset unicorn name
        unicornName = "Unicorn";
    }

    // Transfer funds to the Seller
    function withdraw() public {
        require(msg.sender == seller, "Only the Seller
        can withdraw funds.");
        seller.transfer(address(this).balance);
    }
}
```

Fig. 1 Smart contract code

platform on which it is written and used, we can briefly analyze how autonomous a smart legal contract is on a public and permissionless DLT such as Ethereum's (properly a Blockchain). The word "autonomy" comes from the Greek, "self-nomos," meaning to give oneself the laws and rules one follows in acting. This characteristic can be found exactly in the scripts we are talking about, especially with regard to the "own laws": the autonomy we are talking about, in fact, allows the algorithm—once started—to be independent of the agent who writes or starts it. Subject to the need for external comfort, given by the Oracles, the script (if well written) contains all the information necessary for its own operation and can operate in total autonomy, either as a single one or as part of an interoperable smart contract network. It seems that this feature allows the creation of autonomous orders, which are born, operate, execute, and terminate within an ecosystem which is specifically written and designed for particular needs of the parties. The automatic enforcement of covenants shifts the intervention of the Judge to a moment—eventual and—later: there is no longer a default, but only the possibility of acting in pristine or with an action for unjust enrichment, if a party has benefited from a right that was not his or her due. In this sense the (computer) code becomes law, at least between the parties. It is true that this tool could be used to follow up on illegal transactions (as in the school case of contracting a murderer), however, should this happen one would step out of the legal groove of the smart legal contract and fall into the broader category of the mere script of code executing a (in the example illegal) transaction. In such a case, the contract itself would be null and void, and-if applicable-its perpetrators prosecutable in criminal court: however, it comes hard to imagine a legal instrument that could stem the (technological) advance of the script executed in a permissionless and public context. In this sense, the guardianship of legal practitioners becomes crucial: the correctness of the (computer) code seems to be the dividing line between what can be considered a valid contract and not a mere smart contract [3].

3.3 Smart Contract Life-Cycle

Smart contracts, as pieces of code, need to be written, implemented, and deployed in order to be used.

Typically a smart contract is written in the Solidity language: this is the language used by one of the most common and resilient chains—the Ethereum one—and compatible with any platform that leverages the Ethereum Virtual Machine (EVM). Like any code, the script can be written in a text editor, taking advantage of complex code editors-especially useful for debugging and formatting text-or with online tools, such as Remix, Ethereum's Integrated Developer Environment (IDE).

Once the code is written, it is typically reviewed by one or more subjects: it may be the same programmer who wrote it in the first instance, or a third party. Typically the "third party" route is taken in the case of smart contracts that are complex or that will settle transactions with large volumes of money: in this case we rely on companies whose core business is precisely to audit smart contract code.

Once the code has been audited and secured, the contract can be compiled and deployed: the first action allow to generate to important pieces of code–ABI and bytecode—which are relevant for smart contract interaction. The second action—deployment—is the formal publication of the contract in the chosen blockchain: from now on, the contract is live and can be called—and executed—by anyone, according to its own rules [1].

4 Smart Contract Standards

This section describes a peculiar aspect of smart contracts on the Ethereum blockchain: standards. Generally speaking, the importance of—social, legal, or economic—standards is recognized in any discipline: even more, in a technical context such as software, reaching agreement on common rules is critical to the mass adoption of a given technology. We start with a quick review of what standard contracts are in the Ethereum environment (Sect. 4.1), to then move forward to some practical examples (Sect. 4.2).

4.1 Importance of Standards

It seems to the writer that one of the key factors contributing to the growth and interoperability of Ethereum based applications, based on the above described smart contracts, is the creation of standards known as "ERC," which stands for "Ethereum Request for Comments." As in other industries, the need for standards arose from the rapid expansion of the Ethereum platform and the increasing complexity of smart contract development: as more developers and projects entered the space, it emerged that a common set of rules, code, and interfaces was necessary to ensure compatibility, granting interoperability, and facilitate easy integration between different applications and tokens.

The Ethereum community immediately understood the importance of such approach, which allows developers to create interoperable and easily understandable smart contracts ecosystem: the gencsis of ERC standards can be traced back to the Ethereum collaborative and open-source nature. In fact, ERC generated from Ethereum Improvement Proposals (EIPs), which were introduced specifically as a means for community members to propose changes and improvements to the Ethereum itself. ERC standards are a specific type of EIPs that focus specifically on defining common interfaces and functionalities for smart contracts: they provide developers with a foundation on which to build their applications, reducing complexity and lowering the barriers to entry for newcomers. The compliance to a common set of standards allows developers to create smart contracts and tokens that seamlessly interact with existing tools and interfaces, including wallets, exchanges, and other contracts. Also, using a standard compliant library fastens the writing and deployment of the smart contract and avoids errors in the writing/deployment process. Furthermore, they provide benefits beyond technical compatibility; they contribute to the security and trustworthiness of the Ethereum ecosystem by establishing best practices and common patterns for contract design.

With well-defined standards, developers can more easily audit, review, and verify smart contracts, reducing the risk of vulnerabilities and promoting the adoption of secure coding practices. These standards have become widely recognized and adopted, serving as a blueprint for other blockchain platforms to develop their own token and smart contract standards. In the next section we are going to explore some of the main standards, but it is important to bear in mind that the success of ERC-20, in particular, has led to its emulation on various other blockchain platforms, further fostering interoperability and cross-platform compatibility [1].

4.2 *Example of Useful Standards*

As said in previous sections, smart contracts on the Ethereum blockchain have gained significant popularity, enabling the creation of decentralized applications (DApps) [3] and facilitating the development of the blockchain ecosystem. To promote interoperability and best practices, the Ethereum community has established a set of standards known as ERC (Ethereum Request for Comments) standards.

Here follows some examples:

1. ERC-20, fungible tokens: ERC-20 tokens are interchangeable, meaning one token is equal to another token of the same type. Common functions of these tokens are *balanceOf()*, *transfer()*, and *approve()*, allowing for token transfers and management. This contract supports optional metadata, such as token name, symbol, and decimals, providing additional information about the token [9].
2. ERC-721, Non-Fungible Tokens (NFTs): Each token represents a unique asset, with distinct properties and values. Ownership and transfer: ERC-721 specifies functions like *ownerOf()* and *transferFrom()* to manage ownership and transfer of NFTs. Contracts can store metadata and maintain enumerable lists of NFTs, enabling easy exploration and discovery: metadata are the soft spot of this standard: depending on where they are stored the asset is more or less secure. Typically, the more decentralized is the storage, the more secure is the asset, because it is not prone to modification or deletion [10].
3. ERC-1155, Fungible and non-fungible tokens: ERC-1155 supports both types of tokens within a single contract, allowing for efficient management of diverse token assets. Multiple tokens can be transferred in a single transaction, reducing gas costs and improving scalability, using batch transfers. ERC-1155 introduces a flexible identifier structure, enabling contracts to manage a wide range of token types and instances, thanks to Flexible token IDs [11].

4. ERC-223 introduces a *tokenFallback()* function that prevents accidental loss of tokens by rejecting incompatible transfers, enhancing security in the asset. Gas efficiency: ERC-223 reduces gas costs by eliminating the need for multiple transactions when transferring tokens to contracts. ERC-223 is fully backward-compatible with ERC-20, enabling easy integration with existing token implementations, granting backward compatibility between assets [12].
5. ERC-777 introduces new capabilities, such as hooks for custom token behaviors, improved token control mechanisms, and operators for token management. ERC-777 includes measures to prevent accidental token transfers and enables contract to contract transactions with reduced risks, adding more secure functions. Also, this token is designed to be backward-compatible with ERC-20, ensuring seamless integration with existing ERC-20 token infrastructure, like most of the standards [13].
6. ERC-827 extends ERC-20 by allowing users to attach additional data or instructions when approving token transfers, with specific data transfer approval functions. The additional data feature of ERC-827 enables more complex interactions between contracts, reducing the need for multiple transactions, and of course also this standard allows Interoperability, as its tokens can be easily integrated into existing ERC-20 ecosystems without significant modifications [14].
7. ERC-1337 was specifically developed for the CryptoKitties game [15], defining the interface for managing unique virtual assets like digital collectible cats: this is why it is known as "CryptoKitties standard." It introduces functions such as breed, feed, and sell for managing the unique properties and behaviors of CryptoKitties, and many other projects after. ERC-1337 allows for interoperability between different games or applications built on the standard, facilitating the exchange and use of virtual assets [16].
8. ERC-621 extends ERC-20 to enable token contracts to increase or decrease the total token supply after deployment. The ability to adjust token supplies provides flexibility for scenarios where token minting or burning is required, enabling a dynamic token supply process. Token migration: ERC-621 can be useful when migrating from one token contract to another, allowing for the transfer of existing token balances [17].
9. ERC-6551 is a token standard that was launched on the Ethereum mainnet on May 7th, 2023. This standard aims to enhance the capabilities of ERC-721 NFTs by allowing them to function as their own smart contract accounts. This provides NFTs with increased versatility and functionality. The motivation behind the development of ERC-6551 stems from the growing use of NFTs as a form of on-chain identity. Previously, ERC-721 tokens could not act as agents or own other on-chain assets, which did not align with real-world use cases for non-fungible assets. ERC-6551 addresses this by granting each ERC-721 token the full capabilities of an Ethereum account. This is achieved by deploying unique smart contract accounts, known as "token bound accounts," for each ERC-721 token through a permissionless registry. One interesting aspect of ERC-6551 is that it allows a single NFT to own multiple token bound accounts. This

means that an NFT can easily contain multiple wallets and other resources. The system introduced by ERC-6551 primarily consists of a permissionless registry for deploying token bound accounts and a standard interface for interacting with these accounts. With the implementation of ERC-6551, a single NFT can leverage the full functionalities of a traditional Ethereum wallet. For example, an NFT can store other NFTs or trade ERC-20 tokens: this feature opens up a wide range of possibilities and use cases, such as storing virtual items within an NFT or creating NFT collections that can distribute rewards to NFT holders through their token bound accounts or create NFT game characters that hold—and own—objects in form of NFTs [18].

5 Front-End Necessity and Legal Implications

Once the code of the smart contract has been correctly deployed to the desired decentralized network, a new issue arises: the need for a front-end application which should let users interact with the Smart Contract and execute the operations defined therein in an easy and immediate way.

A front-end application may be considered as a friendly interface between the user and the Smart Contract code. The user may use well-known web components such as buttons, input fields, forms in order to execute Smart Contract functions. From a technical point of view the front-end application is composed by two layers:

- A presentational layer, where the user can view the result of an operation and/or order a new operation.
- A logical layer, where a plenty of functions aimed at data transfer from and to the Smart Contract are defined by the developers.

As we already know the Smart Contract defines some functions. For example, *balanceOf(Address address)*, given an address of an account as input, will return the amount of tokens available for operations. The logical layer of the front-end application discussed above will offer, for example, a function *getBalance(address)* may be used in an event listener applied over an html element (such as a button labeled as "get my balance") in order to return the amount of the current user. Such function, in order to work, shall invoke helpers methods provided by third-party libraries which encapsulate the necessary logic to send requests to the smart contract running on a blockchain.

As an example, we may consider the function in Fig. 2:

This asynchronous function shall perform the following tasks:

1. Read data from function parameters (contract and account). Contract is an instance of a particular object provided by a third-party library (web3.js) which abstracts the logic and the concept of a smart contract.

```
export async function getBalance(contract, account){
/* balance contains the amount of tokens available in the account */
let balance = await contract.methods.balanceOf(account).call();
        return balance
}
```

Fig. 2 Balance function

```
<template>
<button class="button" @click="getMyBalance"> Get Balance </button>
</template>

<script>
    /* we assume that web3.js has been correctly injected into the page
    import { getBalance } from '../helpers/actions.js'
    export default {
        props:['contract','account'],
        methods: {
            getMyBalance(){
                return getBalance(this.contract, this.account)
            }
        }
    }
</script>
```

Fig. 3 HTML code

2. Set a variable named balance which will contain the value returned from another function.
3. The method *balanceOf* of the smart contract provided is called and the *account* is used as function parameter. The meaning of the function call is: *get the balance of the account.*
4. The value returned from the smart contract is then saved into the variable.
5. The value of balance is then returned to the calling function (for example, an event listener defined in the HTML part of the front-end application).

In order to ease clarity of this section, we may draft an HTML component which will fire the action of retrieving the balance of a given account, as per Fig. 3:

In this simple HTML component, which uses the well-known framework *Vue.js*, we defined a method *getMyBalance* which, under the hood, invokes the function *getBalance* (which may interact with the smart contract) with two parameters which are bounded to that component (contract and account).

As a result, whoever wants to write and use a smart contract in a real case scenario has to clearly design a human friendly interface which lets the user interact with the blockchain and, in particular, with smart contract functions.

6 Conclusion

Today, we live in a fast-paced digital world: technology such as blockchain can enable the secure and traceable exchange of value in the global Internet. One of the tools that—leaning on the decentralized ledger—can enable much of these use cases is smart contracts. From a technical point of view, a smart contract is simple code, written and compiled like any software. From a legal point of view, it is interesting to see how this object represents the state of the art of what is called "contract automation."

In this chapter, we have tried to describe, on the one hand, the smart contract as a technical object, and—on the other hand—we wanted to trace the evolution of electronic contract. An evolution that began with changes in the medium—from paper to bit—and has moved in the direction of automation, which is increasingly radical. This has led us to identify three phases: an Instrumental one, where information technology plays a very limited role; a Telematica one, where digital communications and the Web take a preponderant role; and an Autómata one, where contractual automation takes over. At this stage, the front-end with the user becomes critical: the interface is as important for the proper functioning-and use-of smart contracts as it is for ensuring compliance with the different regulations governing contractual relationships in the digital world.

To date, therefore, it is important for both the jurist and the entrepreneur to fully understand the mechanisms of smart contracts in order to build use cases that are secure and usable by the market.

References

1. A.M. Antonopulus, G. Wood, *Mastering Ethereum* (O'REILLY, Springfield, 2018)
2. W.-M. Lee, *Beginning Ethereum Smart Contracts Programming: With Examples in Python, Solidity, and JavaScript* (Apress, New York, 2019)
3. M. Swan, *Blockchain, Blueprint for a New Economy* (O'REILLY, Springfield, 2019)
4. A. Torrente, P. Schlesinger, *Manuale di diritto privato* (Giuffrè Francis Lefebvre, Milano, 2019)
5. G. Finocchiaro, F. Delfini, *Diritto dell'informatica* (UTET, Torino, 2014).
6. *Nick Szabo.* (1996) https://www.fon.hum.uva.nl/rob/Courses/InformationInSpeech/CDROM/Literature/LOTwinterschool2006/szabo.best.vwh.net/smart_contracts_2.html
7. P. De Filippi, A. Wright, *Blockchain and the Law, the Rule of Code* (Harvard University Press, Cambridge, 2019)
8. M. Finck, *Blockchain Regulation and Governance in Europe* (Cambridge University Press, Cambridge, 2019)
9. F. Vogelsteller, V. Buterin, *ERC20* (2015). Available at https://eips.ethereum.org/EIPS/eip-20
10. W. Entriken, D. Shirley, J. Evans, N. Sachs, *ERC721* (2018). Available at https://eips.ethereum.org/EIPS/eip-721
11. W. Radomski, A. Cooke, P. Castonguay, J. Therien, E. Binet, Ronan, *ERC1155* (2018). Available at https://eips.ethereum.org/EIPS/eip-1155
12. Dexaran, *ERC223* (2017). Available at https://eips.ethereum.org/EIPS/eip-223

13. J. Dafflon, J. Baylina, T. Shababi, *ERC777* (2017). Available at https://eips.ethereum.org/EIPS/eip-777
14. A. Lemble, *ERC827* (2018). Available at https://github.com/ethereum/eips/issues/827
15. *Cryptokitties*. Available at https://www.cryptokitties.co/
16. A. LembleKevin Owocki, A. Redden, S. Burke, K. Sea-graves, L. Kacil, Š. Šimec, P. Kosiński, A. Raj, J. Griffin, N. Creswell, *ERC1337* (2018). Available at https://eips.ethereum.org/EIPS/eip-1337
17. A. Miller, *ERC621* (2017). Available at https://github.com/ethereum/EIPs/pull/621
18. D. Downs, R. Huynh, A. Lam, W. Chung, P. Sulli-van, A. Macmillan, J.-F. Schwarz, A. Bukov, M. Melnik, J.W.J. Windle, B. Giang, S. Jang, R. Montgomery, *ERC6551* (2023). Available at https://eips.ethereum.org/EIPS/eip-6551

Designing Future Sustainable Cryptocurrencies: Principles and Expectations

Behzad Esmaeilian, Mark Jamison, Joseph Sarkis, and Sara Behdad

Abstract This book chapter introduces the principles of designing cryptocurrencies and outlines the key characteristics expected from future currencies. The chapter provides an overview of the foundational components of cryptocurrency networks and emphasizes scalability, security, and sustainability as pivotal characteristics for the next-generation currencies. It begins by introducing the fundamental elements involved in designing cryptocurrency networks, which include hash functions, data structure, and digital signatures. It further explains the primary processes necessary for achieving decentralization, and the procedure of mining and verifying transactions. In addition, the chapter describes the environmental sustainability aspects of crypto networks, with a specific focus on three key areas: (1) energy consumption, (2) electronic waste generation, and (3) opportunities for sustainable practices through decentralized transactions. Finally, the chapter highlights the potential for sustainable practices and the social benefits that can be derived from future cryptocurrency technology.

Keywords Cryptocurrency Design · Sustainable Cryptocurrency · Blockchain · Mining · Environmental Sustainability · Digital Currencies · Social Benefits

---

B. Esmaeilian (✉)
Andrew F. Brimmer College of Business and Information Sciences, Tuskegee University, Auburn, AL, USA
e-mail: besmaeilian@tuskegee.edu

M. Jamison
Warrington College of Business, University of Florida, Gainesville, FL, USA
e-mail: mark.jamison@warrington.ufl.edu

J. Sarkis
Worcester Polytechnic Institute, Worcester, MA, USA
e-mail: jsarkis@wpi.edu

S. Behdad (✉)
Environmental Engineering Sciences, University of Florida, Gainesville, FL, USA
e-mail: sarabehdad@ufl.edu

N. El Madhoun et al. (eds.), *Blockchain and Smart-Contract Technologies for Innovative Applications*, https://doi.org/10.1007/978-3-031-50028-2_3

1 Introduction

The number of cryptocurrencies worldwide has surpassed 10,000 as of 2022 [1]. With the proliferation of cryptocurrencies and the growing significance of decentralized networks, it becomes crucial to comprehend the theoretical mechanisms underlying them and integrate sustainability principles into their design. Future crypto networks are anticipated to possess three key attributes: scalability, security, and sustainability. Currently, the open-source development community and practitioners have been at the forefront of cryptocurrency advancements, prioritizing the reduction of costs associated with electronic transaction systems [2]. Nevertheless, the design of robust cryptocurrencies necessitates collaborative efforts involving diverse disciplines such as computer scientists, system designers, finance experts, environmental engineers, economists, and social scientists. By bringing together expertise from multiple fields, the development of capable cryptocurrencies can be realized and more innovative and sustainable solutions in the digital currency landscape can be achieved.

The contribution of disciplines beyond computer science to digital currencies has been hindered by a general lack of understanding regarding the technical aspects of cryptocurrency design and engineering. This chapter aims to bridge this knowledge gap by providing an overview of the fundamental components of cryptocurrency networks, intended to familiarize newcomers with distributed networks, as well as sustainability and scalability concerns.

Figure 1 visually represents the technical elements covered in this chapter, although it is important to note that the discussion here is not exhaustive. Rather, it focuses on a representative selection of studies and adopts a thematic review approach to previous literature, providing a comprehensive yet concise overview.

To identify and discuss the available resources, a qualitative approach inspired by Templier and Pare [3] has been employed. This approach involves six main steps: formulating objectives, collecting relevant studies, screening the results, assessing their quality and relevance, extracting pertinent information, and finally summarizing and documenting the findings. Engineering Village, ScienceDirect, Scopus, and Web of Science were the primary databases used to search for the studies.

By addressing the technical aspects of cryptocurrency design and engineering and offering insights into sustainability and scalability concerns, this chapter seeks to facilitate interdisciplinary contributions to the field and foster a broader understanding of digital currencies beyond computer science.

The remainder of this chapter is organized as follows. Section 2 introduced the technology behind cryptocurrencies. Section 3 describes the essential elements needed for building cryptocurrency networks. Section 4 provides a comprehensive understanding of the process of mining and how to achieve decentralization in such distributed networks. Section 5 explores scalability, invulnerability, and sustainability as the three key characteristics expected from future currencies. Further, it examines the current limitations in estimating the environmental impacts of crypto

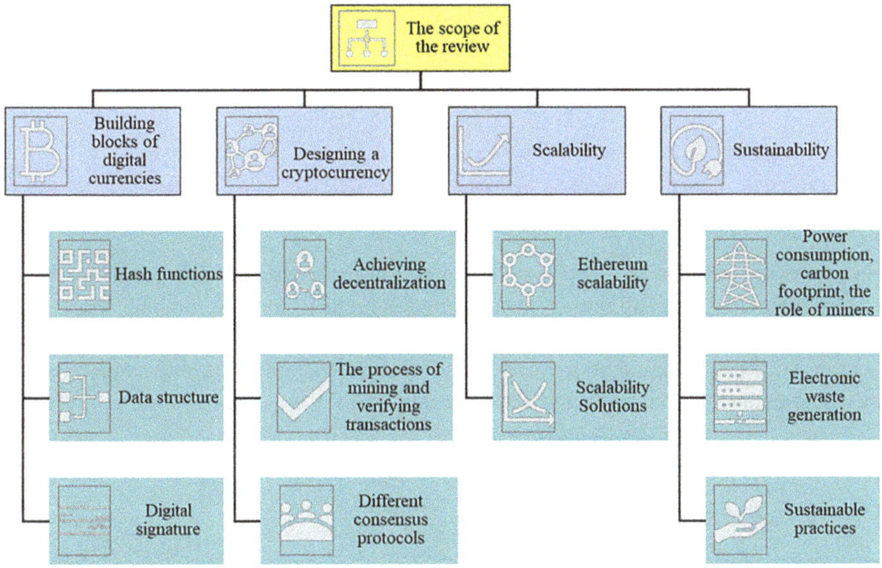

Fig. 1 The scope of the topics included in this study

networks and highlights opportunities for future sustainable practices and the social benefits they can bring.

2 The Technology Behind Digital Money

Cryptocurrencies are the newest forms of digital currencies designed to sustain trust through technology. The technology behind cryptocurrencies is known as distributed ledger technology or blockchain. Blockchain has emerged to enable safe and secure data sharing among entities that otherwise would not trust each other. The tighter integration among mutually untrusted entities creates transformative opportunities for various applications ranging from financial services to supply chains and healthcare.

Blockchain in its abstraction form is a shared, totally ordered, tamper-resistant log of transactions. The log is created via a set of rules written in computer code that is distributed across multiple computers (called computing nodes) and executed by them. Each node holds a copy of the entire ledger. Thus the blockchain infrastructure includes three main elements: (1) a set of rules or computer codes (known as protocol) that govern the way participants transact, (2) a decentralized network of users and computer nodes, and (3) a shared, tamper-resistant, totally ordered ledger that stores the transaction history. The tamper-resistant, distributed nature of blockchain helps make it suitable as a replacement for traditional, centralized financial systems where an entity such as a bank serves as a trusted third party for verifying and recording transactions. In future sections, we describe the elements needed to build such a decentralized distributed ledger.

3 Essential Components of Digital Currencies

This section provides an overview of the key components needed for a decentralized network capable of securely handling financial transactions. The three main elements include the hash function, data structure, and digital signatures (Fig. 2). For clarity, we will describe the concepts using the Bitcoin network as an example and the discussions provided by Narayanan et al. [4].

3.1 Hash Functions

The data stored in blockchain networks are encrypted using hash functions. The hash functions are necessary for ensuring that no one tampers with the data. Hash functions transform input data of varying sizes (like a text message) into a fixed size (such as 256 bits) called a hash value. Encryption converts data into a protected format from which it is nearly impossible to read the original data unless the recipient has the appropriate key.

Cryptocurrency hash functions should have the properties of collision resistance, hiding, and puzzle friendliness. Table 1 summarizes these properties, using for illustrative purposes a hash function H with input X and output $H(X)$. The hash function H is "collision-free" if different values X and Y always give different hash results. This means each $H(X)$ only corresponds to one X. Also, H is "hidden," which means figuring out the input from the output is extremely hard. Moreover, H is "puzzle-friendly" which makes it tough for anyone to pick an input that produces a specific output. In essence, there is no strategy much better than random guessing to solve this [4].

Hash functions are commonly used in cryptography algorithms to verify the authenticity and integrity of data and quickly retrieve and access the data. Hash functions are used in digital signatures and message authentication codes (MACs)

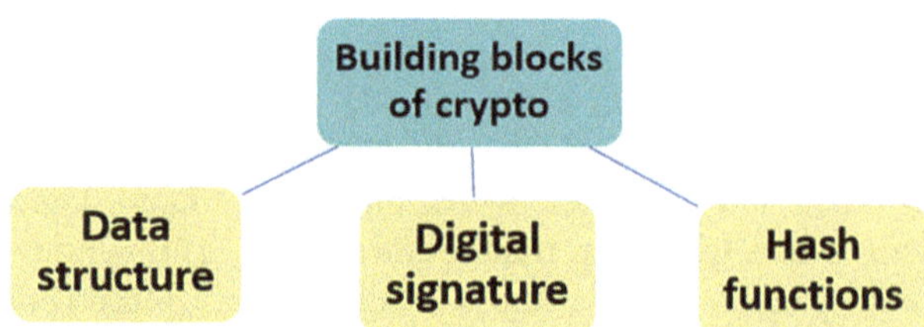

Fig. 2 The essential elements for building digital currencies

Table 1 Properties of hash functions used in crypto design

	Property	Description
1	Collision resistance	It is not possible to locate two different values, *X* and *Y*, for which $x \neq y$, but still have $H(x) = H(y)$
2	Hiding	If $y = H(x)$ represents the output of *H*, there exists no reasonably feasible method to determine the input *x*
3	Puzzle friendliness	For any given output value *y*, if *k* is selected from a distribution with substantial min-entropy, the task of finding an *x* within a reasonable timeframe such that $H(k\|x) = y$ becomes infeasible

Summarized from [4]

to confirm the authenticity of messages. Different types of hash functions exist. For example, Bitcoin uses a hash function named SHA-256.

3.2 Data Structure

In addition to a hash function, we need a data structure that can help identify where the data is located and stored. To design a crypto network, data structures that use pointers such as linked lists or binary search trees can be used. A linked list is a collection of data elements where each contains a piece of information that links it to the next element of the list. A binary search tree is a tree of connected nodes where, for each node, all nodes to its left contain lesser keys and all nodes to its right contain greater keys. In addition to linked lists, to find a way to retrieve the data, we need to hash the value of the data. Therefore, we need a hash pointer. Hash pointers can be used to create different types of data structures. Similar to regular pointers, they help us retrieve the data, and beyond that, they can be used to verify if the information is altered or not. Therefore, hash pointers are necessary parts of the data structure used for building cryptos [4].

There are different types of data structures ranging from arrays and hash tables to graphs, trees, and linked lists (Fig. 3).

Blockchain is a linked list, a data structure in which a series of blocks are linked together using hash pointers. Each block has data and a pointer to the previous block. This pointer is a hash pointer, so each block contains the value of the last block as well as the hash of that value, so it helps to verify that the value of the previous block is not changed. Hash pointers give us where the information is stored and verify that the data is not changed. The hash of the previous block is stored in each block header. That is why blockchain is known as a tamper-evident log; if any data is going to be altered in previous blocks since we stored the last block's hash, we detect the changes [4]. Remember that the hash functions are collision-resistant, so if an adversary tries to alter the data in some block j, the hash in block j + 1 (the hash of the entire block j) does not match up.

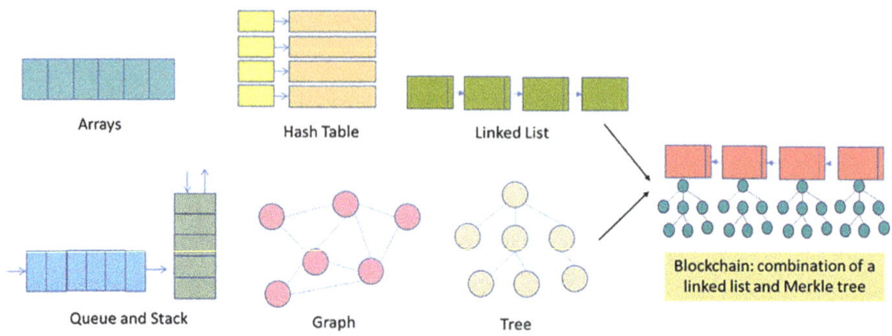

Fig. 3 Blockchain is a combination of a linked list where each block is linked to the previous block and the information within each block can be retrieved using a Merkle tree structure

Table 2 The algorithms needed in a digital signature scheme

Function output	Function	Algorithm type
(sk,pk) := generateKeys (keysize)	Receives a specified keysize and produces a private key (sk) and a corresponding public key (pk)	Randomized
sig := sign (sk,message)	Receives a message and a secret key, then generates a signature	Randomized
inValid := verify (pk, message, sig)	Takes a public key (pk), a message, and a signature as input, and produces a Boolean value of true or false	Always deterministic

Summarized from [4]

3.3 *Digital Signatures*

Besides hash functions and hash pointers, digital signatures are other building blocks of cryptocurrency. Digital signatures should have the same functionalities as handwritten signatures, where only you can create your signature. Still, anyone who sees it can verify it, and also, your signature is attached to a specific document and cannot be reattached or reused in other documents. Thus, a digital signature scheme requires three main algorithms, as listed in Table 2. The three algorithms include (1) getting a key size and generating a private key, and a public key, (2) getting a message and a private key and generating a signature, and (3) taking a public key, a message, and a signature as input, and producing a Boolean outcome – either true or false – to indicate the validity of the signature [4].

Several practical modifications can be made to the digital signature scheme. For example, instead of signing the message itself, we can sign the hash of the message to address the concerns around message size. Moreover, we can sign the hash pointers rather than the hash of the message. This way, the whole data structure

Fig. 4 A simplified overview of how public and private keys work

or the entire blockchain is signed. Bitcoin utilizes a digital signature scheme known as the Elliptic Curve Digital Signature Algorithm (ECDSA), which is a standard by the US government and employs elliptic curves.

The data is encrypted using the public key cryptography method or asymmetric encryption. Two different keys, public and private keys, are needed to encrypt/lock and decrypt/unlock the data. The public key is an identity. The user uses a public key to authenticate their identity electronically or to sign or encrypt data. In Bitcoin, identities are called "addresses." An "address" is just a hash of a public key. The public keys are just addresses that everyone can see, and other people use them to send crypto assets to each other. See [5] for more details on the asymmetric encryption method. Figure 4 shows the general analogy for describing digital signatures and the use of public and private keys.

Suppose that Alice wants to send you a message. She puts the message in a box and encrypts it using her private key. She or anyone with her private key is the only one who can lock the box. She then gives copies of her public key to anyone. Anyone who has Alice's public key can unlock the box and decrypt the message. They can deliver the message to you using your public key. They put the message in your mailbox and lock the box using your public key. Then, you use your private key to unlock the box. To put it simply, if a message is locked with a private key (like Alice's), anyone with her public key can unlock it. But if a message is locked with a public key (like yours), only you can unlock it with your private key.

Now that hash functions and digital signatures are described, we put the information together and see how cryptocurrencies work.

Two properties of cryptocurrencies are that they do not rely on a central authority and they cannot be spent multiple times by the same user. Avoiding the need for a central authority improves the acceptance of the crypto by people suspicious of such authorities and can save costs. Limiting the use of a unit of the crypto to once when in the possession of a user avoids the double-spending problem, that is, the situation where a person uses the same unit of crypto in more than one transaction, resulting

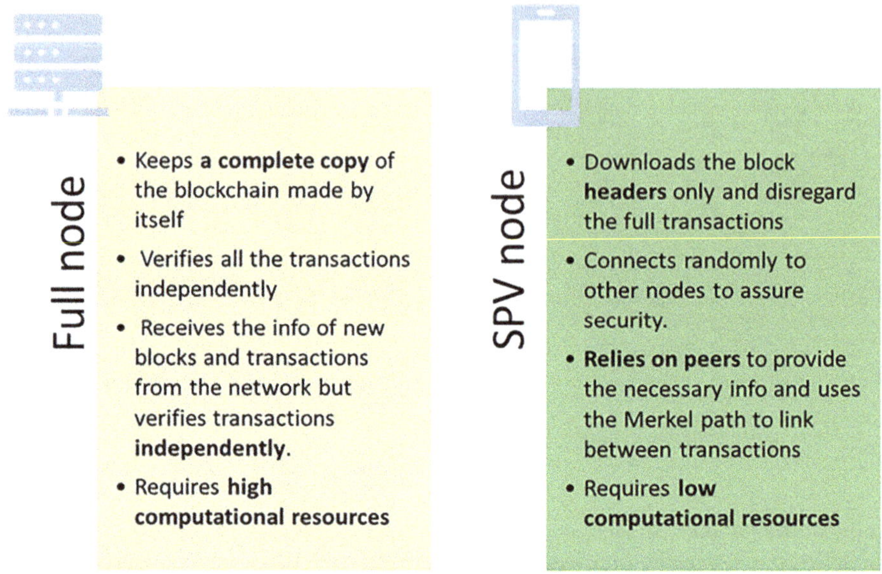

Fig. 5 The comparison of full and SPV nodes

in at least some of the people on the other side of the transactions being cheated. Since blockchain uses hash pointers and users sign the history of transactions or the entire blockchain, users can monitor the history of transactions and identify any sign of double spending of an already-spent coin.

A network consists of two types of nodes: full nodes and Simplified Payment Verification (SPV) nodes. Full nodes verify transactions and create coins by checking the entire history of blocks. SPVs are light nodes that verify transactions without downloading the whole blockchain by using Merkle tree properties, that is, ensuring that every leaf on the tree is labeled. Figure 5 summarizes the basic characteristics of these nodes.

The Merkle trees (binary hash trees) are helpful structures that can be implemented using hash pointers. They are an efficient way of encrypting and storing data needed for verification purposes. The Merkle trees help users prove the integrity and validity of the data. They significantly reduce the amount of memory required as it substantially reduces the amount of data that should be maintained for verification purposes. Users can verify individual parts of a block using the Merkle tree.

4 Designing a Cryptocurrency

A simple cryptocurrency has at least two types of transactions: creating coins and paying coins transaction, which consumes the previously created coins and creates new coins of the same value. Given the use of data structure in recording transactions, the chance of double spending is limited since all transactions are quickly recorded and can be tracked. To eliminate the need for a central authority, digital currencies are often decentralized. To achieve decentralization, we need to determine how all network users can agree upon which transactions have occurred, which transactions are valid, and create coins in a decentralized manner.

4.1 How to Achieve Decentralization?

In Bitcoin, decentralization is achieved through distributed consensus. Every 10 min, each node of the network broadcasts its outstanding transactions and suggests they be included in the next block. A valid block is indicated as the new block if the consensus is achieved among all nodes. This way, if malicious nodes suggest some invalid transactions, they will not be included in the block. Suppose there are some valid transactions by honest nodes that are not included in the block. In that case, they can wait to be included in future blocks once more nodes in the network have those transactions in the list of their outstanding transaction waiting for verification. Nodes of the network should have a consensus on which transactions were broadcasted to the network and the order in which the transactions happened.

In Bitcoin, there is no notion of global time, so the ordering of the transactions cannot be tracked using timestamps. Therefore, nodes need to have a consensus on the order in which transactions have occurred. The consensus is not easy to make since some nodes might crash or act as malicious nodes, and further, the Bitcoin network is not fully connected since not every pair of nodes is connected to others. These points limit the type of consensus algorithms from the "distributed consensus" literature employed in such networks.

Various algorithms have been proposed in distributed consensus, such as Byzantine General Problem, Fischer-Lynch-Paterson, and Paxos protocols. These algorithms discuss situations in which consensus is impossible. For example, in Byzantine General Problem, the consensus is impossible among loyal generals if one third or more of the total number of generals are traitors. Most of these algorithms have analyzed distributed databases that are different from distributed networks such as Bitcoin.

In practice, Bitcoin performs better than in theory. Bitcoin is different from traditional distributed consensus protocols suggested in the literature in two ways: first, Bitcoin introduces incentives. This way, it provides natural mechanisms for nodes to act honestly. Second, Bitcoin's consensus protocol counts on randomization [4].

In Bitcoin, the consensus algorithm works without knowing the node identities. While Bitcoin does not provide complete anonymity since it can link transactions that one node makes, it offers pseudonymity where the nodes do not reveal their actual identity. Due to Bitcoin's pseudonymity, Bitcoin uses a mechanism called implicit consensus to achieve security. In each round, a random node is selected. The selected node broadcasts the next block, which consists of the node's new outstanding transactions. Other nodes in the network either accept (unspent transactions, valid signature) or reject the block. Nodes show their acceptance of a block by including its hash in the next block they create. A transaction will be included in a block once it receives numerous confirmations. While there is no magic number for the number of confirmations for a transaction, six is common.

Bitcoins offer two mechanisms to incentivize users to behave honestly: (2) block reward and (2) transaction fees. In 2015, the value of a block reward was 50 Bitcoins, and it keeps halving every 210,000 blocks or approximately every 4 years. It is a geometric series, meaning that there will be 21 million Bitcoins to create.

The block reward is the only mechanism for generating coins. The way it works is that the node that creates a valid block gets to create a particular transaction in that block called coin-creation transaction, through which the node can send the created coin to his address as the recipient. The block reward becomes zero in 2140. However, Bitcoin still will have another incentivization mechanism: paying transaction fees. The user who creates a transaction can assign a portion of the total value of transactions to whoever creates the block and first puts that transaction into the block. Currently, transaction fees are low. But as the block reward decreases over time, users will need to pay miners more in fees to get their transactions processed quickly.

4.2 The Process of Mining and Verifying Transactions

In previous discussions, we assume that a node is randomly selected to suggest the next valid block. However, the reality is that in the Bitcoin network, all network nodes will compete independently to propose the next block. The process is called proof-of-work in which each node of the network will keep trying to find a number called *nonce* where the hash of that number, together with the hash of the previous hash and the current list of transactions, will be lower than a target value. This process is called solving a hash puzzle. Nodes are competing independently to find the nonce. The node that finds the nonce that satisfies this requirement will be the lucky node to propose the next block. Computationally speaking, nodes that have higher computation power are more likely to find the nonce that satisfies the following equation:

$$H\ (\text{nonce, previous hash, list of current transactions}) < \text{target value}$$

The processing for finding the nonce that satisfies the equation mentioned above is known as Bitcoin mining. For system security, finding this hash value should not be very easy. To adjust the difficulty level in finding the hash function, all network nodes recalculate the target value approximately every 2 weeks or 2016 blocks, meaning that the target level will be redefined. The average time between finding two consequent blocks is 10 min. The 10 min is an average, and in fact, the time between finding blocks follows an exponential distribution with an average of 10 min. The 10 min is not a magical number and can be set to any number (e.g., 6, 5). This time value should not be too small to help maintain system safety. There are disagreements on the best value of the Bitcoin latency, but most people agree that it should be a fixed number [4].

Knowing the average time between finding blocks helps a miner calculate how long it takes to find the next block. To find it, we just need to divide 10 min over the "fraction of hash power of the total network hash power" controlled by the specific miner. For example, if a miner owns 0.2% of the hash power in the network, he will find a block approximately every 5000 min. The block size and the block frequency are the parameters of protocols.

Speaking of the cost of mining, the economy of mining is a complex game theory model for several reasons. The rate of finding blocks by a miner not only depends on the miner hash power but also on the global hash power available in the network. Also, the miner reward, both block reward and transaction fee, is in Bitcoin, not fiat currencies, so the exchange rate at any given point of time influences the economy of Bitcoin mining for a specific miner.

The Bitcoin network needs to have consensus on three things: (1) the value of Bitcoin and the exchange rate of the currency, (2) the status of the system and the number of coins each user owns, and (3) the rules of the systems (e.g., soft forks and hard forks).

Figure 6 summarizes the process of finding blocks in the Bitcoin network.

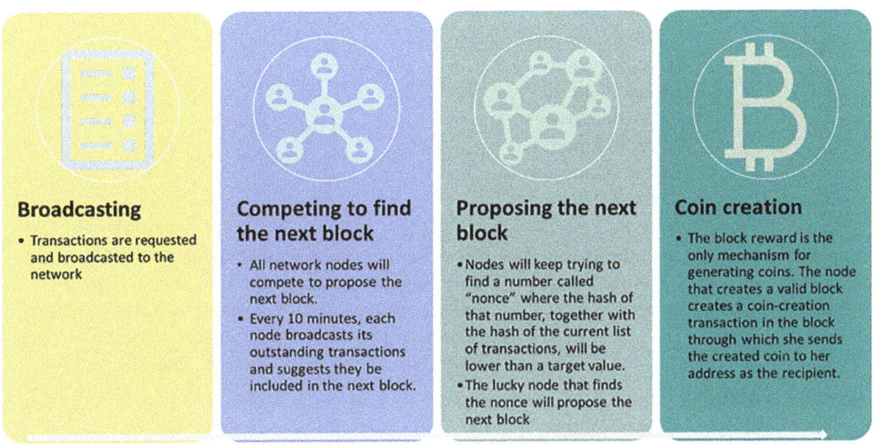

Fig. 6 The process of creating blocks in the Bitcoin network

4.3 Different Consensus Protocols

Since the start of cryptocurrencies, various consensus protocols have been developed to determine the miner who proposes the next block. The consensus protocols can be categorized into computational-based, capability-based, and voting-based protocols. Table 3 provides a summary of available protocols.

A miner is selected based on the computational power s/he has spent in computation-based protocols. In capability-based protocols, miners are chosen based on other factors such as the amount of cryptocurrency they own, the amount of trust the system has on the miner, and the contribution to the community. Finally, in voting-based protocols, a miner is selected using voting mechanisms.

Table 3 Example of current consensus protocols

Type of Protocol	Description	Examples
Computational-Based Consensus Protocols	A miner is selected based on the amount of computational effort she has spent	Pure Proof of Work (PoW), Prime Number Proof of Work (Prime Number PoW), Delayed Proof of Work (DPoW)
Capability-Based Consensus Protocols	A miner is selected based on other capability-related factors such as the miner's cryptocurrency holdings, the level of trust the system places in them, their contributions to the community, or the amount of storage they possess	Proof of Stake (PoS), Delegated Proof of Stake (DPoS), Proof of Stake Velocity (PoSV), Proof of Burn (PoB), Proof of Space (PoSpace), Proof of History (PoH), Proof of Importance (PoI), Proof of Believability (PoBelievability), Proof of Authority (PoAuthority), Proof of Elapsed Time (PoET), Proof of Activity (PoA)
Voting-Based Consensus Protocols	A miner is chosen by using a voting system	Practical Byzantine Fault Tolerance (PBFT), Delegated Byzantine Fault Tolerance (DBFT), Federated Byzantine Agreement (FBA), Combined Delegated Proof of Stake and Byzantine Fault Tolerance (DPoS + BFT), Raft, Federated

Summarized from [6]

5 The Main Features of Future Crypto Networks

The current cryptocurrency networks suffer from several key features such as scalability, vulnerability (full decentralization), and sustainability (Fig. 7). This section provides an overview of two of these challenges and existing solutions to address them.

5.1 Scalability

Cryptocurrency networks face various scalability issues, including problems like communication breakdowns between users, data storage, and the linear recording of transaction history [7]. The challenges with scalability stem from two primary factors: (1) the expansion of blockchain size and transaction volume, and (2) the difficulties of consensus protocols and the maximum block size [8].

A scalable protocol is a protocol in which the waiting time for processing transactions is short, even under high transaction volume [9]. The scalability problem can be identified based on data size, transaction speed, and transaction costs. Several metrics can be used to measure the scalability of a network. Table 4 summarizes some of them.

The upper bound on the transmission of information within a peer-to-peer network can be represented as $L \surd N$, where L represents the link capacity or the number of messages that each node can send or process, and N is the total number of nodes in the network [11]. Note that not all nodes are homogenous and have different capacities as some nodes have potent servers, and some are simple end devices. Bitcoin has a peer-to-peer topology making it among the first class of scalable topologies with $L \surd N$ scaling limit [12].

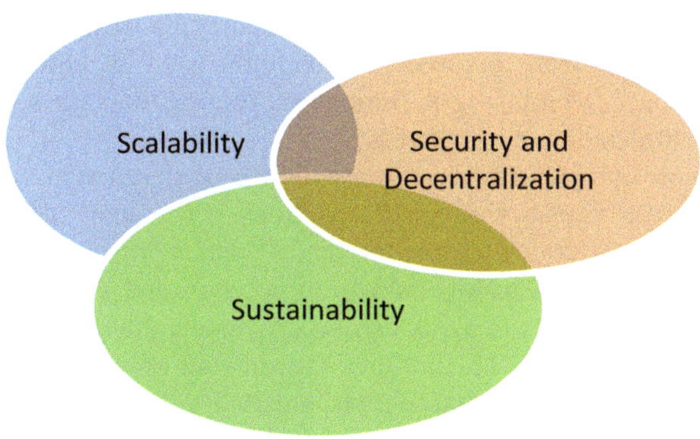

Fig. 7 Three main properties of future cryptocurrencies

Table 4 Key metrics for measuring scalability [10]

Scalability metrics	Definition	Bitcoin
Maximum throughput	The maximum rate of confirming transactions by the network	3.3–7 transactions/s
Latency	Time to confirm a transaction	~10 min
Bootstrap time	Time for a new node to download and process network history	~4 days
Cost per confirmed transaction	Cost consumed by the entire network to confirm a transaction	$1.4–$2.9 (with maximum throughput)

The node's transmission capacity is referred to as the allocated bandwidth which indicates the volume of bandwidth accessible for the node to interact with the network. A higher allocated bandwidth assists miners in swiftly sharing information, as well as receiving and transmitting blocks more speedily [13].

The structure of the network defines the vulnerability and performance of cryptocurrencies. The Bitcoin structure is complicated and is intentionally hidden to preserve users' privacy and network security against denial of service (DoS). Estimating the number of nodes and the full size of the network is difficult as nodes are covered behind firewalls. Moreover, the latency among nodes is unknown. There is often a trade-off between network scalability and the system's security goals.

The current inefficiencies in digital currency networks hinder the potential of blockchain as a valuable technology for businesses [14]. To tackle this challenge, companies are trying to improve the scalability of their networks by enhancing the computational efficiency of mining operations. While more challenging consensus algorithms enhance the security of the blockchain, they also restrict the technology's scalability [8]. The primary purpose of scalability is to increase the transaction speed (faster output) and transaction throughput (transactions per second), without scarifying decentralization or vulnerability.

Bitcoin is slow compared with current credit card transaction systems. At this time, while Bitcoin manages seven transactions per second, the Visa network processes 10,000 transactions per second, and Paypal can handle 100 transactions per second. Each block's total size is 1 million bytes, and given that the average transaction size is 250 bytes, a block can accommodate 4000 transactions in total. Since blocks are found every 10 min, then the network speed is seven transactions per second. In another term, a transaction is called a confirmed transaction once it has been suppressed six blocks into the blockchain. Since it takes 10 min to generate a block, it means, on average, it takes 1 h for your payment transaction to be confirmed. Another limitation is that the choice of cryptographic algorithms

Table 5 Comparison of the crypto protocol parameters

	Bitcoin	Ethereum
Time to find a block	~10 min	10–20 s
Block size	1 MB	Determined by execution fee, called gas

and hash functions in Bitcoin is limited, so the Bitcoin scripting language should be extended to support new cryptographic algorithms.

5.1.1 How Does Ethereum Address Scalability Issues?

Ethereum tries to shorten the time to find a block, provide a fair distribution of rewards among miners, and reduce the incentive for pooled mining. Table 5 compares the time to mine a block and block size in Bitcoin and Ethereum.

In networks with fast block time, such as Ethereum, two miners' chance to find a block simultaneously is higher than the Bitcoin. In Bitcoin, the chance of finding a block at the same time is low, where the time between blocks is approximately 10 min, and it takes relatively 12 s to propagate a block to 50% of the network. Faster networks such as Ethereum generate a large number of stale blocks. Stale blocks, or orphan blocks, are propagated to the network and are verified by some miners as valid blocks but finally do not make it to the longest chain and will end up in forks. Ethereum calls those blocks uncle blocks. To promote decentralization and facilitate fair distribution of reward in the network, Ethereum uses a protocol inspired by the GHOST (Greedy Heaviest Observed Subtree) protocol, where miners of the uncle blocks receive a portion of the block reward.

The GHOST protocol tries to improve Bitcoin's scalability by modifying the selection rule of the chain. In Bitcoin, during a fork, the longest chain with the most proof-of-work becomes the main chain. In GHOST, a node selects the side chain with the most work accumulated in its subtree blocks [15]. Therefore, GHOST improves mining power utilization in the network. Mining power utilization shows how well a protocol uses energy in the network for actual work. It is determined by the fraction of mined blocks that stay in the main chain. In addition, GHOST improves the fairness and security of the system (Fig. 8) since a higher mining power utilization makes it more costly for attackers to launch an attack; therefore, robustness against attacks increases [16].

To have updated information about the number of nodes in the network, Ethereum uses a node discovery mechanism called Kademlia. Kademlia is designed for locating and storing data in a peer-to-peer network. Ethereum uses it to help nodes keep an up-to-date record of their connectedness and their adjacent peers, as nodes may leave or join the network and adjacent nodes may have outdated adjacency information. The node discovery mechanism in Ethereum is UDP-based, a fast and simple protocol; however, the rest of the communication in the network is TCP-based. UDP (User datagram protocol) is quicker and simpler

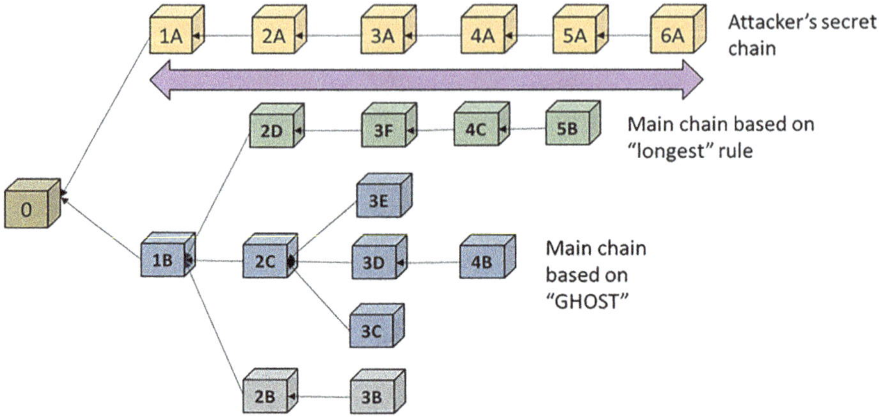

Fig. 8 The comparison of the Bitcoin largest rule protocol with the GHOST protocol. (Redrawn from [17])

than TCP (Transmission Control Protocol), but retransmission of lost data is only possible with TCP standards. After identifying peer connections and determining whether neighbor nodes are responsive using UDP, Ethereum uses TCP to exchange encrypted and authenticated messages [16].

5.1.2 Scalability Solutions

Various scaling methods have been developed in computer science literature to address scalability issues. Some researchers suggested parameter tuning, known as reparametrization of block size and intervals, as a solution for managing scalability; however, it was shown that achieving high load blockchain protocols requires a fundamental rethinking of technical approaches used in designing the network, and parameter tuning is not sufficient [10]. Another solution is using "payment channel networks" which permit users to move funds between intermediaries without needing to record every transaction on the Blockchain [18].

The scaling methods can be categorized into four main groups: (1) on-chain scaling [19], (2) off-chain scaling or using Layer 2 scaling methods [20], (3) compression techniques [21], and (4) secure decentralized randomness techniques [22].

On-chain scaling methods, also known as Layer 1 techniques, improve the throughput of the based layer of the blockchain network by implementing strategies such as increasing blocksize, new address formats, smaller size signatures, database partitioning, and signature aggregation. Off-chain scaling methods, also known as Layer 2 solutions, enhance throughput without touching the main blockchain layer by creating alternative protocols and layers on top of a blockchain; they often require additional software and complexity compared to on-chain scaling

Table 6 The summary of scalability solutions in each layer of blockchain

	Layer	Definition	Example of solutions	Amount of scalability that can be gained
1	Hardware layer	Machines used to run blockchain	Deploy high-end hardware	Up to 5 to 10 times more throughput
2	Network layer	Communications between nodes	Solutions impacting network propagation (e.g., what data is being sent, which transmission method is used)	Up to 5 times more throughput
3	Layer 1, on-chain	On-chain design of Blockchain (e.g., block structure, consensus algorithms, the specific structure of the main chain)	Minor changes (adjusting the block size or the block time interval) or major changes (sharding)	Up to 10 to 20 times more throughput
4	Layer 2 application	Moving computation off the chain	Multi-sided payment channels, payment hubs, complete sidechains	Up to 10,000 to 100,000 more throughput

Summarized from [24]

approaches. Examples of off-chain methods include sidechains, colored coins, and utilizing state channels.

Sharding is an example of a Layer 1 solution for addressing scalability. The transaction throughput is increased by partitioning the database into several shards, horizontal segments, each stored on a separate server, spreading the computational and storage workload across different network users [23]. The information can still be shared among other nodes; however, nodes do not process and store all the data. The nodes should be randomly assigned and reassigned to random shards to assure security while using sharding.

Table 6 provides an overview of different solutions for addressing scalability and examples of each. It is important to note that strategies cannot be compared solely based on the transaction throughput since different methods involve additional security and trust assumptions, among other things [24].

No single solution can guarantee scalability. Multiple solutions can be employed to ensure scalability and reduce the overall burden on the network. Given the scalability methods mentioned above, the application of such methods is still a challenge due to the limited capability of storage and bandwidth of each node, the lack of optimal strategies for placing transactions within each shard, and the efficiency of cross-shard transactions [25].

While scalable protocols may alleviate the energy-intensive nature of verification processes, the main problem still exists since scalable algorithms often compromise security.

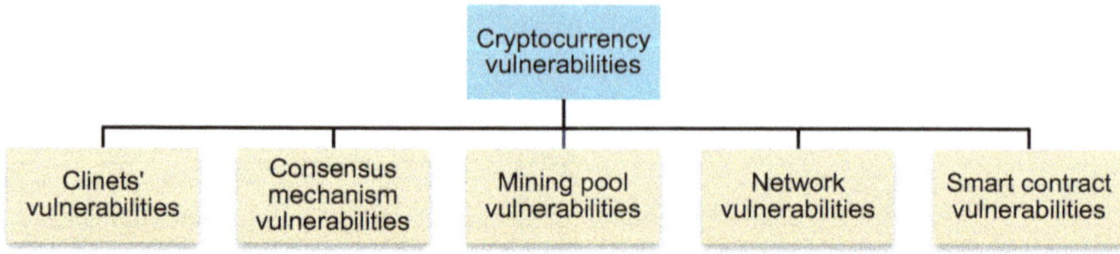

Fig. 9 Five main categories of the vulnerability of crypto networks

Discussing the security of the crypto networks is outside of the scope of this work. However, security and privacy are key requirements of any cryptocurrency. Given the financial nature of cryptocurrencies, crypto networks are an obvious target for adversaries. Various attacks have targeted different parts of the crypto ecosystem, including double spending, transaction malleability, networking attacks, netsplit, and interfering mining processes.

A rich literature exists on the security of blockchain systems. Maleh et al. [26] analyzed 65 different cybersecurity incidents and developed a taxonomy for blockchain attacks. Based on this taxonomy, the threats and vulnerability of crypto networks can be categorized into five main groups as listed in Fig. 9 [26]: (1) clients' vulnerabilities, (2) consensus mechanisms vulnerabilities, (3) mining pool vulnerabilities, (4) network vulnerabilities, and (5) smart contract vulnerabilities.

Digital signature and hash function vulnerabilities are examples of clients' vulnerabilities. Race attacks and 51% vulnerabilities are examples of consensus mechanisms vulnerabilities. In the Race attach, two transactions are generated simultaneously for the same fund, to spend the same fund twice [27]. In the 51% attack, the malicious actor or actors control over 50% of the network computational power, where they will be able to modify the ordering of transactions or prevent confirming certain transactions [28].

Bribery attacks and the Selfish Mining attack are instances of mining vulnerabilities. In a Bribery attack, the adversary bribes miners to maximize their profits, for example, by building on her fork [29]. In the Selfish attack, an adversary pool intentionally keeps discovered block private and does not share it with the public chain [13]. Distributed Denial of Service (DDoS) and Sybil attacks are considered examples of network vulnerabilities. In a DDoS attack, a group of adversary computers floods a targeted server with too much traffic to make it inaccessible to legitimate users [30]. In the Sybil attack, an adversary creates many fake identities to control the network [31].

Finally, Solidity vulnerabilities and Ethereum Virtual Machine (EVM) Bytecode vulnerabilities are cases of smart contract vulnerabilities. Solidity is an object-oriented language for writing smart contracts. When using solidity for developing contracts, developers often use external calls to other functions where those external calls usually do not have enough safety features. EVM is a platform that allows developers to build decentralized applications on Ethereum, which has its limitations [32].

For example, in Bitcoin, an efficient crypto network is expected to have full pseudonymity to satisfy user privacy. Pseudonymity refers to (1) anonymity and (2) unlinkability. The client information should be unidentifiable (anonymous) as well as unlinkable. While Bitcoin provides anonymity since clients use private and public keys, it does not fully guarantee unlinkability, where the relationships between transactions can be inferred by analyzing the transactions broadcasted to the network. Researchers have shown that the linkability of cryptocurrency transactions is possible based on network analysis and transaction propagation analysis [33, 34]. Therefore, one big area for improvement of future crypto networks resides in enhancing their security.

5.2 *Sustainability*

This section describes the sustainability of crypto networks from several perspectives: (1) the carbon footprints of the energy consumed in crypto networks, (2) the e-waste generation rate of mining hardware, and (3) sustainability practices. As can be seen, by these dimensions sustainability focuses primarily on environmental and materials sustainability—although social and economic sustainability may be more broadly related.

5.2.1 Power Consumption, Carbon Footprints, and the Role of Miners

The energy-intensive nature of some crypto networks is probably the most known environmental impact of the crypto networks [35–38]. For example, for Bitcoin, the power consumption and the consequent carbon footprint of mining Bitcoin depend on several factors ranging from the number of miners operating on the network, the type of hardware they use, their geographical region, and the source of electricity they consume—such renewable vs nonrenewable energy sources.

Modeling Energy Consumption

The energy required to verify transactions on blockchain networks depends on factors such as the difficulty of mining, the number of participants, the type of consensus protocol—Proof-of-Work (PoW), Proof of Stake, Proof of Authority—and even strategies set by the mining pools. As previously discussed, most consensus algorithms promoted as energy-efficient scalable algorithms do not have the security level needed for blockchain applications. So far, the most secure algorithm is PoW initially used in the Bitcoin network, which is energy-intensive and is purposely made this way to make cheating costly. The energy consumption of the mining process is measurable in terms of the network hash rates. The hash rates represent the number of calculations performed in seconds—or hashes per second.

Energy consumption is a function of the "difficulty" of mining. The difficulty refers to how difficult is to find a hash—or equivalently, a nonce—below the target value defined by the Bitcoin software (a 256-bit number). The higher the target value, the higher the chance to find a hash lower than the target value, and therefore the lower the mining difficulty and vice versa.

The Bitcoin network controls the rate of discovering Blocks or generating Bitcoins by selecting the target value T. In the Bitcoin network, the target value is noted as the difficulty level [39]:

$$\text{Difficulty} = \frac{T_{\max}}{\text{Target Value}} \qquad D = \frac{T_{\max}}{T}$$

The largest possible target value is $(2^{16} - 1)2^{208} \approx 2^{224}$ which is the original target value used in the Bitcoin network. Therefore, the difficulty compares the target value with its original value (difficulty = original target/target value). The network automatically determines the target value to ensure that the average time to create a block remains equal to 10 min. Therefore, as more miners join the network and start to mine, the difficulty of finding valid blocks, or an acceptable nonce, increases [40].

In the Bitcoin network, the hash function used is $H(S) = \text{SHA256}(\text{SHA256}(S))$, which ensures an even distribution between 0 and $2^{256} - 1$. This means that for any nonce value, we can calculate the probability of the hash value being less than the target value as follows [39]:

$$H(S) = \text{Uniform}\left(0, 2^{256} - 1\right)$$

$$\Pr(H(S) < T) = \Pr(H(\text{Bn}) < T) = \frac{T}{2^{256}} = \frac{T_{\max}}{D2^{256}} = \frac{2^{224}}{D2^{256}} \approx \frac{1}{D2^{32}}$$

The number of attempts needed to acquire a block follows a geometric distribution which implies that the anticipated number of hashes required to find a block is $D2^{32}$. If a system computes hashes at a rate of R, the expected duration for finding a block is [39]:

$$E(t) = \frac{1}{p} = \frac{D2^{32}}{R}$$

The expected time can be used to find the energy needed to discover a block.

The rate at which Bitcoins can be discovered can be controlled by the Bitcoin Network's choice of the value of the target, T. The higher the target value, the lower the mining difficulty (difficulty finding a hash value less than the target value (a 256-bit number). What happens in the network is that once more miners join the network, the rate of creating blocks increases. Therefore, the average mining time decreases. However, since the ideal average mining time is about 10 min, the Bitcoin software increases mining difficulty by setting a lower target value. This reduces the block creation rate, and the average mining time will go back to the expected time, and this cycle repeats.

The above-discussed estimation method relies on several assumptions. For example, in practice, Bitcoin generation is faster than expected since new hash power is determined dynamically by the capabilities of the miner networks, and it can only adjust itself every 2016 block [41]. The accurate estimation of the energy requires relaxing those assumptions, including the number of miners, the speed of mining, how a system calculates hashes (the choice of hardware), and the Bitcoin economics that influence the number of miners.

The Gaps in Current Energy Estimations

Multiple studies have assessed the energy impact of Bitcoin. Mora et al. [42] suggested that if Bitcoin gains popularity at a similar rate as other widely adopted technologies, its CO_2 emissions could contribute to global warming exceeding 2 °C. Krause and Tolaymat [36] estimated the energy demand to produce a US dollar through cryptocurrency mining and contrasted it with the energy used in conventional mining of minerals like aluminum, copper, and gold. They found that crypto mining consumes more energy to generate an equivalent dollar amount. Stoll et al. pointed out that data regarding the geographical locations and IP addresses of miners available through mining pools and websites that disclose pool compositions can aid in translating energy consumption data into metrics for greenhouse gas emissions [43]. Depending on the geographical regions and the main source of electricity generation in each region, the CO_2 footprints can be calculated.

Although some studies aim to discuss the carbon footprint of digital ledger technologies generally, the number of studies on the energy modeling of blockchain networks is limited. While some studies have developed quantitative methods [39, 44], the provided estimates vary significantly among studies since they are based on different sets of assumptions that do not fully capture the reality of the cryptocurrencies market with no consideration of miner behavior. Also, existing studies have not employed the full advantages of available empirical data. Arguments have been made that many projects overestimate near-term Bitcoin CO2 emissions [45]. Finally, since the number of miners operating the network depends on the economy of the Bitcoin, no consideration of the crypto-economics, user adoption, and the incentives offered by the network have been incorporated into existing energy estimation. Given the discrepancies and varying perspectives, public blockchain platforms require greater investigation. Private and permissioned platforms may require significantly less energy and generate future emissions—some blockchain platforms are even selling themselves as greener alternatives. For example, the Signum currency states that it uses less than 0.002% of Bitcoin's energy to drive its blockchain [46].

Blockchain energy consumption does not derive its main cost from the hash rate but from the collective mining process on its corresponding platform. It is essential to understand miners' roles in the digital economy to differentiate between rough estimation and valuable energy reduction innovation. In particular, it is important to understand the interaction between cryptocurrency valuation and Bitcoin adoption (by both clients and miners).

The Concept of Mining Pools

Miners often share their computational recourses over a network by joining some mining pools, where the award is split based on the contribution of each miner. The origin of mining pools occurs because the difficulty of mining has increased to the point that it could take centuries to create a block by one miner—so the solution is to integrate the resources of miners and reward them according to their contributed mining hash power. There are different mining pool methods to reward the miners [47], and currently, there are about 20 major mining pools; most of them—about 81% of the network hash rate—are located in China (e.g., BTCC, Antpool, BW) [48].

While pooled mining is common in the crypto market, previous energy estimations do not consider this pooling situation in energy estimation analysis. The key to overcoming this limitation is developing a framework to attribute blocks to the original miners accurately. Developing an attribution scheme for linking blocks to original miners through support from publicly available datasets helps identify more accurate hash rates used for mining each block and, consequently, a more precise estimation of the energy required for the mining process. We should note that empirically assessing the impact of merged mining on energy consumption modeling is just the first step. The ultimate purpose should be to study the effect of block sharing among mining pools and new constructs such as multi-merged mining on resolving the energy consumption of crypto mining.

5.2.2 Electronics Waste Generation of Mining Hardware

While the energy consumption of Bitcoin has generated significant discussion on the sustainability of cryptocurrency, other crypto network resource consumption with environmental sustainability implications beyond energy implications should be acknowledged. One such resource is materials associated with short-lived hardware infrastructure used for mining processes [44, 49]. Hardware options ranges a spectrum from Central Processing Units (CPUs) and Graphics Processing Units (GPUs) to Field Programmable Gate Arrays (FPGAs), and more recently, Application-Specific Integrated Circuits (ASICs).

In the Bitcoin network, miners use ASICs suitable for performing hashing computations. Once hardware reaches its end-of-use life span, they create a considerable amount of electronic waste that requires proper collection and recovery efforts. De

Vries and Stoll estimated that the life span of Bitcoin mining hardware is 1.29 years which results in up to 30.7 metric kilotons of e-waste equipment annually [44].

The use of special ASIC hardware contributes to e-waste generation and raises concerns about network decentralization, where specific groups of miners with access to mining hardware may control the network and influence its security. The development of ASIC-resistant algorithms is suggested as a solution for enhancing safety. Moreover, minimizing the hardware requirement should be considered a mechanism for improving the sustainability of the technology. Speaking of sustainability, we should note that these ASICS have made computing power greater to save energy per computation, but exponentially greater computation requirements make the energy savings disappear.

Some cryptocurrencies and blockchain platforms have recognized the concerns with e-waste generated by specialized equipment for consensus and mining activities. These greener alternatives state that their actions can be done by using electronic equipment's available computing and disk space. Whether or not this holds actual needs to be proven over the long run if these platforms and cryptocurrencies become more popular.

5.2.3 Sustainability Applications of Cryptocurrencies

Improving efficiency, transparency, and traceability are often highlighted as benefits of blockchain technology, where various use cases would be possible to create more sustainable systems [50]. There is considerable potential for incorporating blockchain into governance efficiency [51], social equity [52], industrial innovation [53], and environmental protection practices [54]. Examples of such practices include real-time monitoring of energy and resource consumption [55], tokenizing carbon credit assets [56], rewarding green actions, tracking emissions [57], addressing information asymmetry to foster resource allocations, establishing market mechanisms to manage resources properly, and financing climate change practices [58].

Proper evaluation mechanisms are needed to measure the exact sustainability consequence of any digital currency precisely. The question arises of what strategies can be employed to green the financial systems and design a low-carbon economic and financial system. There is no consensus on the definition of green financing as it compromises various aspects. It includes financing of green investments in environmentally viable goods and services, investments in the area of prevention, reduction, and compensation of damage to the environment, and also it covers the financing of public policies that promote the implementation of damage mitigation projects or adaptation initiatives [59]. It is expected that future blockchain systems facilitate the implementation of green finance principles due to the decentralized and transparent infrastructure they create. It addresses the mistrust among stakeholders, including donors and recipients of climate change finance.

Blockchain and cryptocurrencies, through these activities, can expand the contribution to sustainability not only to those that can afford it and at the top of the economic pyramid but to all levels and locations of societies. This inclusiveness has great potential to increase equity and inclusion in financial systems [60, 61] which can also be green [62]. These lead to greater potential for social sustainability, as we now discuss.

5.2.4 Social Sustainability

In this section, we highlight three main directions in which cryptocurrency infrastructure influences society: (1) supporting small businesses, (2) empowering the culture of sustainable behavior, and (3) energizing new economic systems.

Supporting Individuals and Small Businesses

Strengthening individuals' innovation and small businesses is a potential advantage of distributed digital ledgers [63] such as NFTs. NFTs, or non-fungible tokens, are special units of data stored on blockchains that are unique and cannot be exchanged for one another. NFTs are tokens associated mainly with digital content such as images, art, songs, and social media posts [64]. NFTs have received attention recently since they facilitate the implementation of a "creator economy" or "attention economy" in which a chain of ownership can be created, and the owner of digital content can authenticate the content, dictate the value, and transfer the ownership to the future buyer with no intermediary.

NFTs allow content creators to connect to their customers directly and provide pricing tiers that give buyers flexibility in paying their desired price. The marketplace for NFTs is growing fast as the concept of "metaverse" is gaining momentum; however, we should also acknowledge that keeping track of the history of the chain of owners does not necessarily reduce market inequality. Still, projects with specific sustainability dimensions as the basic philosophy can contribute in sustainable and inclusive ways. For example, relating NFTs to the United Nation's 17 Sustainable Development Goals to be socially responsible can be a direction for sustainable NFT development and sale. Another aspect that can prove more socially inclusive and sustainable is to use NFT for craft manufacturers in developing nations who use sustainable materials to support their crafts [65].

Encouraging the Culture of Sustainable Behavior

Citizens' behavior in reducing environmental emissions and the drivers of sustainable behavior have long been studied in the literature [66–68]. Moreover, technology has been highlighted as an enabler for promoting sustainable practices and social value creation [69, 70]. For instance, the technology can help monitor consumer behavior, provide feedback to users, reward sustainable behaviors, and consequently help users adjust their behavior towards sustainability.

Blockchain capabilities in measuring, reporting, and verifying practices pave the way for regulating and standardizing habitual user behaviors and tethering society towards an appropriate culture of usage, consumption, and disposal compatible with sustainable principles. Moreover, the technology can be implemented at a larger scale in countries and regions where the international carbon markets and discussions under the Paris Agreement can be regulated [71]. There are also ways of tokenizing plastics, such as ocean plastics, to develop plastic credit within developing countries [72]. Developing general tokenized social credit systems has also been proposed and could be a way to increase effective and flexible philanthropic activities by organizations [73].

New Economic Systems

The crypto peer-to-peer system of exchange is formed around shared ownership and is inherently self-governing, so it has the potential to create new economic structures. For example, one challenge with the concept of the sharing economy as a socioeconomic system is that while online operators and dematerialized organizations have tried to match supply and demand nodes and aggregate the resource of multiple users to provide service to others, they fail to equally distribute the value among all the participants in creating the value. A big fraction of the value often goes to the intermediaries who operate such online platforms. Scientific research is needed to explore how the distributed nature of blockchain makes it possible to address the current imbalanced nature of traditional platforms. The crypto network aggregates the work of disparate groups of miners who can coordinate themselves and run online platforms. The transaction micro fees that users pay to the network will be paid to those who participate in operating the platforms and lessen the inequality issue [74].

The crypto market has its political consequences too. Arguments have been continued regarding whether the crypto networks should maintain pseudonymity to preserve users' privacy or should the users be identifiable to the government as the current implementation of the Know Your Customer (KYC) requirement imposes. Satoshi's intent for creating Bitcoin is not known yet, but there is a possibility that he meant Bitcoin to be an agorist currency [75]. Although Bitcoin has the potential to roll through black and gray markets, its ability to create a free market should not be overlooked. Maintaining privacy, alleviating mistrust among stakeholders, involving all people, and the infrastructure for using unutilized assets of the crowd should be used such that it paves the way for an open free market.

These social and environmental sustainability possibilities for cryptocurrency and blockchain technology require research and monitoring [76]. In each situation, social and environmental outcomes can be both negative and positive. Determining the environments, actions, practices, and decisions to determine whether negative or positive sustainability outcomes occur requires further research [77]. To complete this research knowledge and experience across disciplines, locations, regulatory, and economic systems are needed. The research can be quite substantial and necessary.

6 Conclusion

Financial systems play a critical role in guiding society toward sustainable movements. However, the business community still needs to respond to many concerns about the scalability, vulnerability, and sustainability of crypto networks. This chapter reviewed the basics of developing cryptocurrency systems and elaborates on scalability and sustainability as two main characteristics expected from future cryptocurrency networks.

With a particular focus on sustainability implications, this study provided insights into energy consumption and e-waste generation among the sustainability-related concerns of the crypto networks. It argued the need for opening a new paradigm shift for the design and development of financial systems that provide economic, environmental, and social benefits and provides opportunities for sustainable practices. Further, we have elaborated on the social sustainability of crypto technology and the need for further research on how technology should be employed to empower small businesses, establish a culture of sustainable behavior, and the emergence of new economic models and a free market.

It should be noted that much more work has been accomplished in the computer science community that has not been covered in this chapter. The primary purpose of this study was to give beginners an overview of how the cryptocurrency network functions, what opportunities for social benefits it offers, and what aspects of the technology require further investigation.

References

1. R. de Best, Number of cryptocurrencies worldwide from 2013 to February 2022. (2022), https://www.statista.com/statistics/863917/number-crypto-coins-tokens/
2. A. Klarin, The decade-long cryptocurrencies and the blockchain rollercoaster: Mapping the intellectual structure and charting future directions. Res. Int. Bus. Financ. **51**, 101067 (2020)
3. M. Templier, G. Paré, A framework for guiding and evaluating literature reviews. Commun. Assoc. Inf. Syst. **37**(1), 6 (2015)
4. A. Narayanan, J. Bonneau, E. Felten, A. Miller, S. Goldfeder, *Bitcoin and Cryptocurrency Technologies: A Comprehensive Introduction* (Princeton University Press, Princeton, 2016)
5. P. Vryonis, Explaining public-key cryptography to non-geeks, Mediu.com, 27 Aug 2013
6. L. Ismail, H. Materwala, A review of blockchain architecture and consensus protocols: Use cases, challenges, and solutions. Symmetry (Basel) **11**(10), 1198 (2019)
7. S. Barber, X. Boyen, E. Shi, E. Uzun, Bitter to better—How to make Bitcoin a better currency, in *International Conference on Financial Cryptography and Data Security*, (Springer, Berlin, Heidelberg, 2012), pp. 399–414
8. M. Conoscenti, A. Vetro, J.C. De Martin, Blockchain for the internet of things: A systematic literature review, in *2016 IEEE/ACS 13th International Conference of Computer Systems and Applications (AICCSA)*, (IEEE, Piscataway, 2016), pp. 1–6
9. Y. Sompolinsky, Y. Lewenberg, A. Zohar, SPECTRE: A fast and scalable cryptocurrency protocol. IACR Cryptol. ePrint Arch. **2016**, 1159 (2016)
10. K. Croman, C. Decker, I. Eyal, A.E. Gencer, A. Juels, A. Kosba, A. Miller, P. Saxena, E. Shi, E.G. Sirer, On scaling decentralized blockchains, in *International Conference on Financial Cryptography and Data Security*, (Springer, Berlin, Heidelberg, 2016), pp. 106–125

11. P. Gupta, P.R. Kumar, The capacity of wireless networks. IEEE Trans. Inf. Theory **46**(2), 388–404 (2000)
12. J. Mallett, *Scaling and Consenus in Monetary Systems* (Department of Computer Science, Reykjavik University, Reykjavik, 2020)
13. I. Eyal, E.G. Sirer, Majority is not enough: Bitcoin mining is vulnerable, in *International Conference on Financial Cryptography and Data Security*, (Springer, Berlin, Heidelberg, 2014), pp. 436–454
14. M. Blinder, *Making Cryptocurrency More Environmentally Sustainable*. Harvard Business Review 27 (2018)
15. Y. Sompolinsky, A. Zohar, "*Accelerating Bitcoin's Transaction Processing," Fast Money Grows Trees, Not Chain*. Cryptology ePrint Archive (2013)
16. A.E. Gencer, S. Basu, I. Eyal, R. Van Renesse, E.G. Sirer, Decentralization in Bitcoin and Ethereum networks, in *International Conference on Financial Cryptography and Data Security*, (Springer, Berlin, Heidelberg, 2018), pp. 439–457
17. D. Stone, An overview of proof of work based blockchain consensus protocols (Part 1). (2018), https://medium.com/ [Online]. Available: https://medium.com/@drstone/an-overview-of-proof-of-work-based-blockchain-consensus-protocols-part-1-e04102885093
18. J. Bonneau, A. Miller, J. Clark, A. Narayanan, J.A. Kroll, E.W. Felten, SoK: Research perspectives and challenges for Bitcoin and cryptocurrencies, in *2015 IEEE Symposium on Security and Privacy*, (IEEE, Los Alamitos, 2015), pp. 104–121
19. S. Kim, Y. Kwon, S. Cho, A survey of scalability solutions on blockchain, in *2018 International Conference on Information and Communication Technology Convergence (ICTC)*, (IEEE, Piscataway, 2018), pp. 1204–1207
20. J. Poon, T. Dryja, *The Bitcoin Lightning Network: Scalable off-Chain Instant Payments* (2016) pp. 1–59
21. U. Nadiya, K. Mutijarsa, C.Y. Rizqi, Block summarization and compression in Bitcoin blockchain, in *2018 International Symposium on Electronics and Smart Devices (ISESD)*, (IEEE, Piscataway, 2018), pp. 1–4
22. N. Alsalami, B. Zhang, Uncontrolled randomness in blockchains: Covert bulletin board for illicit activity, in *2020 IEEE/ACM 28th International Symposium on Quality of Service (IWQoS)*, (IEEE, Piscataway, 2020), pp. 1–10
23. L. Luu, V. Narayanan, C. Zheng, K. Baweja, S. Gilbert, P. Saxena, A secure sharding protocol for open blockchains, in *Proceedings of the 2016 ACM SIGSAC Conference on Computer and Communications Security*, (The Association for Computing Machinery, New York, 2016), pp. 17–30
24. Forum, T. E. U. B. O. and F., *Scalability, Interoperability and Sustainability*, Thematic Report, European Union Blockchain Observatory & Forum (2019)
25. Q. Zhou, H. Huang, Z. Zheng, J. Bian, Solutions to scalability of blockchain: A survey. IEEE Access **8**, 16440–16455 (2020)
26. Y. Maleh, M. Shojafar, M. Alazab, I. Romdhani, *Blockchain for Cybersecurity and Privacy: Architectures, Challenges, and Applications* (CRC Press, Boca Raton, 2020)
27. D. Dasgupta, J.M. Shrein, K.D. Gupta, A survey of blockchain from security perspective. J. Bank. Financ. Technol. **3**(1), 1–17 (2019)
28. S. Sayeed, H. Marco-Gisbert, Assessing blockchain consensus and security mechanisms against the 51% attack. Appl. Sci. **9**(9), 1788 (2019)
29. G. Ebrahimpour, M.S. Haghighi, Analysis of Bitcoin vulnerability to bribery attacks launched through large transactions. arXiv Prepr. arXiv2105.07501 (2021)
30. S.T. Zargar, J. Joshi, D. Tipper, A survey of defense mechanisms against distributed denial of service (DDoS) flooding attacks. IEEE Commun. Surv. Tutor. **15**(4), 2046–2069 (2013)
31. B.N. Levine, C. Shields, N.B. Margolin, A survey of solutions to the Sybil attack. Univ. Mass. Amherst. Amherst, MA **7**, 224 (2006)
32. J. Chen, X. Xia, D. Lo, J. Grundy, X. Luo, T. Chen, DefectChecker: Automated smart contract defect detection by analyzing EVM bytecode. IEEE Trans. Softw. Eng. **48**, 2189–2207 (2022)

33. A. Biryukov, S. Tikhomirov, Deanonymization and linkability of cryptocurrency transactions based on network analysis, in *2019 IEEE European Symposium on Security and Privacy (EuroS&P)*, (IEEE, 2019), pp. 172–184
34. L. Zhang, H. Li, Y. Li, Y. Yu, M.H. Au, B. Wang, An efficient linkable group signature for payer tracing in anonymous cryptocurrencies. Futur. Gener. Comput. Syst. **101**, 29–38 (2019)
35. J. Li, N. Li, J. Peng, H. Cui, Z. Wu, Energy consumption of cryptocurrency mining: A study of electricity consumption in mining cryptocurrencies. Energy **168**, 160–168 (2019)
36. M.J. Krause, T. Tolaymat, Quantification of energy and carbon costs for mining cryptocurrencies. Nat. Sustain. **1**(11), 711 (2018)
37. U. Gallersdörfer, L. Klaaßen, C. Stoll, Energy consumption of cryptocurrencies beyond Bitcoin. Joule **4**(9), 1843–1846 (2020)
38. L. Yan, N. Mirza, M. Umar, The cryptocurrency uncertainties and investment transitions: Evidence from high and low carbon energy funds in China. Technol. Forecast. Soc. Change **175**, 121326 (2022)
39. K.J. O'Dwyer, D. Malone, *Bitcoin Mining and Its Energy Footprint* (2014), 25th IET Irish Signals & Systems Conference and China-Ireland International Conference on Information and Communications Technologies (ISSC 2014/CIICT 2014), Limerick, June 26–27.
40. Bitcoin.org, Bitcoin mining. (2019), https://bitcoin.org/en/faq#what-is-bitcoin-mining
41. N. Carter, Digesting 'Quantification of energy and carbon costs for mining cryptocurrencies'. (2018), Medium.com [Online]. Available: https://medium.com/@nic__carter/digesting-quantification-of-energy-and-carbon-costs-for-mining-cryptocurrencies-1f019e10fad4
42. C. Mora, R.L. Rollins, K. Taladay, M.B. Kantar, M.K. Chock, M. Shimada, E.C. Franklin, Bitcoin emissions alone could push global warming above 2 C. Nat. Clim. Chang. **8**(11), 931 (2018)
43. C. Stoll, L. Klaaßen, U. Gallersdörfer, *The Carbon Footprint of Bitcoin* (2018)
44. A. De Vries, C. Stoll, Bitcoin's growing e-waste problem. Resour. Conserv. Recycl. **175**, 105901 (2021)
45. E. Masanet, A. Shehabi, N. Lei, H. Vranken, J. Koomey, J. Malmodin, Implausible projections overestimate near-term Bitcoin CO2 emissions. Nat. Clim. Chang. **9**(9), 653–654 (2019)
46. PRNewswire, Blockchain goes green: Signum – The world's first truly sustainable blockchain steps into the light [Online]. (2021), Available: https://www.prnewswire.com/news-releases/blockchain-goes-green-signum%2D%2D-the-worlds-first-truly-sustainable-blockchain-steps-into-the-light-301320293.html
47. O. Schrijvers, J. Bonneau, D. Boneh, T. Roughgarden, Incentive compatibility of Bitcoin mining pool reward functions, in *International Conference on Financial Cryptography and Data Security*, (Springer, Berlin, Heidelberg, 2016), pp. 477–498
48. J. Tuwiner, Bitcoin mining pools. (2019), https://www.buybitcoinworldwide.com/mining/pools/
49. R.K. Jana, I. Ghosh, M.W. Wallin, Taming energy and electronic waste generation in Bitcoin mining: Insights from Facebook prophet and deep neural network. Technol. Forecast. Soc. Change **178**, 121584 (2022)
50. S. Nandi, J. Sarkis, A.A. Hervani, M.M. Helms, Redesigning supply chains using blockchain-enabled circular economy and COVID-19 experiences. Sustain. Prod. Consum. **27**, 10–22 (2021)
51. W.A. Kaal, Blockchain solutions for agency problems in corporate governance, in *Information for Efficient Decision Making: Big Data, Blockchain and Relevance*, (World Scientific, Singapore, 2021), pp. 313–329
52. A. Whitaker, R. Kräussl, Fractional equity, blockchain, and the future of creative work. Manag. Sci. **66**(10), 4594–4611 (2020)
53. T.L.D. Huynh, E. Hille, M.A. Nasir, Diversification in the age of the 4th industrial revolution: The role of artificial intelligence, green bonds and cryptocurrencies. Technol. Forecast. Soc. Change **159**, 120188 (2020)
54. B. Esmaeilian, J. Sarkis, K. Lewis, S. Behdad, Blockchain for the future of sustainable supply chain management in industry 4.0. Resour. Conserv. Recycl. **163**, 105064 (2020)

55. F. Imbault, M. Swiatek, R. De Beaufort, R. Plana, The green blockchain: Managing decentralized energy production and consumption, in *2017 IEEE International Conference on Environment and Electrical Engineering and 2017 IEEE Industrial and Commercial Power Systems Europe (EEEIC/I&CPS Europe)*, (IEEE, 2017), pp. 1–5
56. J. Woo, R. Fatima, C.J. Kibert, R.E. Newman, Y. Tian, R.S. Srinivasan, Applying blockchain technology for building energy performance measurement, reporting, and verification (MRV) and the carbon credit market: A review of the literature. Build. Environ. **205**, 108199 (2021)
57. D. Patel, B. Britto, S. Sharma, K. Gaikwad, Y. Dusing, M. Gupta, Carbon credits on blockchain, in *2020 International Conference on Innovative Trends in Information Technology (ICITIIT)*, (IEEE, Piscataway, 2020), pp. 1–5
58. G. Dorfleitner, D. Braun, Fintech, digitalization and blockchain: Possible applications for green finance, in *The Rise of Green Finance in Europe*, (Springer, Cham, 2019), pp. 207–237
59. N. Lindenberg, *Definition of Green Finance, Presentation*, German Development Institute / Deutsches Institut für Entwicklungspolitik (2014)
60. T. Conlon, S. Corbet, R.J. McGee, Are cryptocurrencies a safe haven for equity markets? An international perspective from the COVID-19 pandemic. Res. Int. Bus. Financ. **54**, 101248 (2020)
61. J.W. Goodell, S. Goutte, Diversifying equity with cryptocurrencies during COVID-19. Int. Rev. Financ. Anal. **76**, 101781 (2021)
62. A. Rijanto, Business financing and blockchain technology adoption in agroindustry. J. Sci. Technol. Policy Manag. **12**, 215 (2020)
63. H. Mora, M.R. Morales-Morales, F.A. Pujol-López, R. Mollá-Sirvent, Social cryptocurrencies as model for enhancing sustainable development. Kybernetes **50**, 2883 (2021)
64. L. Kugler, Non-fungible tokens and the future of art. Commun. ACM **64**(9), 19–20 (2021)
65. N.O. Madichie, R.E. Hinson, The future of Africa's creative industries, in *The Creative Industries and International Business Development in Africa*, (Emerald Publishing Limited, Bingley, 2022)
66. T. Wastling, F. Charnley, M. Moreno, Design for circular behaviour: Considering users in a circular economy. Sustainability **10**(6), 1743 (2018)
67. E.L. Amel, C.M. Manning, B.A. Scott, Mindfulness and sustainable behavior: Pondering attention and awareness as means for increasing green behavior. Ecopsychology **1**(1), 14–25 (2009)
68. D. McKenzie-Mohr, *Fostering Sustainable Behavior: An Introduction to Community-Based Social Marketing* (New Society Publishers, Gabriola, 2011)
69. R. Tao, C.-W. Su, B. Naqvi, S.K.A. Rizvi, Can Fintech development pave the way for a transition towards low-carbon economy: A global perspective. Technol. Forecast. Soc. Change **174**, 121278 (2022)
70. L.T.Q. Nguyen, T.G. Hoang, L.H. Do, X.T. Ngo, P.H.T. Nguyen, G.D.L. Nguyen, G.N.T. Nguyen, The role of blockchain technology-based social crowdfunding in advancing social value creation. Technol. Forecast. Soc. Change **170**, 120898 (2021)
71. J. Truby, Decarbonizing Bitcoin: Law and policy choices for reducing the energy consumption of blockchain technologies and digital currencies. Energy Res. Soc. Sci. **44**, 399–410 (2018)
72. C. Liu, X. Zhang, F. Medda, Plastic credit: A consortium blockchain-based plastic recyclability system. Waste Manag. **121**, 42–51 (2021)
73. R. Ajwani-Ramchandani, J. Sarkis, Time to consider circular and social credits exchanges? Resour. Conserv. Recycl. **175**, 105860 (2021)
74. P. De Filippi, What blockchain means for the sharing economy. Harv. Bus. Rev. **15**, 1–6 (2017)
75. W. McElroy, Bitcoin markets: Black and gray, white and red. (2016), Bitcoin.com
76. F. Mustafa, S. Lodh, M. Nandy, V. Kumar, Coupling of cryptocurrency trading with the sustainable environmental goals: Is it on the cards? Bus. Strateg. Environ. **31**(1), 1152 (2021). https://doi.org/10.1002/bse.2947
77. C.A. Bai, J. Cordeiro, J. Sarkis, Blockchain technology: Business, strategy, the environment, and sustainability. Bus. Strateg. Environ. **29**(1), 321–322 (2020)

Part II
Surveys on Blockchain and Smart-Contract Technologies

A Taxonomy on Blockchain-Based Technology in the Financial Industry: Drivers, Applications, Benefits, and Threats

Andry Alamsyah and Syahputra Syahrir

Abstract Blockchain technology, renowned for its decentralization capabilities, has led to numerous innovations across various sectors. One of the most heavily regulated sectors, the financial industry faces a significant challenge as decentralized finance (DeFi) advantages are increasingly recognized and adopted. The decentralized nature of blockchain technology has two primary facets: improved operational efficiency and product innovation. The first facet concerns transparent recordkeeping, establishing trust in business processes, enhancing efficiency, and reducing stakeholder coordination efforts. The second aspect involves the creation of innovative digital financial products that circumvent intermediaries, ensure security, and maintain privacy. However, most practical implementations and academic resources primarily focus on the centralized paradigm within the financial industry. This chapter, therefore, proposes a taxonomy that outlines how the aspects above can be realized. Our research methodology includes a literature review and interviews with domain experts. We delineate four dimensions of the proposed taxonomy: drivers, benefits, applications, and threats. Moreover, we discuss potential solutions to address these emerging threats.

Keywords Financial Industry · Decentralized Finance · Financial Technology · Blockchain Technology · Review

1 Introduction

As of 2022, the financial industries are slightly trailing behind the technological sector in terms of the world's largest industries in market capitalization, valued at \$7 trillion compared to \$10 trillion [1]. Technological advancements have significantly impacted the financial sector, leading to Financial Technology

A. Alamsyah (✉) · S. Syahrir
School of Economic and Business, Telkom University, Bandung, Indonesia
e-mail: andrya@telkomuniversity.ac.id

N. El Madhoun et al. (eds.), *Blockchain and Smart-Contract Technologies for Innovative Applications*, https://doi.org/10.1007/978-3-031-50028-2_4

(FinTech) companies. FinTech encompasses financial organizations that employ primarily Internet-based technology to develop and innovate products/services [2]. Such technology facilitates the introduction of novel products characterized by enhanced security, efficiency, flexibility, and opportunities. For instance, leveraging technology for credit card payments via smartphones and tablets [3] improves payment efficiency by enabling access to users' devices without needing a physical card.

The development of the financial industry has been profoundly influenced by the evolution of FinTech, which consists of three digitalization phases. The first phase, dubbed "internal" digitalization, focuses on streamlining financial institutions' internal operations by automating products and services. ATMs, which provide customers access to banking services without visiting a branch, exemplify this phase. The second phase, known as provider-oriented digitalization, targets the standardization of processes and applications across various organizational functions, aiming to establish a cohesive business process and reduce in-house production degrees. The third and final phase, customer-oriented digitalization, prioritizes customer needs and fosters the development of a new financial ecosystem, enabling users to conduct peer-to-peer transactions without third-party involvement.

Blockchain technology holds the potential to disrupt business processes and catalyzes innovations. As a secure ledger comprising a chain of blocks protected by cryptographic techniques to ensure integrity [4], blockchain finds applications in both financial and nonfinancial sectors [5]. The transformative impact of blockchain on organizational structures is embodied by Decentralized Autonomous Organizations (DAOs), which offer a decentralized system as opposed to traditional pyramid hierarchies. Decentralization also stimulates the innovation of novel products by obviating intermediary functions. For example, Bitcoin, proposed by Satoshi Nakamoto in 2008, is underpinned by blockchain technology, providing a decentralized mechanism for peer-to-peer payments without third-party interference [6]. The blockchain relies on two critical drivers: Smart Contracts and Distributed Ledger Technology (DLT). Immutable by nature, Smart Contracts necessitate thorough scrutiny by both parties prior to signing. At the same time, DLT securely records transactions by sharing records among multiple nodes and employing a consensus algorithm for data validation, ensuring all nodes possess up-to-date information.

Blockchain's decentralization concept fosters new product innovations, enabling stakeholders to engage directly without central authority intervention. In the financial domain, this suggests that activities such as savings, lending, and borrowing can occur between individuals. DeFi, which empowers users to manage their virtual assets instead of relying on traditional financial institutions, arises from this decentralization principle. Centralized Finance (CeFi), another blockchain-based financial product, does not embody blockchain's decentralization essence, as it represents an institution managing user assets. Consequently, our analysis focuses on DeFi and its derivatives.

This chapter aims to offer a classification of blockchain's integration into finance, encompassing the financial industry and blockchain background with a focus on DeFi, a literature review, a taxonomy outlining dimensions of key drivers, benefits, products, and threats, potential solutions to address these threats, and a conclusion. We organize the chapter as follows: Section 1 discusses the chapter motivation and FinTech phenomenon, including blockchain technology impact. Section 2 discusses the taxonomy construction; it contains the methodology explanation and several critical points in the literature review, such as blockchain and decentralization idea, decentralized vs. traditional finance, tokenization, technological issues and regulations. Section 3 explains the constructed taxonomy and its explanations. Section 4 offers approaches to mitigate threats relating to blockchain adoptions. Section 5 summarizes and discusses the whole chapter idea, while Sect. 6 provides the conclusions and suggestions for future research.

2 The Taxonomy Construction

This chapter employs a dual-method research approach: a literature review and interviews. The literature analysis thoroughly examines numerous studies related to blockchain technology and its various applications. For example, Zhang and Chan's work [7] explores energy consumption in blockchain consensus algorithms, while Tarr's research [8] investigates blockchain use cases in the insurance sector. Despite its potential for widespread use across multiple industries, blockchain technology is still relatively new, with only a limited number of organizations adopting it thus far. Our literature analysis encompasses 73 research articles from various sources such as Elsevier, Emerald, MDPI, IEEE, ACM, and Taylor & Francis. We used keywords like blockchain technology, application, and implementation, as well as other terms specific to financial industry blockchain products. We have found five articles that cover existing DeFi or blockchain in financial industry taxonomy and compare them to this chapter. The comparison is shown in Table 1.

Moreover, we have examined several well-known DeFi/crypto project whitepapers, their respective websites, and source code from an industry perspective. These projects include Ethererium, MakerDAO, Uniswap, Compound, Aave, Yearn.finance, Curve Finance, Balancer, Sushiswap, and Synthetix. Additionally, we have compiled data on Central Bank Digital Currency (CBDC) from cbdctracker.org, a portal committed to monitoring CBDC progress, along with publications produced by global institutions such as the Bank for International Settlements (BIS), incorporating the views of financial regulatory bodies.

The interview component consists of engaging with experts and practitioners in blockchain and finance to comprehend the ramifications of blockchain technology on the financial sector. Interviewees comprise notable individuals from various backgrounds, including programmers, policymakers, and researchers. Questions are formulated based on two aspects: the technological disruption in the financial industry and the elucidation of blockchain features and potential to support taxonomy construction. The interview encompasses inquiries about blockchain definitions and features, its adoption and development within the financial industry, comparisons with conventional technology, financial industry business processes, regulatory

Table 1 Comparison of existing DeFi or blockchain in financial industry taxonomy

Publication	Scope
A Taxonomy of Decentralized Finance. *Puschmann* et al. *(2023)* [9]	Specifically, explore DeFi operations from conceptual and empirical approaches, including business process, price mechanism, token type, and value proposition
Managing Risk in DeFi. *Jensen* et al. *(2021)* [10]	Present the taxonomy of agents, roles, and risks in open finance DeFi. It presents four critical sources of complexity and risks for managers, scholars, and practitioners to engage with this new breed of financial applications
Finance Infrastructure through Blockchain-Based Tokenization. *Tian* et al. *(2020)* [11]	Provide the role of tokenization in improving infrastructure assets liquidity, transaction efficiency, and transparency. The study also found the potential of tokenization has not yet been fully realized because of the limited technical infrastructures, regulation uncertainties, volatilities in the token market, and absence of the public sector
Automated market makers (AMM) and decentralized exchange: a DeFi primer. *Mohan. (2022)* [12]	Organize AMM development as a neoclassical black box that converts inputs (tokens) to outputs (prices). Governed tokenization by the technology of the AMM summarized by an "exchange function." various types of AMMs are examined. The paper also looks at the impact of introducing concentrated liquidity in an AMM
The future of FinTech: Integrating finance and Technology in Financial Services. *Nicoletti. (2017)* [13]	Provides an introduction to the state of the art in FinTech. Covers innovation in products, processes, and business models. Includes coverage of regulation in FinTech. Features case studies of key FinTech products and e-business models. Includes a range of social funding platforms and virtual currencies
This chapter	Our objective is to construct a broader taxonomy aspect of blockchain implementation in a digestible approach to the public. We discuss the following points: Scope: A broader range of implementation, from enhancing existing business processes (speed, transparency, security, auditing, etc.) to support new products or innovations. It could involve both centralized and decentralized systems Control: We discuss blockchain utilization under centralized finance (traditional finance or TradFi) and decentralized finance (DeFi). DeFi products/services will be extensively discussed in the applications dimension in the taxonomy Financial instruments: DeFi aims to recreate TradFi products controlled by smart contracts to operate on a peer-to-peer basis, while TradFi does not necessarily use blockchain technology for decentralized objectives Interoperability: By using popular blockchain platform such as Ethereum, DeFi is looking for interoperability to support composability features, like creating "money lego." TradFi, on the other hand, does not intend to create this level of interoperability Risk and regulation: DeFi's decentralized and permissionless nature potentially creates certain risks do not present in TradFi. We discuss broader aspects of how blockchain could enforce risk management that might occur in financial industry practice

principles and frameworks, opportunities and challenges, and risk mitigation. The interviews and literature review yield several significant aspects elaborated upon in subsequent paragraph discussions.

2.1 Blockchain and Decentralization Idea

Blockchain technology can be divided into permissioned and permissionless blockchains based on member participation [14]. Permissioned blockchains restrict access to information and system participation to specific authorized parties, whereas permissionless blockchains permit universal access and oversight of system information. Depending on their requirements, financial institutions can opt for one or both blockchains. For instance, an insurance company may implement a permissioned blockchain to safeguard privacy and prevent public data exposure. Blockchain technology is driven by two critical components: Smart Contracts and DLT. Smart Contracts represent legal contracts that can be executed and expressed in software [15], essentially constituting directly coded, self-executing agreements between buyers and sellers [16]. Smart Contracts function by incorporating code into the blockchain, which, once deployed to the network, is immutable and can only be executed once the stipulated conditions are satisfied. DLT comprises multiple transaction copies distributed among numerous participants and updated via consensus among the involved parties [17]. This technology ensures information synchronization across multiple nodes, guaranteeing each node possesses the latest ledger update.

A decentralized system serves as a significant distinguishing factor between blockchain and traditional technologies. Centralized systems store records on a single server while maintaining backups on additional devices. In contrast, decentralized systems employ consensus algorithms to synchronize records across multiple devices concurrently. Various consensus algorithms exist, with Proof of Work (PoW) and Proof of Stake (PoS) being two of the most prominent. PoW rewards miners based on their capacity and speed in solving mathematical problems provided by the network, with miners possessing greater computational power having increased chances of receiving network rewards. Bitcoin, the most valuable cryptocurrency, utilizes PoW as its consensus algorithm. The underlying principle of PoW necessitates that no miner has more than 50% of the network's total computational power, preventing any one miner from controlling the entire network. Conversely, the PoS system distributes rewards to miners based on the quantity and duration of coins held within the network [18], making it more challenging for malicious actors to seize control of the network, as this would require substantial financial and temporal investments. PoS is regarded as a more secure alternative to PoW, as it can address the 51% attack issue. PoS is also more energy efficient than PoW, as it selects network leaders based on their stake and time commitment, thereby reducing the energy consumption required for block mining compared to PoW, which relies on high computational power [19].

Blockchain technology's characteristics have the potential to transform business processes within FinTech companies, encompassing aspects of security and governance. The term "security" pertains to several properties, including immutability, transparency, traceability, data records, and tokenization. For example, FinTech companies can facilitate improved procurement agreements with other organizations through Smart Contracts. The second characteristic involves the capacity to modify a company's governance system, decentralizing the organizational structure to achieve more efficient business processes. In the blockchain governance system, the relationships between nodes are characterized by values of voluntariness, equality, and mutual benefit.

2.2 Decentralized Versus Traditional Finance

Conventional financial systems, also referred to as Traditional Finance (TradFi) or CeFi, have historically depended on intermediation mechanisms, encompassing intermediaries, due diligence processes, and regulatory frameworks to bolster the security of stakeholders engaging in transactions. Nevertheless, this approach incurs a compromise in optimal market efficiency and constrains opportunities for asset proprietors. Market effectiveness and prospects are curtailed when participants are reluctant to incur fees associated with rent-seeking intermediaries, are precluded from market access by centralized authorities, or harbor reservations about engaging in a system where their assets are subject to third-party control. The centralized nature of these systems forms the basis for their designation as CeFi.

DeFi exemplifies the core principles of blockchain within the FinTech sector. We focus on DeFi's discussion, as it epitomizes the essence of blockchain: decentralization. DeFi enables users to manage their financial assets, while CeFi denotes a financial institution that controls its users' financial assets [20]. DeFi encompasses elements: DLT, Smart Contracts, disintermediation, and open banking [21]. Open banking supports inclusivity, and supported by mechanisms such as permissionless, trustless, transparent, programmable, and censorship-resistant are already in place. In reality, we are far from the decentralization nature of everything in the digital realm. For example, we still need to exchange assets to enter the DeFi space. The exchange entity is centralized by nature; they provide swap services for different formats and types of assets. Some assets need support and meticulous curation from the exchange to be promoted in the exchange list and adopted by a wider audience. From CeFi/exchanges such as *Binance, Coinbase, and Uphold*, we can see how they accumulated cost compared to fully decentralized like Defi.

Figure 1 demonstrates how TradFi can incur higher costs compared to decentralized finance (DeFi). TradFi experiences elevated intermediation costs, while DeFi has minimal costs (limited to gas fees for transactions or agreements). These significant cost differences result in higher transaction thresholds or restrictions for TradFi, whereas DeFi allows unrestricted access and even supports fractionalized

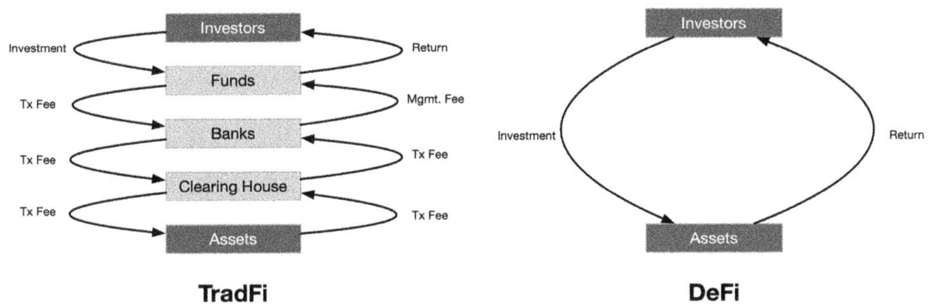

Fig. 1 Different process of TradFi vs. DeFi

ownership. Additionally, DeFi enables instant finality and transparent order flow, while TradFi exhibits longer settlement times and private order flows.

2.3 *Token and Tokenization*

Tokenization is a process of transforming traditional assets or real-world assets (RWA) into digital assets. The digital token as a representation of a digital asset is written on blockchain or DLT in general. The process is critical due to its role in becoming the portal for RWA, including financial products such as securities, equities, bonds, commodities, real estate, stocks, insurance, and other financial products. In principle, there are two categories of asset tokenization [22]: security tokens that are native to the blockchain with no physical presence and tokenized securities that are not native to the blockchain and have a physical presence. Tokenized securities, any RWA, such as bonds, stock, real estate, etc., are put in custody, tokenized, and a digital representation on a blockchain is created. The digital natives token on a blockchain, like Bitcoin, inspired the conversion of RWA to digital financial markets based on blockchain technology.

The tokenization aspect of blockchain allows users to manage their digital assets in a more decentralized manner compared to traditional methods. Tokenization refers to the digitalization of assets, enabling fractional investment and ownership [23]. These tokens can then be traded or transferred, akin to conventional securities [16]. In an economy fueled by tokenization, technical protocols supplant the role of intermediaries [24], thereby facilitating streamlined tracking of assets' origins, authenticity, and rights [25]. Tokenized assets are not restricted to physical items but can also extend to intangible assets, such as Non-Fungible Assets (NFTs). For example, John owns 30% of a digitized yellow dog NFT. As the owner, John has the right to sell his NFT; however, this is limited to his percentage, equating to 30% of the total asset.

Tokenization could be treated as information bridging, transferring information regarding an asset's economic value and ownership to the on-chain (to be stored

Table 2 A different proposal for asset tokenization

Proposal	Asset coverage	Features
ERC-20	Anything	Most popular standard in tokenized assets. Wide DeFi support. Lacks legal specificity
ERC-721	Anything	Commonly used to represent unique goods, e.g., art/NFT. Some adoption in DeFi to represent proof of ownership
ERC-3643	Securities	Specific to assets that are deemed as securities. Has encoded compliance
ERC-2222	Dividend bearing	The standard for assets that are claims on future cash flows. No legal compliance
ERC-4626	Yield bearing	The standard for assets that use DeFi's yield-bearing vaults. No legal compliance

within the blockchain ledger). There have been numerous proposals to create a DeFi standard for asset tokenization, as shown in Table 2. It requires external data to portray the value of assets that lives off-chain accurately. Oracle is the mechanism to pull data from centralized or decentralized sources so that assets on-chain can mimic its market value off-chain.

2.4 *Technological Issues and Regulations*

Despite the numerous benefits provided by blockchain technology, it is not without its imperfections. Blockchain may give rise to new threats originating from both internal and external factors. If not adequately addressed, these threats could potentially harm financial institutions. However, researchers have endeavored to tackle these risks, which will be outlined in Sect. 4. We have identified five specific threats to blockchain implementation in the financial industry: scalability and speed, security, high operational cost, regulation, environment and energy.

Limited scalability and speed present significant obstacles to blockchain implementation. In contrast to Visa, which can accommodate approximately 2000 transactions per second, blockchain systems such as Bitcoin can only process seven transactions per second [20]. Furthermore, a blockchain typically generates a mere 1 MB of data every 10 minutes [21], constraining the number of transactions. Storage capacity is another concern, as data produced by multiple devices on the blockchain must be processed and stored back on the network. Networking also poses challenges for scalability and speed, as transactions must be transmitted to each node twice [26], leading to block propagation delays and substantial network resource consumption.

Despite its advantages for FinTech, blockchain remains susceptible to security threats. The PoW algorithm gives rise to a threat called the 51% attack, which involves seizing control of the entire network by commanding more than 50% of the total computational power. Upon achieving 51%, malicious parties can control the blockchain and impose rules as they see fit. Such rules may include halting the verification of transaction blocks, preventing miners from mining the remainder of available blocks, and altering data containing all transactions. Another security threat is phishing [27], in which attackers place fraudulent wallets on various popular apps, leaving victims unable to withdraw funds from these fake wallets since the private key already belongs to the scammer.

Although blockchain can reduce intermediary costs through decentralization, it introduces a new problem. The limited size of a blockchain system prompts miners to impose higher mining fees, with larger transactions incurring higher fees. Blockchain miners, particularly those on Ethereum, receive payment in the form of gas rather than fiat currency. The gas price refers to the fee required to perform computational effort [28], while the total transaction fee in Ethereum is calculated by multiplying the gas limit by the gas price [29]. The gas limit signifies the maximum amount of gas the creator is willing to expend, and the gas price indicates the amount of Gwei the creator is prepared to pay [30]. It is worth noting that 1 Ether equals 1^0.18 Wei, and 1 Gwei equals 1,000,000,000 Wei [29]. As blockchain's popularity and demand surge over time, gas costs continue to rise, necessitating Ethereum's management of millions of daily transactions.

Inadequate regulations may heighten the risk of criminal activity. For example, criminals might utilize blockchain-based assets like cryptocurrency for money laundering and obfuscating the origins of money. Blockchain-based digital assets are vulnerable to money laundering for two reasons [31]. First, global anti-laundering agencies primarily focus on illegal financial flows and transactions and whether they align with standard theoretical money. Fiat money and blockchain-based digital assets receive similar treatment in combating money laundering. Second, digital assets' innovative methods for verification, execution, and transaction publication occur almost instantaneously.

Blockchain technology potentially consumes more energy than conventional technologies. For context, mining 1 Bitcoin demands energy equivalent to powering one US household for 2 years [32], illustrating blockchain's enormous energy consumption and its consequent carbon footprint. Consensus algorithms like PoW require significant power, as miners receive rewards when discovering a new block in the network [33]. The increased work rate consumes energy, which ultimately harms the environment. Blockchain's energy consumption has been widely criticized due to its environmental impact. As the title implies, this chapter concentrates on the influence of blockchain on the financial industry and does not emphasize its implementation in other sectors. However, we provide an overview of how blockchain technology generally operates, to familiarize readers with the broad workings of blockchain before examining its application in the financial industry.

3 The Taxonomy

Based on the discourse in the taxonomy construction, we construct the taxonomy into four main dimensions: drivers, benefits, applications, and threats. Figure 2 illustrates a broad view of the taxonomy.

An in-depth examination of each of these four dimensions is conducted, ranging from drivers to threats. Furthermore, we investigate the relationships between dimensions to comprehend how one may influence another. The bold line signifies the branching of the original dimensions. For instance, the Smart Contract comprises three branches: Decentralized Autonomous Organizations (DAO), Decentralized Applications (DApps), and Automated Market Makers (AMM). The dotted line represents the characteristics/traits empowered by the originating dimension. For example, Distributed Ledger Technology (DLT) attributes foster transparency, immutability, data consistency, and traceability.

3.1 *Drivers*

Blockchain technology adoption in the financial industry could be influenced by two main factors: the environment and technology capabilities. In the environmental setting, a disruption effort by the public to have less centralized financial authority learning from the downside of the current system, which shows a wider gap between rich-poor, inefficient financial activity, low financial inclusions, less control over financial products, privacy, and other factors. At the same time, the technology

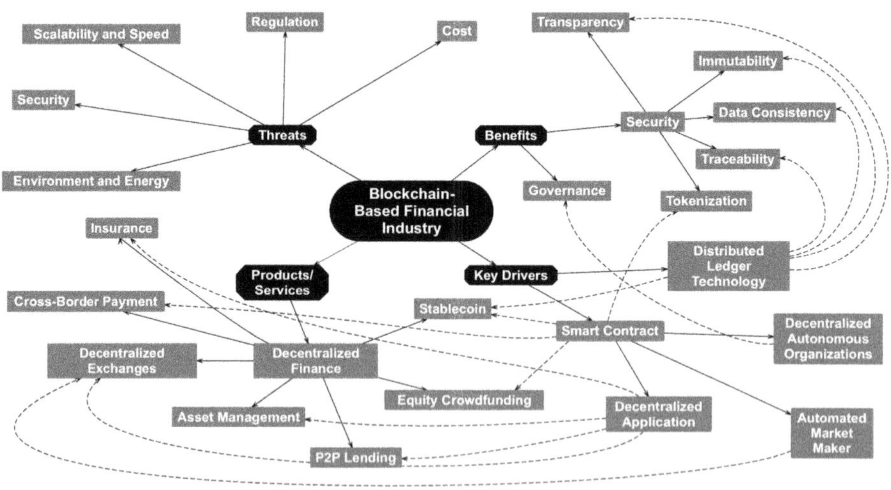

Fig. 2 A taxonomy of blockchain-based technology in the financial industry

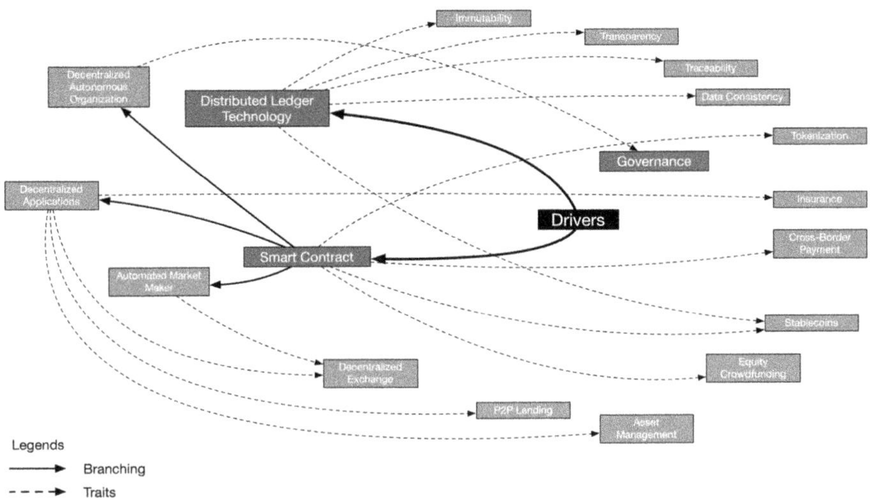

Fig. 3 Driver dimension

capabilities refer to the ability of blockchain itself to represent digital assets with tempting features, as we mentioned in blockchain technology properties. We focus the taxonomy on technological capabilities. They are Smart Contracts and Distributed Ledger Technology (DLT). Figure 3 focuses on the driver dimensions and their relations. The subsequent subsections provide a comprehensive discussion of each driver:

3.1.1 Smart Contract

A Smart Contract is defined as a computer programming code embedded within the blockchain data structure. As it is integrated into the blockchain, it inherits characteristics such as immutability, tamper resistance, and transparency. Once established, a Smart Contract can be executed on the blockchain network and applied to any participant within the network. Smart Contracts can function as an alternative to intermediaries in the financial industry, facilitating decentralized transactions [34]. This allows users to engage in direct transactions with one another, bypassing the need for a financial institution to dictate the terms of trade. The peer-to-peer interaction characteristic within the financial industry is commonly referred to as Decentralized Finance (DeFi). Moreover, Smart Contracts foster a financial ecosystem where users maintain control over their assets. Although financial companies continue to operate within the DeFi space, they serve primarily as platform providers to enable asset trading among users without directly holding any assets themselves.

The Smart Contract allows for unalterable and globally enforced agreements between parties, as it maintains a synchronized record. The code within a Smart Contract is self-executing once the agreed-upon conditions are met. For example, a company may utilize a Smart Contract to manage employee payroll, with the sole requirement being the arrival of the 25th day of each month. In this case, factors such as performance are not incorporated into the contract. Consequently, employees will automatically receive their salaries on the 25th day of the month, irrespective of their performance or other negative factors. Smart Contracts serve as the foundational principle for various technologies, including Decentralized Autonomous Organizations (DAO), Decentralized Applications (DApp), and Automated Market Makers (AMM), which will be further elaborated in the subsequent subsections.

Decentralized Autonomous Organization (DAO)

DAO is a blockchain-based mechanism enabling individuals within an organization to govern and coordinate themselves according to a set of predefined rules [35]. DAO can transition an organization's hierarchical structure into a decentralized one. Traditional corporate governance presents two fundamental challenges [34]: individuals within the organization may not consistently adhere to the rules, and others may not always agree with the established regulations. However, DAO possesses the capacity to address these issues through the following attributes [36]:

1. The absence of a hierarchical structure and central authority signifies that connections between nodes are not governed by managerial associations but by voluntariness, equality, and mutual benefit principles.
2. The existence of distributed organizations, wherein stakeholders define all rules and collaboration patterns within the DAO. As a result, implementing DAO can enhance trust while reducing communication and transaction costs.
3. The reliance on Smart Contracts bolsters DAO's transparency and openness concerning participants' responsibilities, authorities, operational rules, penalties, and rewards.

Decentralized Applications (DApps)

DApps are a practical manifestation of the functioning of DAO. These applications are characterized by a central entity's lack of control or authority; instead, control rests with the users. DApps rely on Smart Contracts to execute transactions automatically and transparently without the need for intermediaries. Generally, there are three primary types of applications: centralized, distributed, and decentralized. Centralized applications are governed by a single central authority, while distributed applications entail the distribution of information across multiple nodes. Decentralized applications, however, lack any central authority. DApps possess four key features [37]:

1. Firstly, they are open-source applications, allowing for network transparency. Still, it also renders them susceptible to plagiarism as the content is accessible to anyone on the network.
2. DApp developers participate in economic activities by developing a coin that they can regulate the distribution and supply of, and its worth is determined by public perception.
3. To validate transactions in DApps, the implementation of blockchain technology and a consensus algorithm is essential. Transactions are recorded immutably in the blockchain, guaranteeing their history, and preventing malicious parties from altering them.
4. DApps are designed to avoid the impact of server shutdowns by ensuring that another node takes over the application's functions if one node fails.

Automated Market Maker (AMM)

AMM is a novel method of conducting trade that involves the utilization of a third party to operate the order book. AMM is essentially an algorithm or Smart Contract allowing participation in an electronic market [38]. The primary advantage of employing AMM is that it provides liquidity, which is the ability to convert assets into cash without causing significant changes to the market price. AMM functions as a single-purpose algorithm that matches trading orders and determines execution prices [39]. Through a liquidity pool, it offers superior liquidity compared to a conventional order book. The liquidity pool serves as a "vault" or "pool" in which the liquidity provider (investor) deposits one or more tokens for the trader's use in activities such as lending, borrowing, derivatives, insurance, and swapping [40]. The trader pays a service fee for conducting a transaction in the liquidity pool, which is then transferred to the liquidity provider as a reward for supplying liquidity. AMM is employed by various blockchain-based FinTech firms such as *Uniswap, Balancer, PancakeSwap, Osmosis,* and *1inch*.

3.1.2 Distributed Ledger Technology (DLT)

DLT stores and shares data across a network of computers, making it more resilient compared to a centralized ledger [41]. This characteristic ensures that nodes are synchronized with the latest information updates across the network. Thus, if one node fails to receive updated information, another node can notify it. For example, a centralized system may back up data hourly, recording it an hour later. The system records a transaction at 07.00 AM and backs up the data at 08.00 AM. However, if a system malfunction occurs at 7.30 AM, it could permanently alter all data recorded from 07.00 to 07.30 AM. On the other hand, a decentralized system synchronizes

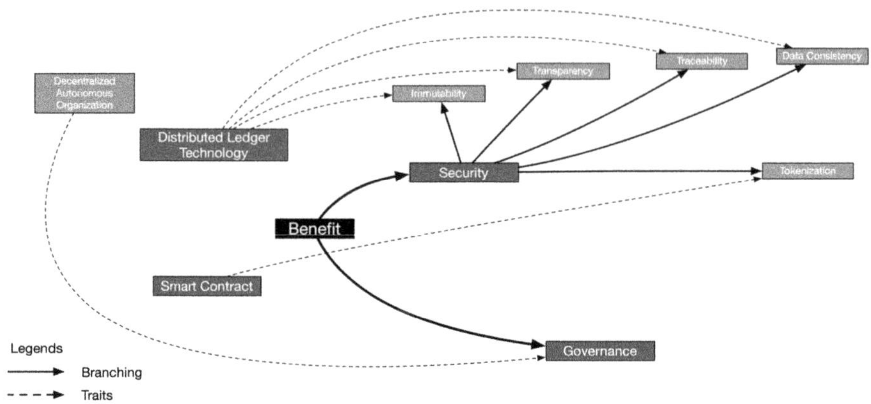

Fig. 4 Benefit dimension

transaction records simultaneously to every node in the system. Several important blockchain features, such as transparency, data consistency, immutability, and traceability, are enabled by DLT.

DLT comprises three primary layers [42]:

1. Foundation: This refers to the entire DLT system's underlying framework, which describes the rules governing how the network distributes information.
2. Network: This layer comprises interconnected nodes or actors sharing, storing, and processing data.
3. Data: This layer represents the data stored in the DLT system as records.

3.2 *Benefits*

Developing innovative technologies primarily aims to offer advantages beyond traditional methods. In the context of incorporating blockchain into FinTech firms, a thorough analysis has identified key components for inclusion in the classification system. The core benefits of blockchain implementation include enhanced security and governance, which are vital for the financial industry. Other advantages, such as transparency, cost reduction, faster transactions, improved efficiency, data integrity, accessibility, simplified compliance, and tokenization, are also significant but fall under different feature dimensions. The focus remains on security and governance, as the financial sector requires secure representation of wealth/assets and welcomes innovative transaction governance, particularly in P2P transactions. Figure 4 exhibits details of the benefit dimension and its relations.

3.2.1 Security

Security constitutes the foremost advantage offered by blockchain technology. Enhanced security enables financial institutions and their clientele to engage in transactions without concerns about asset theft or information compromise. Key features of blockchain, such as Smart Contracts and DLT, play pivotal roles in bolstering the security landscape of FinTech. Smart Contracts facilitate agreements between multiple parties and are activated only upon fulfilling specified conditions. DLT, conversely, concurrently updates records across all authorized nodes. The confluence of attributes such as immutability, transparency, traceability, data consistency, tokenization, and governance contribute to the appeal of blockchain for maintaining security and privacy. Within the financial sector, this technology empowers FinTech firms to introduce fully digital products/services, which assure performance comparable to traditional offerings while capitalizing on the flexibility inherent in digital frameworks.

An additional security feature integrated within the blockchain is the Public Key Infrastructure (PKI). PKI essentially employs public key cryptography to safeguard transactions and ensure privacy and authentication. Each user possesses a public key for receiving transactions and a private key for signing them. Upon initiating a transaction, users sign it with their private key before broadcasting it to the network. Subsequently, other nodes on the network verify the transaction utilizing the sender's public key. Digital signatures offer authenticity and transaction integrity, generated by encrypting transaction data with the sender's private key, and can be verified using the sender's public key.

In summary, blockchain security is bolstered through three primary mechanisms:

1. Hashing and encryption: Hashing connects nodes to achieve a harmonized record, with each node encrypted using private or public keys.
2. Consensus algorithm: This algorithm determines the blockchain participants and limits transaction validation to selected entities. Various blockchain systems may employ different consensus algorithms.
3. Network consensus pertains to the community's collective decision-making process to establish the most recent consensus state, effectively disregarding any questionable transactions.

Immutability

Immutability signifies a state where transactions remain unmodified once verified and logged within the system [43]. Maintaining distributed records across the network makes it increasingly difficult for attackers to manipulate the truth, as they would need to replace each record with their version. As the network expands, such record replacement becomes virtually impossible. Immutability ensures information security and influences the reliability of Smart Contracts, as they inherit these immutable characteristics. Consequently, all parties must carefully review Smart Contract terms before consenting, as these terms cannot be altered once introduced into the system. Immutability fosters trust between parties, continually eliminating the need to verify adherence to the agreed-upon terms.

Immutability affects several positive features, including data integrity, trust, transparency, and security. Ensuring unalterable financial records strengthens data accuracy and trustworthiness. The transparent, verifiable audit trail simplifies regulatory tracking and fraud detection. By preventing unauthorized changes, immutability safeguards against cyber threats and enhances dispute resolution. Streamlining financial processes eliminates manual reconciliation, reduces human error, and saves costs. Immutable records foster trust, ensure regulatory compliance, and minimize legal risks for financial institutions.

Transparency

Transparency represents a critical aspect of blockchain security, leading to enhanced traceability, consistency, and immutability. Blockchain facilitates sharing a replicated ledger, serving as a reliable record system and the definitive source of truth for activities or transactions [44]. The DLT employed by blockchain allows specified parties or anyone to monitor transaction activities. The network can determine data access rights by opting for permissioned or permissionless blockchains. Data transparency is crucial for building trust among stakeholders, particularly in DeFi, where peer-to-peer transactions are promoted. Transparency is fundamental for fostering trust and establishing stakeholder reputations in the absence of a central authority to validate stakeholder legitimacy, preventing untrustworthy stakeholders from gaining endorsement in DeFi. Data transparency covers both state and history on the ledgers.

In the financial industry context, transparency empowers consumers to make informed decisions based on accessible product information. It fosters trust between consumers and institutions by demonstrating honesty regarding product features, fees, and risks. Market-wide transparency enables efficient resource allocation as consumers compare products and promotes fair competition, leading to better pricing, innovation, and customer service. Transparent products allow both consumers and institutions to manage investment risks, promoting market stability and reducing the likelihood of financial crises. Regulators can monitor institutions more effectively, contributing to the financial system's integrity. Furthermore, transparency helps protect consumers from misleading or fraudulent practices by facilitating informed choices.

Traceability

Blockchain bolsters data traceability within a system, which pertains to the capacity to track and trace the history of transactions, interactions, or movements. The DLT employed by blockchain is a key factor in facilitating easy information tracking, as each block contains data from the preceding block. The recorded history is essential for stakeholders to obtain a comprehensive view of any item, such as digital assets representing financial products/services. Consequently, traceability reveals the flow of information stemming from DLT's decentralized nature, where the information within the system is visible to all participants [45, 46]. Each node in the DLT encompasses a complete transaction record, thereby augmenting the traceability of an asset's movement.

There are several advantages of traceability in the financial industry contexts; traceability ensures that all parties involved in a transaction access the complete history of a financial product, fostering trust among stakeholders and reducing information asymmetry. Regulatory bodies often require financial institutions to provide detailed information about transactions and assets. Traceability facilitates compliance by providing an auditable trail of transactions and asset movements. The ability to trace the history of financial products makes it easier to detect and prevent fraudulent activities, as suspicious transactions or patterns can be identified and investigated more efficiently. Traceability allows financial institutions to better understand the origin and movement of assets, enabling them to make informed decisions about risk exposure and implement appropriate risk mitigation strategies. By providing a comprehensive overview of financial products and their associated transactions, traceability streamlines the reconciliation process, reduces manual intervention, and improves overall operational efficiency. In case of disagreements between parties, traceability serves as an unbiased source of truth, helping to resolve disputes by providing a clear and verifiable record of transactions and asset movements. Traceability enhances consumer confidence in financial products by providing an increased level of transparency, accountability, and security.

Data Consistency

Data consistency refers to the ability of blockchain technology to ensure that all copies of information on each stakeholder are identical and up-to-date. This requires record synchronization throughout the operation. It implies that all nodes in the network share the same information, ensuring the information's consistency and accuracy throughout the network. The improved data consistency originated from DLT implementation, where the transactions are recorded and distributed simultaneously to every node. Hence every node has the latest transaction update. Moreover, participants in the system can notice the double-spending issue [47], which refers to a problem where money or asset is spent more than once. Overall, the advantage of data consistency in the financial industry enforces the benefit that we already mentioned in immutability, transparency, and traceability traits.

Tokenization

Tokenization refers to the process of converting real-world assets into digital ones. By default, the financial instrument and products are the abstractions of the idea; thus, it would be logical to transform them into digital assets. Blockchain technology often ensures these digital assets' security, robustness, and privacy protection. The tokenized asset's ownership is implanted in the system; therefore, the owners can oversee the transaction records and identify the asset's seller and buyer. In general, tokenized assets originate from RWA, both tangible and intangible assets. These assets include artworks, precious stones, airplanes, intellectual property, company equity, etc. We have provided an early discussion and literature review in Sect. 2.3. Moreover, the term Token Economy [48] is a unit of storage or medium of exchange in a particular digital economy ecosystem.

There are several prime benefits of tokenizing an asset [49]:

1. Fractional ownership. It enables the participation of retail and small investors.
2. Efficiency enhancement. It comes from the automated execution of transactions, contracts, interests, dividends, and limiting intermediaries' involvement.
3. Borderless asset exchange.
4. Improved transparency.

3.2.2 Governance

In the decentralization setting, governance holds the most important aspects of how rules, processes, and decision-making mechanisms of the operation and management of organizations, communities, and crowds are arranged. Governance in the blockchain is complex; every piece of information must be disseminated and go through a consensus mechanism process to be accepted as finality. As the opposite of centralized and hierarchical operations, each stakeholder has their share proportions of the organization's decision based on the agreed-upon preliminary rules or protocol. Therefore, societies and interactions are governed by various actors' networks [47]. Blockchain-based governance aims to balance accountability and decentralization to ensure the organization's long-term sustainability. Blockchain governance consists of on-chain and off-chain governance [50]. The on-chain governance explicitly defines the governance arrangement in the protocol, allowing stakeholders to vote or make changes to the proposal. Off-chain governance refers to the external governance structure protocol. DAO represents the implementation of decentralized corporate governance, which utilizes tokenized tradable shares to give network participants the weight of voice in governing the organization. It could function to provide dividends to shareholders [35].

DeFi inherits the blockchain governance model, allowing people to make transactions directly with each other without a central authority (P2P transactions). It revolutionizes and challenges TradFi operations. Large-scale P2P financial activity such as DeFi would have no problem being supported with this kind of governance.

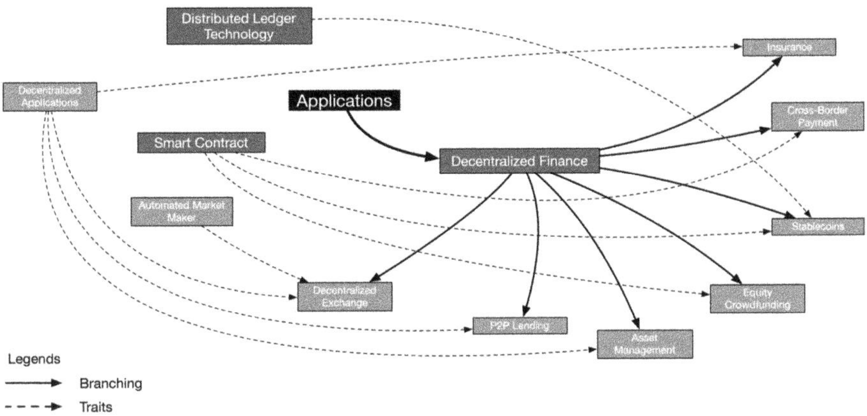

Fig. 5 Applications dimension

The crucial point of financial product innovation using blockchain technology is how financial transactions can organically follow crowd interaction, such as in social media.

3.3 Applications

Recent technological advancements have led to the development of innovative products offering various features and benefits. In the financial industry, the core idea of enabling peer-to-peer transactions has led to the creation of decentralized finance, decentralized exchange, P2P lending, equity crowdfunding, stablecoins, and many more product and concept innovations. Our taxonomy shows that innovations are possible to develop due to drivers, features, and benefits offered by technology. Figure 5 presents the application dimension and its relations.

3.3.1 Decentralized Finance (DeFi)

DeFi, a blockchain product in the financial industry valued at over 50 billion US dollars, challenges the fundamentals of TradFi by removing intermediaries [51]. Smart Contracts replace these intermediaries, acting as transaction supervisors. Though financial institutions exist in the DeFi ecosystem, they primarily serve as platforms for user activities. TradFi institutions' drawbacks, such as complex service policies, high costs, and limited transparency, are mitigated through DeFi, enabling peer-to-peer transactions, reduced bureaucracy, lower costs, and a transparent ecosystem. In short, we present Fig. 6 about DeFi characteristics traits [52]. The explanation of

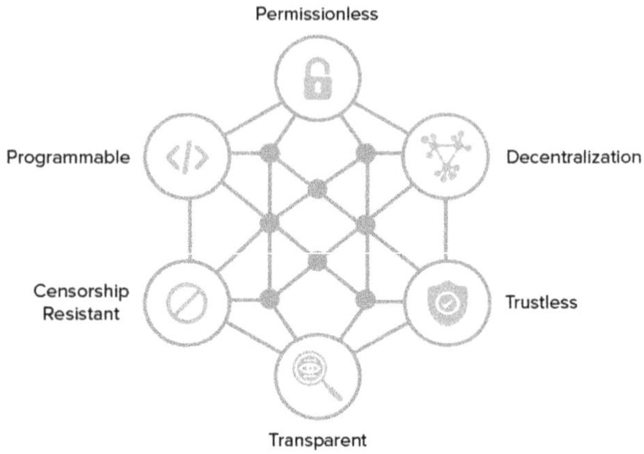

Fig. 6 DeFi characteristics traits

Table 3 TradFi vs. DeFi operational comparisons

TradFi	DeFi
1. Rely on extensive intermediation – Limited hours, high latency, manual error	1. Asset move through an open, secure settlements network that runs 24/7
2. Infrastructure is closed and proprietary	2. Infrastructure is open + open liquidity
3. High compliance burden	3. Global protocol and agnostic to jurisdictional regulations ⟶ compliance at the app layer
4. New entrants require licenses/partnerships with banks	4. Startup costs are low
5. Corporations own the services + extract value from users	5. Users own the services + capture value

each trait is followed. From the trait list, we can derive the operational comparison between TradFi and DeFi in Table 3.

1. Decentralization: Refers to the capacity for conducting peer-to-peer transactions in DeFi, thus eliminating intermediaries' roles.
2. Trustlessness: Smart Contracts automate user agreements, removing the need for manual checks or middlemen.
3. Transparency: DeFi's DLT system permits authorized users to monitor transaction activities, promoting data traceability, consistency, and immutability.
4. Censorship Resistance: Users can make transactions without being stopped or altered by middlemen.
5. Programmability: Smart Contract users can set their own agreement terms, with middlemen only providing the platform and not controlling terms.
6. Permissionlessness: DeFi allows everyone to use the platform for financial activities without limiting who can join.
7. Modularity: Contracts can work with other contracts, like combining savings and insurance, letting users mix and match based on their needs and preferences.

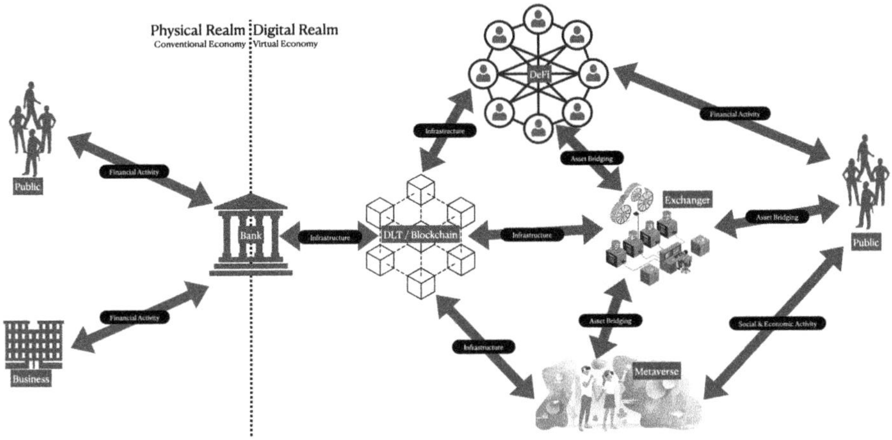

Fig. 7 General CBDC workflow in supporting physical and digital economic activity

The recent DeFi projects and ideas surge has prompted regulators, financial institutions, and banks to reassess their practices. The COVID-19 pandemic accelerated the digital economy, increasing online interactions and transactions. Digital natives or younger generations often seek opportunities for value creation among peers. Regulators, such as central banks, are adapting to the digital economy and peer-to-peer financial activities like DeFi by introducing Central Bank Digital Currencies (CBDCs). CBDCs enable effective monetary policy implementation, protect the public from the risks of malicious financial activities in DeFi, and support innovation [53]. As mentioned above, it is hoped that CBDCs could increase financial inclusion, enable faster and cheaper transactions, improve transparency, and reduce fraud. Notably, the most important regulators could maintain control over their monetary policy in the face of threats from the shadow economy.

We construct the general CBDC workflow using several sources [54, 55]. Figure 7 highlights the existence of two economic realms: physical and digital. Traditionally, regulators focus only on the physical aspect or actual economic activity, while a fully digital economy needs to catch up. Digital money, or CBDC, enables citizens to participate in the digital economy directly without the need for exchanges, efficiently promoting economic activity growth. On the physical side, banks serve retail and business clients. In the digital realm, banks could implement CBDCs in the form of DLT, which can then be transacted in DeFi activities or metaverse spaces, either directly or via exchangers.

Figure 8 illustrates the role of a commercial bank in promoting the use of CBDCs by authenticating customer identities through a Know Your Customer (KYC) procedure [56]. Once verified, customers can employ a digital wallet as a medium to interact with DLT/Blockchain technology in the bank's central or commercial back office. This wallet also allows users to communicate, conduct transactions, collaborate, generate content, and participate with their peers in a fully

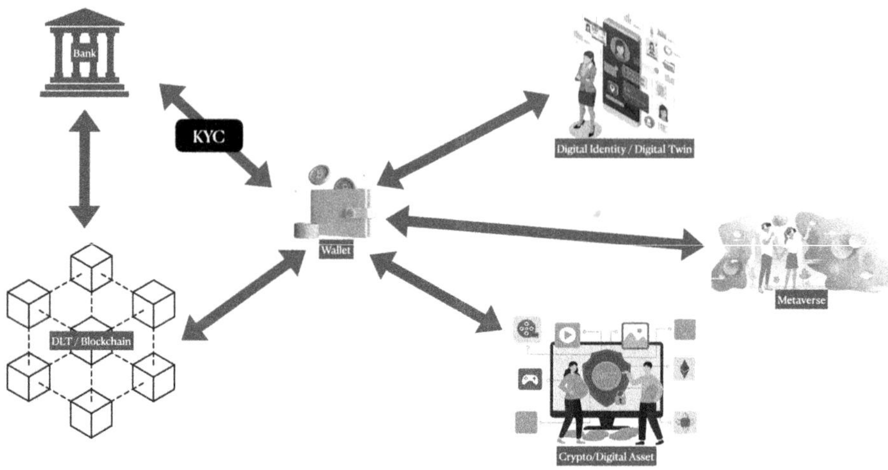

Fig. 8 A commercial bank mechanism to distribute CBDC and its usage

digital environment, such as the Metaverse. Furthermore, the digital wallet offers secure and private digital identity protection against harmful Internet activities and acts as a repository for crypto/digital assets, including NFTs, cryptocurrencies, and other assets.

Decentralized Exchange (DEX)

A decentralized exchange, or DEX, is a platform that allows the trading of assets without the need for an intermediary [57]. This contrasts with centralized exchanges (CEX), where a financial institution supervises trading activities. The primary advantage of DEX over traditional exchanges lies in its liquidity [58]. Here, liquidity pertains to the ease of converting one type of cryptocurrency to another without significantly impacting the market price. This characteristic of DEX is dependent on AMM. At present, cryptocurrencies are the most common assets traded on DEXs. However, the platform may soon facilitate trading fiat currencies, stocks, or other commodities. For example, *Uniswap*, a prominent DEX, enables users to seamlessly swap between Bitcoin and Ethereum based on the stored ratio of Bitcoin to Ethereum within the Smart Contract [59].

P2P Lending

P2P lending is a business model enabling individuals to borrow money through profit-oriented online platforms, circumventing TradFi institutions as intermediaries. P2P lending solutions cater to two types of users: general platforms for individuals or small businesses, and professional platforms for specific application domains. Blockchain technology can revolutionize P2P lending through the following steps:

1. A transaction request is broadcasted to P2P nodes.
2. Nodes authenticate transactions and the user's status.
3. Confirmed transactions are consolidated into a new data block for the ledger.
4. The new block is added to the existing blockchain.

Several prominent decentralized P2P lending platforms have emerged, such as:

1. *Aave*: A decentralized platform enabling lending and borrowing various cryptocurrencies without intermediaries, utilizing smart contracts for P2P lending.
2. *Compound*: A decentralized platform for lending and borrowing various cryptocurrencies, leveraging smart contracts to execute loans in a trustless and secure manner.
3. *MakerDAO*: A decentralized platform allowing users to borrow a stablecoin called DAI using other cryptocurrencies as collateral, employing an intricate smart contract system to maintain DAI token stability.
4. *BlockFi*: Initially a centralized lending platform, BlockFi has expanded into decentralized lending with its BlockFi Interest Account (BIA), enabling users to earn interest on their cryptocurrencies by lending them to others.

Equity Crowdfunding

Equity crowdfunding is a specialized business model aimed at obtaining outside financing for emerging companies [60]. Distinct from other crowdfunding methods, it primarily links entrepreneurs to investors interested in specific enterprises [61]. Blockchain technology presents two major benefits for equity crowdfunding [62]:

1. Safe transfer of business ownership: Blockchain's consensus algorithm and public key encryption ensure the secure and tamper-resistant exchange of company ownership, providing added trust and security for both entrepreneurs and investors.
2. Task automation: Blockchain enables the automation of various functions, such as interest payments and dividend distributions, via smart contracts. This automation removes the need for middlemen, streamlines the process, and minimizes potential inefficiencies or transaction delays, simplifying the investment process and lowering associated expenses for both entrepreneurs and investors.

Republic, StartEngine, and *Wefunder* are decentralized equity crowdfunding platforms that facilitate fundraising for startups and small businesses from a wide range of investors. *Republic* permits investments as low as $10 in company equity, *StartEngine* presents investment opportunities across sectors, and *Wefunder* has assisted over 600 companies in raising over $200 million, enabling anyone to support the startups they believe in.

Cross-Border Payment

Cross-border financial institutions facilitate transactions between countries with diverse currencies. Nevertheless, cross-border payments have been affected by economic volatility arising from events such as the Argentine financial crisis (2001–2002), the global financial crisis (2008), the Russian financial crisis (2014–2016), the US-China trade war (2018–2020), and the Covid-19 pandemic. The Society of Worldwide Interbank Financial Telecommunications (SWIFT) plays a significant role in this field, but its centralized nature leads to extended transaction durations and ambiguous fees [63]. Blockchain technology offers benefits such as lower costs, faster settlement times, increased transparency, and enhanced security.

Ripple, a blockchain-based initiative, tackles these challenges by implementing a decentralized peer-to-peer network that provides quicker transactions and lower fees than SWIFT [64]. *Ripple* employs its consensus algorithm, the Unique Node List (UNL), where node operators choose a roster of "unique" nodes they endorse. Nodes within the UNL are deemed valid only if they receive approval from other "unique" nodes, thereby safeguarding the network from external threats and preventing majority votes from jeopardizing nodes within the UNL [65]. Additional services, like *Stellar*, utilize their native digital currency, Lumens (XLM), as a bridge between various currencies, facilitating seamless and cost-effective money transfers.

Stablecoin

Stablecoins represent a form of digital currency engineered to retain a consistent value by pegging it to another asset, like gold or fiat money [66]. They offer a more dependable and stable method of exchanging and preserving value compared to other cryptocurrencies, which can experience significant price fluctuations. The creation of stablecoins addresses the market's desire for a dependable digital currency that mimics fiat money, which is typically overseen by regulatory bodies or central authorities.

Cryptocurrencies like Bitcoin showcase blockchain technology, but their high volatility and susceptibility to regulatory and informational factors have caused investor concerns [67]. In contrast, stablecoins demonstrate low volatility and are increasingly employed as an alternative to fiat currency in digital markets [66]. There are three main types of stablecoins based on their stabilization techniques [68]:

1. Asset-backed: Stablecoins tied to and supported by assets held in private banks, such as *Tether*, which has value pegged to the USD.
2. Crypto-collateralized: Stablecoins whose value is anchored to other cryptocurrencies. The value of the underlying cryptocurrencies is greater than the issued crypto-collateralized coins, like *DAI*, which the USD backs, and other cryptocurrencies, such as Ethereum's ETH.

3. Algorithmic: A stablecoin type that does not depend on another asset for support. Instead, it utilizes Oracle price feeds, algorithmic stabilization, and user engagement (trading) to maintain its peg, as *LUNA* was developed by *Terra*, a community-driven blockchain platform. However, algorithmic stablecoins are often perceived as failures due to their residual volatility, limited adoption, and vulnerability to unforeseen events that impact supply and demand.

Asset Management

The goal of an asset management company is to document and transfer asset ownership based on a contractual agreement [69]. Through a decentralized system, blockchain streamlines asset transactions and manages asset registries. This decentralized approach to asset management accelerates the settlement process, as central authority validation is unnecessary, and offers increased transparency since monitoring occurs via Smart Contracts. Asset management involves two types of assets [70]:

1. Fungible: These assets, like currency, can be exchanged with any other object. For instance, ten one-dollar bills hold the same value as a single ten-dollar bill.
2. Non-fungible: These are unique assets that have distinct attributes despite their similarities to others. For example, diamonds are categorized by their clarity, cut, color, carat, and certification. Each diamond varies, even though they are all called "diamonds."

Asset management firms can leverage permissioned or permissionless blockchains, or both, to enhance their operations [71]. Government officials can manage permissioned blockchains to confirm asset registration, prevent double-spending, and guarantee the legality of traded assets. Conversely, permissionless blockchains symbolize the asset trading marketplace. Asset management firms might opt for permissioned blockchains to prioritize scalability and security, while permissionless blockchains could be chosen for decentralization and accessibility.

We present three examples of decentralized asset management platforms. The first, *Melon Protocol*, is a platform enabling users to create, manage, and invest in digital asset portfolios using smart contracts for secure and transparent management. The second, *Balancer*, is a platform allowing users to develop custom cryptocurrency portfolios, utilizing an automated market maker algorithm to maintain portfolio value balance. Lastly, *Set Protocol* is a platform that lets users create, manage, and invest in tokenized cryptocurrency portfolios, employing smart contracts to automate portfolio management tasks and ensure compliance.

Insurance

Fraud is a major issue in the traditional insurance industry, with the FBI estimating losses in the USA to be over $40 billion, or around $120 per person. Coupled with challenges like insufficient customization, bureaucratic complexities, limited transparency, and customer dissatisfaction, innovation is needed. Blockchain technology can significantly improve the efficiency of the insurance sector in several ways:

1. Fraud prevention: Firms like *Everledger* use blockchain to establish a worldwide registry for assets like gemstones, making fraudulent activities more challenging.
2. Streamlined claim management: Integrating claim data improves visibility for insurers, brokers, and reinsurers.
3. Reinsurance: Smart contracts enable transparent and open systems between reinsurers and insurers.
4. Transparency: Blockchain enhances claim tracking, product verification, and service authentication transparency.
5. Improved customer management: Blockchain allows for tailored insurance products that cater to individual client requirements.
6. Data analysis: Insurance companies can make informed business decisions by examining customer data within the blockchain network, including credit information, policy specifics, and accident-related data.

We present three examples of blockchain-based insurance platforms. *Nexus Mutual* is a decentralized insurer providing coverage for smart contract failures on the Ethereum network. *Etherisc* is a decentralized platform that lets individuals and businesses create and buy insurance products using smart contracts, covering various needs like flight delays and crop insurance. *InsurAce* is a decentralized platform offering coverage for a variety of DeFi products and services, utilizing both on-chain and off-chain data for policy pricing and claim evaluations.

3.4 Threats

We have provided drivers and benefits of blockchain technology implementation in the financial industry. Several key applications are also presented that might disrupt how financial industries operate. Nevertheless, it is essential to recognize that certain inherent limitations associated with this technology warrant close scrutiny. Failure to adequately address and manage these limitations may lead to unfavorable consequences for organizations that adopt blockchain technology.

A critical aspect to consider when discussing the adoption of blockchain technology is the so-called blockchain trilemma. This concept highlights the challenges faced when trying to achieve the ideal combination of decentralization, security, and scalability within a blockchain network. To better understand and visualize the intricate relationships between these three components, Fig. 9 provides three features of the blockchain trilemma, which serves to elucidate the trade-offs and decisions that must be made when designing and deploying blockchain solutions in the financial sector.

The trilemma posits that achieving all three characteristics in a blockchain system is difficult [72]. These characteristics include decentralization, security, and

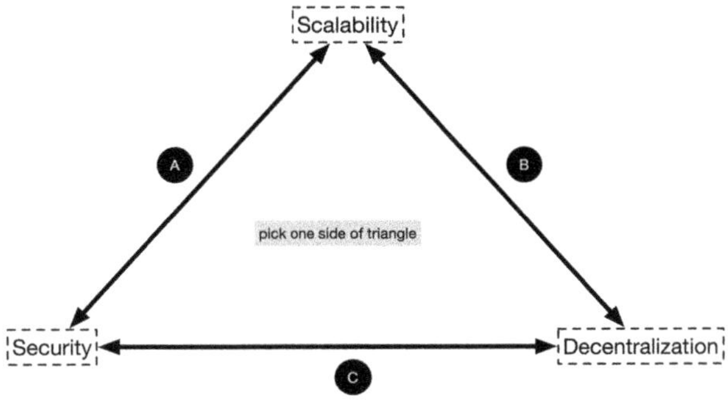

Fig. 9 The blockchain trilemma is concerned with finding the best balance between three aspects, which cannot all be achieved simultaneously

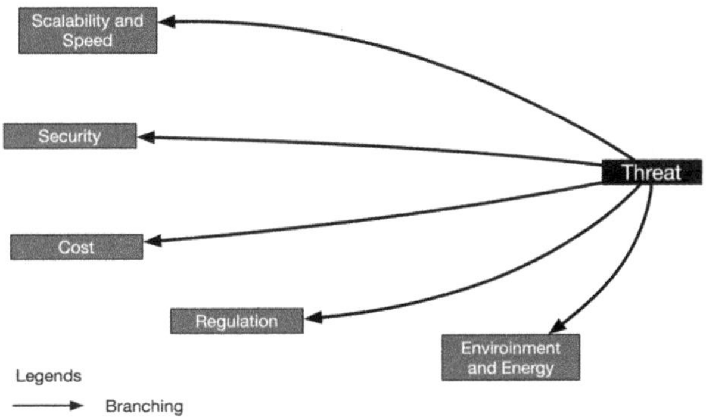

Fig. 10 Threat dimension

scalability. Decentralization involves distributing the network across multiple nodes, avoiding centralization. Security ensures the network is protected from tampering and attacks. Scalability pertains to the network's ability to quickly process a high volume of transactions. Currently, a blockchain network can only excel in two of these characteristics, posing a challenge for developers to find the right balance among the three according to a product or service's requirements. Blockchain trilemma inherits several threats or risks related to implementing blockchain in the financial industry. Figure 10 shows the threat dimensions, and we discuss the dimension in the following subsection.

3.4.1 Scalability and Speed

Scalability and speed are two interconnected challenges in blockchain implementation. Scalability pertains to the capacity of a blockchain to manage a growing volume of tasks, while speed refers to the time it takes for a system to perform a specific task. These challenges are linked, as a system is considered scalable if it can accommodate an expanding workload within a given time frame. In many instances, blockchain networks might struggle to cope with the growing workload, which in turn slows down the entire network. The system becomes less efficient as the number of transactions on the blockchain network increases. However, researchers and platform developers are working on solutions to tackle these challenges, aiming to make blockchain more practical and effective in handling the rising workload. According to the blockchain trilemma, the general principle is that to improve a blockchain's speed and scalability, it must compromise either decentralization or security.

3.4.2 Security

Multiple factors contribute to security risks in the digital realm. One such factor is the vulnerability of consensus algorithms to manipulation, such as PoW. Attackers can take over the system by acquiring 51% of its computational power, allowing them to impose their own rules. However, achieving this level of control will be difficult if we have a large number of network participants. This is why blockchain networks like Bitcoin are considered safe since global collaboration cannot be easily defeated. A Sybil attack is another significant security risk involving creating multiple false identities to gain network control. If successful, the attacker can exploit the network without consequences. Phishing is another security threat; fraudsters trick victims into investing in fake cryptocurrency wallets. Once victims provide their information to the deceptive wallet, they are unable to access their funds without the private key.

3.4.3 Cost

While blockchain has the potential to lower transaction costs, FinTech users must pay service fees to miners instead of a financial institution, with fluctuating costs based on the required effort and network traffic. Fees motivate miners to incorporate transactions into the upcoming block, with the amount determined by the demand and supply of block space. Increased demand results in higher fees, which in turn contribute to escalating costs. The costs also factor into expenses related to hardware, bandwidth needs, and network synchronization complexities. These elements have impeded mass adoption, but emerging innovations such as Layer 2 (L2) solutions provide a swifter and more cost-effective implementation. The L2 is a secondary framework or protocol built on top of an existing blockchain network called Layer 1 or L1 to enhance scalability, speed, and efficiency, including transaction costs.

3.4.4 Regulation

Inadequate regulations can adversely affect the introduction of new products, leading to chaos and increased criminal activities. As such, it is crucial to maintain a balance between regulation and innovation. It is worth mentioning that regulatory levels may vary across countries, and weak regulations combined with a limited user understanding of blockchain can make it easier for criminals to engage in illicit activities. Regulators have a critical role in formulating regulations for blockchain-based digital assets to mitigate negative consequences.

If authorities do not regulate blockchain effectively, several negative outcomes may arise. First, it could enable money laundering due to the anonymity of blockchain transactions, allowing criminals to conceal the source of their funds. Second, customers might be susceptible to blockchain scams because of the technology's complexity and absence of safeguards. Thus, regulators should establish suitable rules to protect citizens. Finally, the lack of standardization and regulation might cause interoperability issues among different blockchain networks, obstructing or even preventing transactions between companies.

Blockchain regulations differ between developed and developing countries. Developed countries generally have a more transparent legal framework for blockchain and a higher adoption rate than developing countries. Moreover, developed countries tend to have clearer tax regulations, while developing countries may still be in the process of defining tax procedures. According to World Economic Forum Global Future Council on Cryptocurrencies, no internationally coordinated regulation of cryptocurrencies – though international bodies have been working on assessing risks and appropriate policy responses to the rise of the crypto industry. Globally, central banks and regulators already share a common objective – stabilizing monetary systems and spurring innovation and economic growth [73]. Table 4 shows different approaches from different countries.

3.4.5 Environment and Energy

The energy consumption associated with blockchain technology, particularly the PoW mechanism, has raised concerns about its adverse environmental effects. In fact, a single Bitcoin transaction can consume up to 1499 kWh of electricity due to intense competition among miners to solve computational problems and receive

Table 4 Regulatory approaches for the crypto industry from different countries

Country	Cryptocurrency	Cryptocurrency exchanges
USA	Not legal tender	Legal, regulated
Canada	Not legal tender	Legal, registration needed
Singapore	Not legal tender	Legal, registration needed
Australia	Legal, treated as property	Legal, registration needed
Japan	Legal, treated as property	Legal, registration needed
South Korea	Not legal tender	Legal, registration needed
China	Not legal tender	Illegal
India	Not legal tender	Regulation being considered
UK	Not legal tender	Legal, registration needed
Switzerland	Legal, accepted as payment to some extent	Legal, regulated

rewards [74]. This heightened electricity demand has resulted in increased fossil fuel consumption, leading to higher carbon emissions and pollution. The prevalent use of PoW as a consensus mechanism in cryptocurrencies exacerbates these environmental problems, highlighting the need for alternative consensus algorithms that support environmental sustainability.

Moreover, the issue of hardware waste significantly impacts environmental sustainability, as blockchain companies often need to continuously acquire new hardware with a limited life span. Additionally, the substantial energy requirements for mining activities have caused a concentration of mining operations in areas with low energy costs or favorable tax policies. This results in centralization, which goes against blockchain technology's decentralized principles. Therefore, addressing these environmental concerns is essential to ensure blockchain technology's sustainable growth and development.

4 Approaches to Mitigate Blockchain Threats

Numerous proposals, including new blockchain protocols and solution layers, have been presented as potential solutions to mitigate the threats described in the preceding section (Sect. 3.4). However, it is important to recognize that these proposed solutions are not the sole means of addressing these risks. Financial institutions can adopt the recommended strategies or seek out more effective alternatives as they arise. As a result, financial institutions must stay alert and adaptable in their security efforts, enabling them to combat the ever-changing landscape of these threats effectively. Financial institutions should consistently evaluate their security protocols and invest in groundbreaking methods that offer optimal protection against cyber threats. Adopting this forward-thinking stance will ensure they stay one step ahead in the continuous struggle against cyber criminals. By prioritizing cybersecurity, financial institutions can more effectively shield their clients' confidential data and uphold their reputation for trustworthiness and dependability.

4.1 Scalability and Speed

Researchers have proposed solutions to tackle scalability and speed issues in blockchain systems, specifically Layer 1 (L1) platforms like Bitcoin and Ethereum, which are difficult to modify. A Layer 2 (L2) solution has been devised to surmount these constraints, which functions by consolidating transactions before committing them to L1. A variety of L2 techniques have been developed, such as Simplified Payment Verification (SPV), which minimizes data consumption compared to conventional methods; *Lightning Network,* which employs micropayment channels for continuous balance updates [75]; and sharding, a method that distributes data across multiple nodes. Sharding segments the network into smaller sections [76], with each responsible for handling a portion of transactions, which eases the load on individual nodes and boosts transaction processing speeds. Moreover, sharding provides advantages such as sublinear communication, increased resilience, swift committee consensus, secure setup, rapid cross-shard verification, and decentralized bootstrapping. Notably, platforms like *Rapidchain* and *Ethereum 2.0* have embraced sharding as part of their solutions.

4.2 Security

To reduce the risk of a 51% attack, alternative consensus algorithms such as Proof of Stake (PoS), Unique Node List (UNL), Proof of History (PoH), and Proof of Authority (PoA) can be employed in place of the conventional Proof of Work (PoW). For instance, PoS minimizes the likelihood of a malicious actor gaining control over the network by making a 51% attack nearly unfeasible. *Peercoin* is an example of a blockchain project that implements the PoS mechanism [77]. UNL is another approach that defends the network against external attacks by preventing malevolent entities from entering the network without being identified as the node operator. PoH, a consensus algorithm introduced by *Solana*, assigns a unique hash and count to every transaction, functioning as a real-time record that miners can use to reconstruct transactions based on historical data, enhancing the network's resistance to potential attacks [78]. PoA is a different consensus model where miners stake their reputation, instead of coins, to be acknowledged as validators [79]. Unlike PoW and PoS, they have not operated anonymously, making it simpler to detect colluding parties and increasing resistance to 51% of attacks.

However, security threats can also stem from a lack of user awareness rather than system vulnerabilities. Various solutions exist to prevent such attacks. Firstly, users should avoid dubious financial investments and opt for registered financial institutions, which are less likely to engage in scams. Increasing user awareness of different scamming techniques, such as phishing, pop-ups, and spoofing, is another way to prevent attacks [27].

Over several years, the Financial Action Task Force (FATF) has introduced numerous policy changes, culminating in the issuance of Recommendation No. 15. This recommendation requires cryptocurrencies and associated financial institutions to disclose information about the parties and beneficiaries involved in digital asset transactions, often referred to as the travel rule [80]. However, as financial institutions, Virtual Asset Service Providers (VASPs) have faced challenges when trying to implement the travel rule for blockchain-based digital assets. The primary problem is that VASPs have difficulty managing cryptographic keys linked to their customers since digital assets on a blockchain are directly controlled via a private-public key mechanism. To tackle these issues, there are two possible approaches [77]:

1. Noncustodial: This method employs a secure system where both the private and public keys are held by the parties participating in the asset transfer, such as the originators and beneficiaries, while the VASP only has access to their public keys. This limits the VASP's involvement in overseeing the transaction without intermediation. By adhering to the travel rule, this approach allows financial institutions to validate and verify transactions before they are carried out, thereby improving the security and stability of the financial system.
2. Custodial: This involves a custody-centric approach, where the VASP takes on the responsibility of securely storing both the public and private keys. Neither the originator nor the beneficiary possesses direct control over the keys. This technique enables financial institutions to oversee the transfer of their client's assets and is commonly employed by blockchain-based exchanges such as Binance, Coinbase, and Uphold. By utilizing a robust custody mechanism, financial institutions can ensure the security of their client's assets and protect them from potential threats or losses.

4.3 Cost

Ethereum 2.0 offers a promising solution to tackle the high expenses linked to blockchain technology. By employing PoS and sharding, *Ethereum 2.0* diminishes the computational demands of blockchain transactions, reducing transaction costs. The integration of the Beacon Chain, an innovative consensus layer, with the existing PoW layer in Ethereum 2.0 represents a novel strategy that eliminates the necessity for conventional PoW consensus, leading to quicker processing times and substantially decreased transaction costs [78]. Miners can optimize gas consumption to cut costs by enhancing code efficiency and bundling multiple transactions into a single batch.

However, addressing the cost-related challenges in the blockchain sector requires a multifaceted approach combining technological solutions and market reforms. As blockchain innovation continues to advance, it creates opportunities for the emergence of new platforms with lower costs than their predecessors, thereby making the industry more accessible and widely adopted. The development of various new blockchain platforms is driven by the demand for cost reduction, the majority of which are L2 solutions, as discussed in Sect. 4.1.

These L2 solutions, such as the *Lightning Network* for Bitcoin or *Optimistic Rollups* for Ethereum, function as an additional layer on top of the base blockchain, enabling faster and cheaper transactions. These solutions process transactions off-chain and then periodically batch and settle them on the main blockchain, effectively reducing network congestion and transaction fees. By embracing these L2 solutions and continually exploring new cost-effective alternatives, the blockchain industry can become more accessible and experience broader adoption across various sectors.

4.4 Regulation

Regulating blockchain technology is a complex issue that poses challenges due to the varying regulatory landscapes across different jurisdictions. However, several potential solutions could be employed worldwide to tackle this challenge. The first solution entails governments prioritizing the establishment of blockchain regulations that emphasize data privacy, anti-money laundering, and consumer protection. By implementing such regulations, governments can ensure blockchain operations comply with legal frameworks and protect stakeholder interests.

Secondly, developing regulatory sandboxes provides a controlled environment for blockchain companies to test regulatory approaches in partnership with regulators. This method facilitates a balance between fostering innovation and maintaining regulation [81]. Thirdly, international collaboration among countries is crucial, given the borderless nature of blockchain technology. Governments can work together to create global standards and coordinate regulatory efforts, encouraging innovation, accountability, and consumer protection on a worldwide scale.

Lastly, adopting Central Bank Digital Currencies (CBDCs) offers a potential solution for regulating blockchain technology within the financial sector. By issuing a regulated form of cryptocurrency, central banks can oversee and enforce compliance with regulations, enhancing transaction transparency and mitigating threats such as money laundering.

However, striking a balance between fostering innovation and upholding regulation is essential to address potential risks and support the industry's growth. This balance ensures that the blockchain ecosystem can develop without compromising the security, privacy, and compliance required for widespread adoption and integration into various sectors of the economy.

4.5 Environment and Energy

Several approaches can be employed to tackle blockchain technology's environmental and energy concerns. The first approach involves adopting the PoS algorithm, which eases the mining difficulty for each node and accelerates the mining process based on the quantity and age of coins held by the node [7]. This leads to fewer participating nodes compared to PoW and reduces overall energy consumption.

The second approach involves using permissioned blockchains, which restrict the number of participants in the network, ultimately lowering energy consumption. Another potential solution is to shift mining operations towards renewable energy sources like solar, hydro, and wind power, which can help decrease the carbon footprint of the mining process.

Furthermore, Blockchain-as-a-Service (BaaS) can be harnessed as a way to reduce energy consumption. This approach allows companies to outsource their blockchain networks to third-party providers, eliminating the need for businesses to set up and maintain their own energy-intensive blockchain networks.

Raising awareness and educating people about blockchain technology's environmental impact and energy sustainability concerns is crucial. By taking proactive measures, individuals and organizations can minimize their environmental footprint and promote energy sustainability through their use of blockchain technology. A multifaceted approach, which blends both technological and social solutions, can help mitigate blockchain technology's negative consequences on the environment and energy sustainability. This comprehensive strategy can pave the way for a more eco-friendly and efficient infrastructure, fostering the technology's adoption and growth while minimizing its environmental impact.

5 Discussion

We have presented a taxonomy of blockchain-based technology in the financial industry. It discloses four critical dimensions and their branches. The intricate relationship between entities in the form of branching and traits is also revealed, leading to a comprehensive understanding of the complexity of blockchain-based products/services innovation and operational capabilities. Blockchain technology has recently attracted considerable attention, especially in the financial sector, due to its numerous potential benefits. These include low-cost financial services, enhanced security, and privacy, efficient transactions, high returns on investment, and transparency. Decentralized systems built on DLT and Smart Contracts are key driving forces behind blockchain technology. The decentralized nature of blockchain can foster financial inclusion by offering affordable services to unbanked and underbanked populations. This particularly impacts developing countries where TradFi services can be expensive or inaccessible. Blockchain technology also ensures high security and privacy for users' sensitive financial information, while still maintaining transaction transparency. It facilitates efficient transactions and traceability, such as remittances, with faster processing times and lower fees compared to TradFi institutions. Additionally, the programmable aspect of blockchain allows for the development of new financial instruments and business models through programmable assets and currencies.

Despite these advantages, blockchain technology faces several challenges in its adoption within the financial sector. One major issue is the system's complexity, which demands significant technical expertise for implementation and maintenance, For example, how to transform the business process into the blockchainable one [82]. A decentralized mindset like DeFi products/services, which gives end users much power to control their assets, is still not within the grasp of the majority of people who used to deal with centralized systems (TradFi). Associated with the mindset, there are questions on how the industry can adapt and create new business models related to the decentralized perspective. The lack of standardization and regulation further complicates its adoption. Other significant concerns include cost, speed, and scalability issues, which have been extensively discussed in the threat and mitigation effort sections.

To address these challenges, educational and awareness programs are needed to equip individuals with the skills necessary to implement and maintain blockchain technology. Standardization and regulatory efforts can ensure blockchain operates within a legal framework protecting all parties involved. It is essential to recognize that the problems and solutions discussed may not apply universally, as each country has its own rules and regulations governing blockchain technology. However, cross-border regulation may become necessary as blockchain continues to evolve and expand globally. Regulators must strike a balance that guarantees all parties' safety and security while preventing emerging negative consequences.

Future Research might dive deeper into several streams as follows. The first stream is looking at the technological aspect, especially the taxonomy map of the existing blockchain platform. There are still questions on what are the optimal configurations between scalability, decentralization, and security for the financial industry. Each platform offers the ideal pricing structure for different type industry such as banking, insurance, security, and more. The second stream considers product and service type, and potential features that can be combined, constructed, or modified to create innovation. While the third stream focuses holistically on sensible regulation, which can protect the public from malicious actors in the crypto industry. At the same time, it still gives room for a new idea to grow, which is essential in the digital economy sphere.

6 Conclusion

We have carefully designed a taxonomy consisting of four dimensions, which we believe are crucial for comprehending and categorizing the blockchain-based financial sector. We offer in-depth explanations for each dimension, discussing blockchain technology's primary drivers, advantages, products, and risks. Furthermore, we examine the interplay between these dimensions to grasp their mutual impact better. It is worth mentioning that our taxonomy's dimensions may not be all-encompassing as the blockchain-based financial landscape continually changes. Nonetheless, our chapter delivers a thorough analysis of the circumstances at the time of writing and emphasizes blockchain technology's most significant aspects in finance.

A major benefit of blockchain is its potential to enhance business process efficiency and foster innovation in the financial industry. However, it is vital to develop adaptive regulations to ensure that users and financial institutions securely and responsibly employ blockchain technology. Given the fast-paced evolution of this technology, our chapter may not cover all facets of the blockchain-based financial sector. As a result, future research could focus on exploring these gaps to gain a deeper understanding of blockchain's role in FinTech companies. In summary, our chapter strives to offer valuable insights into the blockchain-based financial industry and its potential to disrupt TradFi systems.

References

1. Market capitalization of leading 100 banks worldwide from 1st quarter 2016 to 3rd quarter 2022. Statista. https://www.statista.com/statistics/267682/market-capitalization-of-the-largest-banks-in-investment-banking-in-2010/. Accessed 29 Apr 2023
2. T. Puschmann, Fintech. Bus. Inf. Syst. Eng.: Int. J. Wirtschaftsinformatik. **59**(1), 69–76 (2017). https://doi.org/10.1007/s12599-017-0464-6
3. P. Gomber, J.A. Koch, M. Siering, Digital Finance and FinTech: Current research and future research directions. J. Bus. Econ. **87**(5), 537–580 (2017). https://doi.org/10.1007/s11573-017-0852-x
4. R. Zhang, R. Xue, L. Liu, Security and privacy on blockchain. ACM Comput. Surv. **52**(3) (2019). https://doi.org/10.1145/3316481
5. M. Nofer, P. Gomber, O. Hinz, D. Schiereck, Blockchain. Bus. Inf. Syst. Eng. **59**(3), 183–187 (2017). https://doi.org/10.1007/s12599-017-0467-3
6. S. Nakamoto, Bitcoin: A peer-to-peer electronic cash system, (2008). [Online]. Available: https://www.bitcoin.org
7. R. Zhang, W.K. Chan, Evaluation of energy consumption in block-chains with proof of work and proof of stake. J. Phys. Conf. Ser. **1584**(1) (2020). https://doi.org/10.1088/1742-6596/1584/1/012023
8. J.A. Tarr, Distributed ledger technology, blockchain and insurance: Opportunities, risks, and challenges. Insur. Law J. **29**(3), 254–268 (2018)
9. T. Puschmann, M. Huang-Sui, A taxonomy for decentralized finance. SSRN (2023). https://doi.org/10.2139/ssrn.4360067
10. J.R. Jensen, O. Ross, Managing risk in DeFi. SSRN (2021). https://doi.org/10.2139/ssrn.3745568.2021
11. Y. Tian, Z. Lu, P. Adriaens, A. Caithness, J. Woo, Finance infrastructure through blockchain-based tokenization. Front. Eng. Manage. **7**, 485–499 (2020)
12. V. Mohan, Automated market makers and decentralized exchanges: A DeFi primer. Financ. Innov. **8**, 20 (2022)
13. B. Nicoletti, The future of Fintech: Integrating finance and technology in financial services, in *Palgrave Studies in Financial Services Technology*, (2017). https://doi.org/10.1007/978-3-319-51415-4
14. C.V. Helliar, L. Crawford, L. Rocca, C. Teodori, M. Veneziani, Permissionless and permissioned blockchain diffusion. Int. J. Inf. Manag. **54** (2020). https://doi.org/10.1016/j.ijinfomgt.2020.102136
15. W. Zou et al., Smart contract development: Challenges and opportunities. IEEE Trans. Softw. Eng. **47**(10), 2084–2106 (2021). https://doi.org/10.1109/TSE.2019.2942301
16. Y. Tian, Y. Zhang, R.E. Minchin, A. Asutosh, C. Kan, An innovative infrastructure financing instrument: Blockchain-based tokenization, in *Constr. Res. Congr. 2020 Infrastruct. Syst. Sustain. - Sel. Pap. from Constr. Res. Congr. 2020, no. October*, (2020), pp. 731–740. https://doi.org/10.1061/9780784482858.079
17. J. Frankenfield, Distributed ledger technology definition and how it works. *Investopedia*, March 2023, [Online]. Available: https://www.investopedia.com/terms/d/distributed-ledger-technology-dlt.asp. Accessed 29 Apr 2023
18. E. Deirmentzoglou, G. Papakyriakopoulos, C. Patsakis, A survey on long-range attacks for proof of stake protocols. IEEE Access **7**, 28712–28725 (2019). https://doi.org/10.1109/ACCESS.2019.2901858
19. C.T. Nguyen, D.T. Hoang, D.N. Nguyen, D. Niyato, H.T. Nguyen, E. Dutkiewicz, Proof-of-stake consensus mechanisms for future blockchain networks: Fundamentals, applica-

tions and opportunities. IEEE Access **7**, 85727–85745 (2019). https://doi.org/10.1109/ACCESS.2019.2925010
20. J. Xie, F.R. Yu, T. Huang, R. Xie, J. Liu, Y. Liu, A survey on the scalability of blockchain systems. IEEE Netw. **33**(5), 166–173 (2019). https://doi.org/10.1109/MNET.001.1800290
21. E. Georgiadis, How many transactions per second can bitcoin really handle? Theoretically. Cryptol. ePrint Arch., 1–6 (2019)., [Online]. Available: https://eprint.iacr.org/2019/416
22. KPMG, Asset tokenization and its implications on securities markets, December 2020. [Online]. Available: https://lhoft.com/en/insights/asset-tokenization-and-its-implications-on-securities-markets/. Accessed 29 Apr 2023
23. G. Sazandrishvili, Asset tokenization in plain English. J. Corp. Account. Financ. **31**(2), 68–73 (2020). https://doi.org/10.1002/jcaf.22432
24. A. Sunyaev et al., Token economy. Bus. Inf. Syst. Eng. **63**(4), 457–478 (2021). https://doi.org/10.1007/s12599-021-00684-1
25. L. Lotti, The art of tokenization: Blockchain affordances and the invention of future milieus. Media Theory **3**(1), 287–320 (2019)
26. I.C. Lin, T.C. Liao, A survey of blockchain security issues and challenges. Int. J. Netw. Secur. **19**(5), 653–659 (2017). https://doi.org/10.6633/IJNS.201709.19(5).01
27. A.A. Andryukhin, Phishing attacks and preventions in blockchain based projects, in *Proc. - 2019 Int. Conf. Eng. Technol. Comput. Sci. Innov. Appl. EnT 2019*, (2019), pp. 15–19. https://doi.org/10.1109/EnT.2019.00008
28. Y. Faqir-Rhazoui, M.J. Ariza-Garzón, J. Arroyo, S. Hassan, Effect of the gas price surges on user activity in the DAOs of the ethereum blockchain, in *Conf. Hum. Factors Comput. Syst. - Proc.*, (2021), pp. 4–10. https://doi.org/10.1145/3411763.3451755
29. M. D'Aliessi, How does Ethereum work?. *Medium*, [Online]. Available: https://medium.com/@micheledaliessi/how-does-ethereum-work-8244b6f55297. Accessed 29 Apr 2023
30. D. Carl, C. Ewerhart, Ethereum gas price statistics. SSRN Electron. J. **7041**(373) (2021). https://doi.org/10.2139/ssrn.3754217
31. M. Campbell-Verduyn, Bitcoin, crypto-coins, and global anti-money laundering governance. Crime Law Soc. Chang. **69**, 283–305 (2018)
32. The New York Time, Bitcoin uses more electricity than many countries. How is that possible? 2021/09/03 edition. Accessed 29 Apr 2023
33. J. Sedlmeir, H.U. Buhl, G. Fridgen, R. Keller, Recent developments in blockchain technology and their impact on energy consumption, https://arxiv.org/abs/2102.07886, (2021). https://doi.org/10.1007/s00287-020-01321-z
34. C. Jentzsch, Decentralized autonomous organization to automate governance. SlockIt, 1–30 (2016) [Online]. Available: slock.it/dao.html. Accessed 29 Apr 2023
35. S. Hassan, P. De Filippi, Decentralized autonomous organization. Internet Policy Rev. **10**(2), 1–10 (2021). https://doi.org/10.14763/2021.2.1556
36. S. Wang, W. Ding, J. Li, Y. Yuan, L. Ouyang, F.Y. Wang, Decentralized autonomous organizations: Concept, model, and applications. IEEE Trans. Comput. Soc. Syst. **6**(5), 870–878 (2019). https://doi.org/10.1109/TCSS.2019.2938190
37. S. Raval, *Decentralized Applications: Harnessing Bitcoin's Blockchain Technology*, 1st edn. 1230 (O'Reilly Book, California, USA, 2016). https://www.oreilly.com/library/view/decentralized-applications/9781491924532/
38. A. Othman, D.M. Pennock, D.M. Reeves, T. Sandholm, A practical liquidity-sensitive automated market maker. Proc. ACM Conf. Electron. Commer. **1**(3), 377–386 (2013). https://doi.org/10.1145/1807342.1807402
39. J. Aoyagi, Liquidity provision by automated market makers. SSRN Electron. J. (2020)., [Online]. Available: https://papers.ssrn.com/sol3/papers.cfm?abstract_id=3674178. Accessed 29 Apr 2023
40. J.R. Jensen, O. Ross, Managing risk in DeFi. SSRN Electron. J. (2021). https://doi.org/10.2139/ssrn.3745568. Available: https://papers.ssrn.com/sol3/papers.cfm?abstract_id=3745568. Accessed 29 Apr 2023
41. M.J.M. Chowdhury et al., A comparative analysis of distributed ledger technology platforms. IEEE Access **7**, 167930–167943 (2019). https://doi.org/10.1109/ACCESS.2019.2953729

42. M. Rauchs et al., Distributed ledger technology systems: A conceptual framework. SSRN Electron. J. (2019). https://doi.org/10.2139/ssrn.3230013. https://papers.ssrn.com/sol3/papers.cfm?abstract_id=3230013. Accessed 29 Apr 2023
43. E. Politou, F. Casino, E. Alepis, C. Patsakis, Blockchain mutability: Challenges and proposed solutions. IEEE Trans. Emerg. Top. Comput. **9**(4), 1972–1986 (2021). https://doi.org/10.1109/TETC.2019.2949510
44. E. Bertino, A. Kundu, Z. Sura, Data transparency with blockchain and AI ethics. J. Data Inf. Qual. **11**(4) (2019). https://doi.org/10.1145/3312750
45. A. Alamsyah, N. Hakim, R. Hendayani, Blockchain-based traceability system to support the Indonesian halal supply chain ecosystem. Economies **10**(6) (2022). https://doi.org/10.3390/economies10060134
46. F.R. Yu, J. Liu, Y. He, P. Si, Y. Zhang, Virtualization for distributed ledger technology (vDLT). IEEE Access **6**, 25019–25028 (2018). https://doi.org/10.1109/ACCESS.2018.2829141
47. A. Zwitter, J. Hazenberg, Decentralized network governance: Blockchain technology and the future of regulation. Front. Blockchain **3**(March), 1–12 (2020). https://doi.org/10.3389/fbloc.2020.00012
48. S. Voshmgir, *Token Economy: How the Web3 Reinvents the Internet*, 2nd edn. (Token Kitchen, Berlin, Germany, 2020). https://token.kitchen/token-economy
49. Y. Tian, Z. Lu, P. Adriaens, R.E. Minchin, A. Caithness, J. Woo, Finance infrastructure through blockchain-based tokenization. Front. Eng. Manag. **7**(4), 485–499 (2020). https://doi.org/10.1007/s42524-020-0140-2
50. D.W.E. Allen, C. Berg, C., Forthcoming in the journal of the British Blockchain Association blockchain governance: What we can learn from the economics of corporate governance. SSRN Electron. J., 1–14 (2020)., [Online]. Available: https://ssrn.com/abstract=3519564
51. D.A. Zetzsche, D.W. Arner, R.P. Buckley, Decentralized finance. J. Financ. Regul. **6**(2), 172–203 (2020). https://doi.org/10.1093/jfr/fjaa010
52. The 2020 Year in Review. Blockchain Capital. [Online]. https://blockchain.capital/2020-year-in-review/. Accessed 29 Apr 2023)
53. A. Grym, Lessons learned from the world's first CBDC, (2020). [Online]. Available: https://publications.bof.fi/. Accessed 29 Apr 2023
54. CBDC Tracker. [Online]. https://cbdctracker.org/. Accessed 29 Apr 2023
55. T. Mancini-Griffoli et al., Casting light on central Bank digital currency, in *Book Chapter on Cryptoassets: Legal, Regulatory, and Monetary Perspectives*, (Oxford Academic, 2019). https://doi.org/10.1093/oso/9780190077310.003.0012
56. T.H. Chen, Do you know your customer? Bank risk assessment based on machine learning. Appl. Soft Comput. J. **86**, 105779 (2020). https://doi.org/10.1016/j.asoc.2019.105779
57. V. Mohan, Automated market makers and decentralized exchanges: A DeFi primer. Financ. Innov. **8**(1) (2022). https://doi.org/10.1186/s40854-021-00314-5
58. L. Heimbach, Y. Wang, R. Wattenhofer, Behavior of liquidity providers in decentralized exchanges. (2021). [Online]. Available: http://arxiv.org/abs/2105.13822
59. C.A. Parlour, Decentralized exchanges. SSRN Electron. J. (2021). https://doi.org/10.2139/ssrn.3905316
60. G.K.C. Ahlers, D. Cumming, C. Günther, D. Schweizer, Signaling in equity crowdfunding. Entrep. Theory Pract. **39**(4), 955–980 (2015). https://doi.org/10.1111/etap.12157
61. K. Mochkabadi, C.K. Volkmann, Equity crowdfunding: A systematic review of the literature. Small Bus. Econ. **54**(1), 75–118 (2020). https://doi.org/10.1007/s11187-018-0081-x
62. J. Roth, F. Schär, A. Schöpfer, The tokenization of assets: Using blockchains for equity crowdfunding, in *Theories of Change: Change Leadership Tools, Models and Applications for Investing in Sustainable Development*, (2021), pp. 329–350
63. S.V. Scott, M. Zachariadis, The Society for Worldwide Interbank Financial Telecommunication (SWIFT): Cooperative governance for network innovation, standards, and community (2013)
64. T. Qiu, R. Zhang, Y. Gao, Ripple vs. SWIFT: Transforming cross border remittance using blockchain technology. Procedia Comput. Sci. **147**, 428–434 (2019). https://doi.org/10.1016/j.procs.2019.01.260

65. P. Todd, Ripple protocol consensus algorithm review, (2015), [Online]. Available: https://raw.githubusercontent.com/petertodd/ripple-consensus-analysis-paper/master/paper.pdf. Accessed 29 Apr 2023
66. L. Ante, I. Fiedler, E. Strehle, The impact of transparent money flows: Effects of stablecoin transfers on the returns and trading volume of Bitcoin. Technol. Forecast. Soc. Chang. **170**(April), 120851 (2021). https://doi.org/10.1016/j.techfore.2021.120851
67. E.L. Sidorenko, Stablecoin as a new financial instrument, in *2019 ISCDTE International Scientific Conference "Digital Transformation of the Economy: Challenges, Trends, New Opportunities"*, (Springer International Publishing, New York, USA, 2019). https://link.springer.com/chapter/10.1007/978-3-030-27015-5_75
68. A. Kahya, B. Krishnamachari, S. Yun, Reducing the volatility of cryptocurrencies – a survey of stablecoins, (2021), [Online]. Available: http://arxiv.org/abs/2103.01340
69. Q. Lu et al., Integrated model-driven engineering of blockchain applications for business processes and asset management. Softw. - Pract. Exper. **51**(5), 1059–1079 (2021). https://doi.org/10.1002/spe.2931
70. U.W. Chohan, Non-fungible tokens: Blockchains, scarcity, and value. SSRN Electron. J. (2021). https://doi.org/10.2139/ssrn.3822743
71. V. Zakhary, M.J. Amiri, S. Maiyya, D. Agrawal, A. El Abbadi, Towards global asset management in blockchain systems, (2019), [Online]. Available: http://arxiv.org/abs/1905.09359
72. Q. Zhou, H. Huang, Z. Zheng, J. Bian, Solutions to scalability of blockchain: A survey. IEEE Access **8**, 16440–16455 (2020). https://doi.org/10.1109/aCCESS.2020.2967218
73. Cryptocurrency Regulations Around The World. [Online]. Available: https://complyadvantage.com/insights/cryptocurrency-regulations-around-world. Accessed 29 Apr 2023
74. M. Platt et al., The energy footprint of blockchain consensus mechanisms beyond proof-of-work, in *Proc. - 2021 21st Int. Conf. Softw. Qual. Reliab. Secur. Companion, QRS-C 2021*, (2021), pp. 1135–1144. https://doi.org/10.1109/QRS-C55045.2021.00168
75. J. Poon, T. Dryja, The bitcoin lightning network scalable off-chain instant payments. 2016. [Online]. https://lightning.network/lightning-network-paper.pdf. Accessed 29 Apr 2023
76. M. Zamani, M. Movahedi, M. Raykova, RapidChain: Scaling blockchain via full sharding. Proc. ACM Conf. Comput. Commun. Secur., 931–948 (2018). https://doi.org/10.1145/3243734.3243853
77. Hardjono, T., Lipton, A., and Pentland, A.: Wallet attestations for virtual asset service providers and crypto-assets insurance. pp. 1–35, (2020), [Online]. Available: http://arxiv.org/abs/2005.14689
78. M. Neuder, D.J. Moroz, R. Rao, D.C. Parkes, Low-cost attacks on Ethereum 2.0 by sub-1/3 stakeholders. pp. 1–11, (2021), [Online]. Available: http://arxiv.org/abs/2102.02247
79. J.P. Aumasson, D. Kolegov, E. Stathopoulou, Security review of ethereum beacon clients, (2021), [Online]. Available: http://arxiv.org/abs/2109.11677
80. T. Hardjono, Development of trust infrastructures for virtual asset service providers, in *European Symposium on Research in Computer Security*, (2020), pp. 74–91. https://doi.org/10.1007/978-3-030-66504-3_5
81. Y. Guo, C. Liang, Blockchain application and outlook in the banking industry. Financ. Innov. **2**(1) (2016). https://doi.org/10.1186/s40854-016-0034-9
82. A. Alamsyah et al., Blockchain traceability model in the coffee industry. J. Open Innov. Technol. Mark. Complex. **9**(1), 100008 (2023). https://doi.org/10.1016/j.joitmc.2023.100008

Scientific Workflows Management with Blockchain: A Survey

Tiphaine Henry and Sara Tucci-Piergiovanni

Abstract In science, a key requirement for conducting a scientific experiment is that it must be reproducible, meaning that a different team should be able to carry out the same or similar experiment and obtain similar results. For this to be possible, it is important to maintain the traceability of the end-to-end workflow used to conduct the experiment. However, partners may be responsible for specific process segments in large projects involving multiple laboratories or collaborating partners. As a result, maintaining the provenance of results requires keeping track of datasets and computing tools at each step of the workflow in a cross-partner fashion. While centralized solutions exist for tracking scientific workflows, they can sometimes be met with distrust. Blockchain-based scientific workflow management can help overcome this issue by offering integrity, transparency, and accessibility. This chapter aims to provide a better understanding of blockchain uses for scientific workflow management. To do so, it examines the existing literature on the topic, discusses the associated opportunities and challenges, and highlights future research topics.

Keywords Blockchain · Survey · Patterns · Scientific workflows · Data traceability · Experiments reproducibility

1 Introduction

Scientific workflows support the management and execution of complex and data-intensive scientific experiments. They allow for the integration and automation of multiple steps and tasks involved in a scientific study, such as data acquisition, processing, analysis, and visualization [1, 2]. However, the credibility and useful-

T. Henry (✉) · S. Tucci-Piergiovanni
Université Paris-Saclay, CEA, List, Palaiseau, France
e-mail: tiphaine.henry@cea.fr; sara.tucci@cea.fr

N. El Madhoun et al. (eds.), *Blockchain and Smart-Contract Technologies for Innovative Applications*, https://doi.org/10.1007/978-3-031-50028-2_5

ness of the results obtained from scientific workflows depend on their accuracy and reliability, which in turn rely on the traceability of the workflow process [3].

Traceability involves keeping track of all the steps involved in the workflow, including the input data, assumptions, algorithms used, users, certifications, and the final results. Tracking this information allows for transparency and accountability in the workflow process, allowing future practitioners to verify, validate, and reuse a former experiment. Indeed it allows for a clear understanding of the underlying assumptions and methods used in the workflow [4]. For example, researchers could accurately and reliably track the origins and history of a mechanical experiment, from environmental constraints definitions to model execution, analysis, and validation. This provenance history can provide valuable information about the quality and integrity of the experiment and any potential contamination or degradation that may have occurred during the workflow process.

However, maintaining traceability in scientific workflows is a challenging task due to the complexity and dynamism of the workflow process, which requires a significant amount of effort and resources to maintain a complete and accurate record of all the steps involved in the experiment process: the traceability of an experiment may not always be given sufficient consideration during the design and execution of the experiment, leading to incomplete or inadequate traceability information. There are also limitations in the current methods and tools used for traceability, which may need to be revised to ensure the accuracy and reliability of the results obtained from scientific workflows. Indeed, it may be challenging to redo an experiment after some time and encounter the same results: data may have changed, as long as software versions, etc. Hence, a need for liability, reliability, integrity, and trust in scientific experiments arises.

Moreover, scientific workflow systems shall be flexible to support the rerun of experiments. The design includes the ability to easily modify and extend the workflow components, as well as the ability to integrate and automate these components dynamically and flexibly. As experiments are often complex and dynamic, the traceability information may need to be constantly updated and refined to reflect the latest assumptions and algorithms used. Additionally, the workflow system should support storing and accessing data and results conveniently and efficiently. However, keeping available and accurate input data and other parameters used in the experiment may also be challenging. The workflow management system must carefully manage experiments' traceability to ensure the rerun's reliability.

These challenges highlight the need for effective strategies and tools to facilitate the rerunning of experiments with traceability while supporting scientific research's dynamic and evolving nature. Preserving the integrity of data and its provenance in a trustworthy fashion appears as a requirement for the reproducibility of scientific experiments [4–7].

In recent years, blockchain technology has emerged as a promising solution for improving traceability in scientific workflows by providing a decentralized and secure platform for storing and accessing the traceability information [8–

11]. A blockchain is a distributed database that allows multiple parties to record data without the need for a central authority securely. It enforces a commonly agreed-upon single source of truth in a trustless environment. In scientific workflows, a blockchain could store traceability information in a tamper-proof manner, allowing for transparency and accountability in the workflow process for auditing purposes [12]. It could ease scientific workflow monitoring, auditing, and dispute resolution as stakeholder trust improves [13]. Additionally, using smart contracts on a blockchain could enable automated verification of the input data for quality norms and check the integrity of data or code used for a computation task, further enhancing the reliability of the results.

Research has been undertaken on the best ways to ensure the traceability of scientific workflows with blockchains. Several proofs-of-use keep track of datasets and code in an open science context [14, 15]. Other works track assets in a more privacy-preserving fashion between a set of known actors, such as with BIM [16], or computing code in a distributed manner, such as with federated learning [17]. However, to the best of our knowledge, there is no comprehensive summary of the use cases and challenges linked to the use of distributed ledger technologies (DLTs) for cross-collaboration scientific workflows. The main objective of this survey is to give a comprehensive overview of the use of the blockchain in cross-collaboration scientific workflow management systems. We first propose a categorization framework regrouping the use cases published in the literature. We then present the main challenges of developing blockchain-based scientific workflow management systems. We finally compare the identified use cases with these research challenges and suggest future research directions.

The remainder of this chapter is structured as follows. Section 2 introduces key concepts related to scientific workflows and blockchain. Section 3 presents the method used to build this survey. Section 4 presents the blockchain-based scientific workflow management systems published in the literature. Section 5 summarizes the applications and use cases found in the reviewed papers. Section 6 analyzes the challenges that emerge from analyzing the retrieved papers. Section 7 presents an evaluation of the proofs-of-use in light of the mentioned challenges. Section 8 discusses these findings and concludes the chapter with a summary of the results and some considerations for future work.

2 Background

2.1 DLTs and Blockchains

A DLT is a shared immutable ledger based on a peer-to-peer network, i.e., a ledger spread across multiple nodes or servers [12]. DLTs have originally emerged to keep track of digital transfers in financial applications, but they can be used to store any kind of data. Data recorded in a DLT is replicated at each peer of the blockchain network. In order to add new data to the ledger a user will issue a so-called transaction containing the data. The transaction is then collaboratively validated by blockchain network peers, through a consensus mechanism, before being added to each copy of the ledger.

DLT systems store data transactions hierarchically using blocks. Each block holds (1) the reference to the previous block, (2) a tamper evident digest of the transaction history to attest the integrity and the ordering of the blocks, and (3) the list of the transactions to record [18]. The type of cryptography block-linkage chosen gives birth to different ledgers. In the case of linear linkage, we talk about blockchains.

The trusted behavior of blockchain systems builds upon the consensus protocol used to update the chain of blocks. This protocol ensures a secured and continuous growth of the ledger and fosters the DLT resilience to node failures [19, 20]. Deterministic scripts called smart contracts can be implemented on top of DLTs to propose a given set of services to the users (e.g., accept a transaction under a set of programmatic conditions) [21–23].

Several identity management and access control schemes characterize DLTs [24]. Public or permissionless blockchains do not perform access control; users can both submit and validate transactions by following an agreed-upon consensus protocol. This protocol states the conditions for a block of transactions to be valid, as well as the set of rules to insert new blocks in the ledger. Permissioned blockchains set identity guards. In fully private blockchains, only white-listed participants can submit and mine transactions, while in consortium blockchains, only a subset of trusted nodes is responsible for mining [19].

2.2 *Scientific Workflows*

Scientific workflows abstract the procedural steps of a scientific experiment through a set of interdependent activities and patterns, likely repeatable. These workflows are used to automate and manage the execution of complex and time-consuming scientific tasks and experiments (e.g., the processing, transformation, and analysis of data) [25], by defining clearly the participants, the activities, and the information flow between them.

Workflow provenance refers to all the information used and produced during the execution of a scientific workflow [11, 26]. The collected information includes information on data, its usage, its modification, the identity of the modifier, the purpose, date, and time, and its lineage (e.g., by linking input data to produced data outputs, and the data generation context, e.g., specific computation resources).

Data provenance of scientific workflow helps keep track of its history for auditing times and ensure its integrity. Indeed, data provenance can help invalidate data segments if an error is identified a posteriori. Data provenance can also support practitioners regarding data replication and reconciliation [4, 27]. It can finally help practitioners detect different patterns during workflow redesign [28].

3 Methodology

To conduct a survey [29] on the theme of trustworthy scientific workflow management systems using blockchain, we first retrieve the associated literature by applying a subjective search string to the main databases related to the topic of scientific workflow management [30]. We also combine synonyms for blockchain and smart contracts and synonyms for traceability, data anchoring, data integrity, and trustworthiness in scientific workflows into a Boolean query.

The output search string is *("distributed ledger technology" OR "blockchain-based systems" OR "decentralized ledger technology" OR "smart contracts") AND ("traceability" OR "provenance tracking" OR "data provenance" OR "provenance management" OR "workflow traceability" OR "scientific workflow traceability") AND ("data anchoring" OR "data integrity" OR "data verifiability" OR "data immutability") AND ("trustworthiness" OR "reliability" OR "security" OR "integrity").*

We query the main databases related to scientific workflow management with this search string. We remove duplicates and apply the following inclusion criteria: (1) is it a peer-reviewed article? (2) does it mention both blockchain and scientific workflows? We then recursively pursued the related work references included in these papers. Table 1 outlines the number of papers found in each database before and after the inclusion criteria filtering. We get 27 papers after removing duplicates. A backward search led to the retrieval of two more papers. After removing duplicates, the subset of papers kept for further analysis comprises 29 papers. Table 2 lists the selected peer-reviewed conference or journal articles.

We classify each work using a concept matrix presented by Webster and Watson [29] to systematically collect and analyze the different blockchain-based scientific workflow management systems. Table 7 gathers the results of the analysis.

Table 1 Search query results

Database	Initial queries	Filtered queries
ACM Digital library	15	12
AISel	4	2
IEEE	6	5
ScienceDirect	54	9
SpringerLink	41	8
Wiley	5	1
Total	125	27
Total after backward search		29

4 Scientific Workflows Management with Blockchain: A Synopsis

This section presents existing works on scientific workflow management using blockchain for traceability purposes.

Distributed scientific experiments require the trustworthy management of a set of data elements (e.g., test data, results, reports, or code). Indeed, data should be referenced and traced on-chain, alongside a set of metadata. Additionally, once referenced on-chain, data may be shared with other blockchain participants or reused for new experiments (possibly in a privacy-preserving fashion). Finally, archived data may become deprecated, as well as its data dependencies. Hence, the blockchain system should also support the deprecation of data experiments. Figure 1 proposes an abstraction of the blockchain-based data management lifecycle that is leveraged during scientific experiments.

In the following section, we present a synopsis of patterns related to system architectures, notarization strategies, privacy strategies, and auditing choices appearing in the surveyed papers that aim at managing the scientific experiments data lifecycle.

4.1 System Architecture Choices

In this subsection, we review the on-chain scientific workflows architectures presented in the literature. We look at the distributed ledger architectures, the data storage strategies, and conclude with a focus on off-chain data storage support. Table 3 compares the surveyed papers based on their choice of distributed ledger architecture and data storage location strategy.

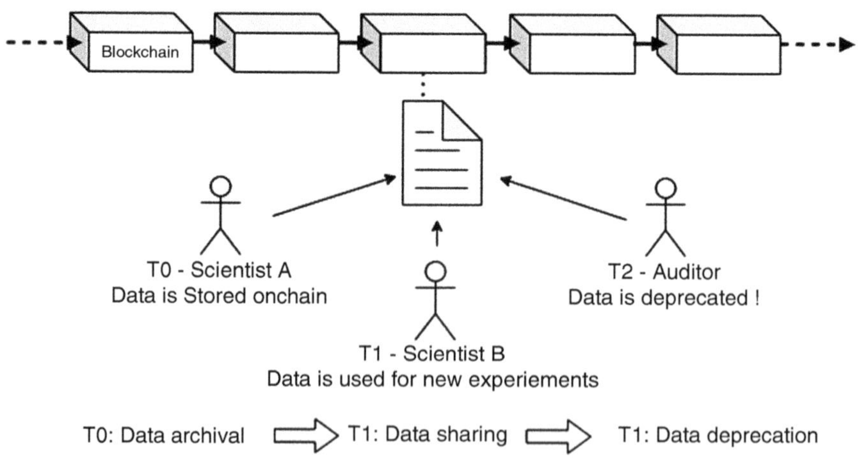

Fig. 1 Illustration of the three steps of the archived data management lifecycle

Table 2 List of papers analyzed for the survey after the filtering stage

Ref.	Publisher	Type[a]
Bell et al. [42]	IEEE/ACM 12th International Workshop on Software Engineering for Science	C
Bose et al. [38]	26th Asia-Pacific Software Engineering Conference, APSEC	C
Chenli et al. [34]	Journal of Parallel and Distributed Computing	J
Chen et al. [11]	IEEE International Conference on Big Data (Big Data)	C
Coelho et al. [26]	Journal of Grid Computing	J
Coelho et al. [43]	HICSS 54	C
Costa et al. [44]	IEEE 10th International Conference on e-Science	C
Daidone et al. [45]	IEEE International Conference on Web Services (ICWS)	C
Demichev et al. [35]	Journal of Grid Computing	J
Fernando et al. [8]	International Conference on Collaboration and Internet Computing (CIC)	C
Gipp et al. [46]	ACM/IEEE Joint Conference on Digital Libraries (JCDL)	C
Hoffman et al. [41]	The Semantic Web: 15th International Conference, ESWC	C
Honar et al. [33]	Journal of Big Data	J
Kim et al. [31]	Intelligent Systems in Accounting, Finance and Management	J
Liang et al. [47]	IEEE/ACM International Symposium on Cluster, Cloud and Grid Computing (CCGRID)	C
Martinez et al. [32]	Information Processing & Management	J
Moller et al. [36]	Innovation Through Information Systems: Volume III: A Collection of Latest Research on Management Issues	J
Nandigam et al. [39]	Practice and Experience in Advanced Research Computing (PEARC)	C
Neisse et al. [48]	12th international conference on availability, reliability and security	C
Pradeep et al. [7]	Automation in Construction	J
Radha et al. [37]	Blockchain: Research and Applications	J
Ramachandran et al. [49]	ACM Conference on Data and Application Security and Privacy, CODASPY	C
Raman et al. [54]	IEEE International Conference on Blockchain and Cryptocurrency (ICBC)	C
Ruan et al. [50]	VLDB	C
Sivagnanam et al. [40]	Practice and Experience in Advanced Research Computing (PEARC)	C
Tang et al. [5]	ACM/SIGAPP Symposium on Applied Computing	C
Wang et al. [51]	Big Scientific Data Management: First International Conference, BigSDM	C
Wen et al. [52]	Big Scientific Data Management: First International Conference, BigSDM	C
Wittek et al. [53]	IEEE International Conference on Blockchain and Cryptocurrency (ICBC)	C

[a] J = journal, C = conference

Table 3 Comparison of the blockchain-based architectures of the surveyed papers

Ref.	DLT architecture	Data storage location
Tang et al. [5]	Custom	Ciphered data stored on-chain
Chen et al. [11]	Custom	Merkle anchor
Daidone et al. [45]	Custom	File location
Liang et al. [47]	Custom	Merkle anchor + Anchoring
Wittek et al. [53]	Custom	Anchoring
Martinez et al. [32]	Custom (tuned Hyperledger)	Anchoring
Ruan et al. [50]	Custom (tuned Hyperledger)	Merkle anchor
Gipp et al. [46]	Bitcoin	Anchoring
Möller et al. [36]	Corda	Anchoring
Bose et al. [38], Wang et al. [51], Hoffman et al. [41]	Ethereum	Anchoring—IPFS
Kim et al. [31], Radha et al. [37]	Ethereum	–
Pradeep et al. [7], Neisse et al. [48], Ramachandran et al. [49], Fernando et al. [8]	Ethereum	Anchoring
Wen et al. [52]	Ethereum	Ciphered data stored on-chain
Bell et al. [42]	Ethereum + Hyperledger	Anchoring
Costa et al. [44]	Early work	NA
Chenli et al. [34]	Hyperledger	On-chain
Honar et al. [33]	Hyperledger	Anchoring
Coelho et al. [26]	Hyperledger	Anchoring
Demichev et al. [35], Nandigam et al. [39]	Hyperledger	File location
Sivagnanam et al. [40], Coelho et al. [43]	Hyperledger	Anchoring
Raman et al. [54]	Hyperledger	On-chain

4.1.1 Distributed Ledger Architectures

In the retrieved articles, three main choices of blockchain architecture appear: (1) public blockchains, (2) private/permissioned blockchains, and (3) customized blockchains.

In the surveyed papers, several systems leverage public blockchains, such as Ethereum and Bitcoin, to provide open access to the transactions' history to future auditors. A majority leverage Ethereum, building on smart contract features, in [7, 8, 31, 37, 38, 41, 42, 48, 49, 52], while Gipp et al. [46] leverages Bitcoin for scientific manuscripts management [46].

Private/permissioned distributed ledgers, such as Hyperledger and Corda, are also used to offer a controlled environment with limited access to members. Access to such a network is conditioned by the approval of a designated authority or of all participants. Hence, access to the ledger is limited to a set of predefined users to

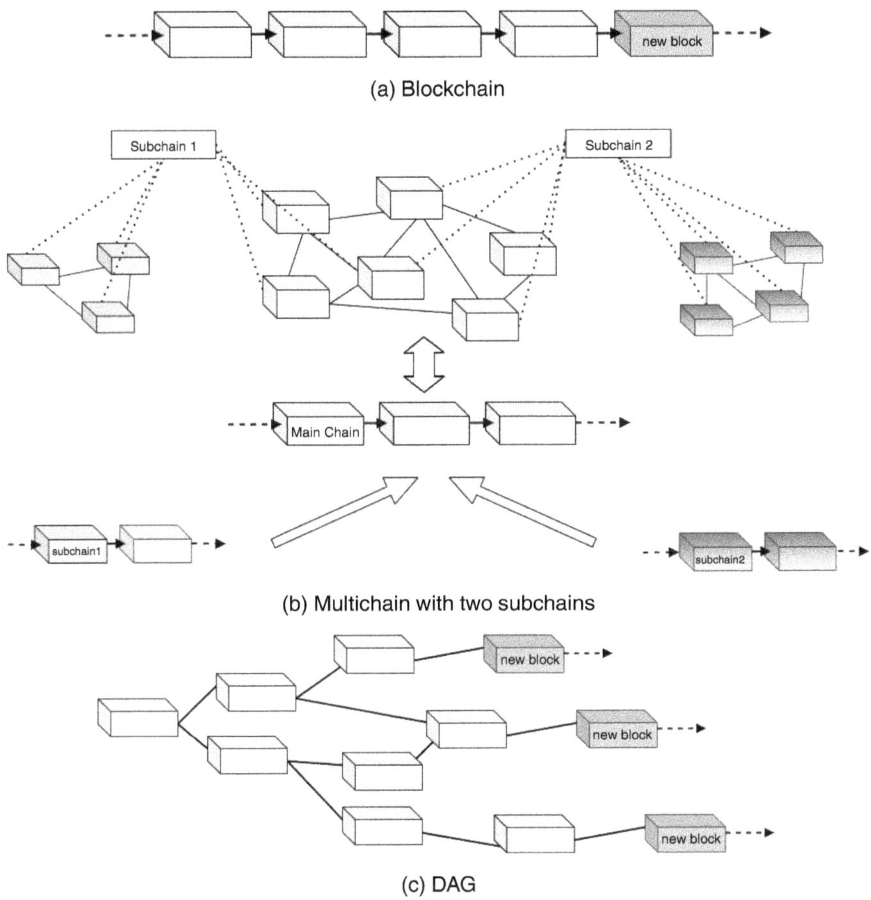

Fig. 2 Illustration of the three distributed ledger architectures used in surveyed papers

protect sensitive data [8, 26, 35, 48]. A majority of the retrieved articles builds on the Hyperledger blockchain [26, 33–35, 39, 40, 43, 48, 54]. Moller et al. [36] uses Corda to build multichains [36]. Figure 2b illustrates this architecture: subchains are created between blockchain peer subsets requiring confidential data provenance. This architecture prevents external users from accessing the workflow execution logs by storing only the beginning and the end of the side chain in the main chain.

Finally, several works propose customized blockchains that enable efficient provenance tracking and tracing [5, 11, 32, 45, 47, 50, 53]. Among these initiatives, Wittek et al. [53] builds on a tuned version of Ethereum with a proof-of-authority consensus [53]. Tang et al. [5] develops a blockchain architecture supporting privacy-preserving data linkages [5], while Chen et al. [11] and Liang et al. [47] elaborate dedicated blockchains designed for efficient provenance queries [11, 47]. At last, directed acyclic graphs (DAG) are used in two works. Martinez et al. [32]

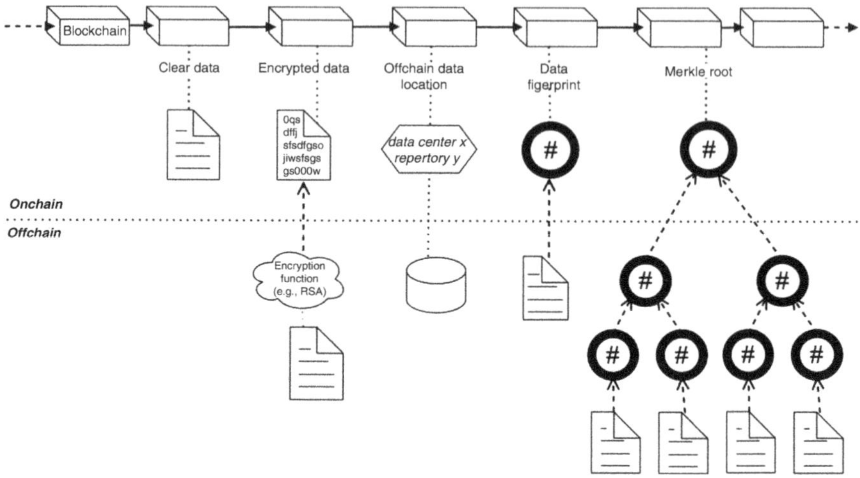

Fig. 3 Data storage location strategies retrieved from papers

uses a DAG for IoT-based workflows to improve the solution's scalability [32]. Ruan et al. [50] customizes the Hyperledger storage structure using a Merkle DAG with a data indexing adapted to efficient queries (a deterministic skip list indexing) [50]. This data structure supports (1) explicit data versioning to track data derivation history dynamically instead of off-line, (2) fork semantics to let users work on independent data copies for data derivation analysis, which is especially relevant in the case of concurrent transactions.

4.1.2 Data Storage Strategies

Figure 3 presents the strategies identified in the retrieved papers to store scientific workflow data on-chain.

Two strategies rely on fully on-chain data storage. The first strategy is to store data and its provenance directly on-chain [34, 54]. For example, Raman et al. [54] uses a smart contract to keep track of the computation results in multi-party computation schemes by storing model parameters with the intermediate results of computation steps on-chain [54]. If a dispute arises, computation steps can be re-audited in a trustworthy fashion by the participants. The second strategy consists into storing encrypted data for confidentiality purposes. Recipients share deciphering keys off-chain with participants that wish to access the archived data [5, 52].

To avoid storing data directly on-chain, other strategies rely on a hybrid on/off-chain storage for privacy or efficiency concerns. Several papers use a location pointer of the data on-chain alongside other metadata (e.g., file owner id, file type). The actual scientific experiments data is stored securely off-chain. Examples of this strategy are found in Nandigam et al. [39], Daidone et al. [45], and Demichev et al. [35]. Another strategy is data fingerprinting, where a reference of the hashed data is stored on-chain, while data is kept off-chain. The hashing strategy classically leverages sha256 [7, 8, 26, 32, 33, 36, 40–43, 46–49, 51, 53]. Compared to referencing the storage location on-chain, this method enables users to verify the integrity of the off-chain data by comparing the hash of this data element to the hash stored on-chain. The link to the external storage is also stored on-chain alongside the data fingerprint in a subset of articles [8, 33, 39, 42, 43, 53]. Finally, Merkle tree structures have been leveraged for scalability concerns and storage efficiency [11, 47, 50]. The leaves of the Merkle tree each correspond to the hash of the tracked data. The generated hashes are computed into a parent node until reaching the root of the Merkle tree. This Merkle root is stored on-chain. Future auditors will assess the integrity of a given file by recomputing the Merkle root using the off-chain storage and comparing it to the tamper-proof root stored on-chain. The integrity of the data history is safeguarded on-chain in an efficient fashion for a dataset. However, with Merkle trees, the integrity check is not at the level of a single file: if the integrity check is negative it is not possible to determine which file has been corrupted precisely based on the Merkle root. An off-chain storage having recorded the Merkle tree construction is necessary, which may lower trust in the system due to the dependency to a centralized component.

4.1.3 Off-Chain Data Storage Support

Modular designs can be noticed regarding the corresponding off-chain storage of data. Figure 4 summarizes the off-chain storage strategies identified in the retrieved set of papers.

First, in the context of academic publishing, data is stored in a centralized database administered by the publisher in Gipp et al. [46] or by trusted third party cloud providers in [7, 8, 26, 32, 33, 43, 47, 49]. These models preserve the privacy of data, as access control schemes can be enforced by data owners. However, long-term auditing is challenging in these schemes as data may become obsolete or be erased off-chain. Additionally, trust issues may arise for use cases relying on cloud services as they rely on trusted third parties to manage off-chain storage. Thirdly, distributed storage approaches are used in a set of papers: several articles rely on several data storage locations [7, 8, 36, 44]. For example, participants may notarize emails or models stored in the cloud (Pradeep et al. [7]) or in data owner personal storage locations (Moller et al. [36]). Moreover, two papers use a distributed storage to support efficient auditing queries based on hash retrieval (Fernando et al. [8] and Costa et al. [44]). Distributed storage is used as a way to alleviate the dependency on trusted third parties. Though long-term storage and data control issues arise in these schemes too, the risk is shared between the different data storage locations. Finally, Hoffman et al. [41] and Wang et al. [51] leverage IPFS, a decentralized information storage platform, to store data off the blockchain in clear to address

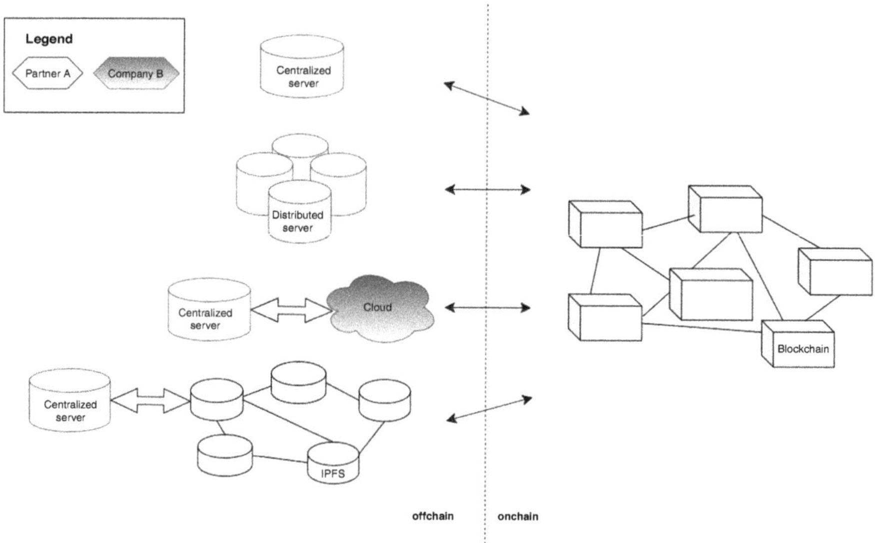

Fig. 4 Off-chain storage choices identified in the retrieved papers

blockchain scalability issues [41, 51]. IPFS is mentioned in Neisse et al. [48] but not implemented. This choice fosters trust in off-chain data, as it mimics the blockchain architecture: it offers a trustworthy and long-term off-chain data storage, without relying on a trusted third party. However, these use cases do not ensure data privacy as data is stored in clear in IPFS.

4.2 Archiving Data On-Chain: Overview of Notarization Strategies

We investigate smart contract-based patterns presented in the literature related to data notarization. We first present literature patterns related to the data notarization lifecycle. We then overview patterns related to the quality assessment of notarized data.

4.2.1 Smart Contract Based Notarization Lifecycle

We investigate hereinafter the on-chain notarization patterns appearing in the literature that support the data management lifecycle (i.e., data archival, data sharing, and data deprecation).

Table 4 Data file notarization—elements retrieved from the literature

Element notarized	References
Type (file/stream/indexed data/message)	[7, 47]
Id	[47]
Name	[47]
Data/file description	[7]
Data hash	[5, 7, 8, 26, 33, 34, 39–43, 48, 52, 53]
Data location pointer (anchoring strategy = certifying file existence at some point in time only)	[8, 33, 39, 42, 43, 47, 53]
Timestamp	[7, 33, 40, 42, 46, 47, 49]
Attribution (ownership/copyright)	[41, 42, 52]
Annotations/feedbacks	[40, 41]

Data Archival

First, a set of smart contract operations aims at the registration of new data items [32, 33, 38, 47]. Table 4 presents the elements notarized on-chain in the papers analyzed for this study. Data content may be stored directly or indirectly using the data anchoring technique. In retrieved papers, this technique involves notarizing the hash of the data and the data location pointer on-chain. Additionally, associated metadata may be registered on-chain. Data type, id, name, and description may be stored [7, 47], sometimes alongside the notarization timestamp [7, 33, 40, 42, 46, 47, 49] and the data owner [41, 42, 52]. Finally, a data field dedicated to annotations/feedback is proposed in two papers for future data sharing [40, 41].

Data Sharing/Dissemination

Once the data is registered on-chain, users may consult the smart contract to query the data (Honar et al. [33], Bose et al. [38], Liang et al. [47]). If the data owner accepts to share it, then the smart contract may provide access to the data. In Wittek et al. [53] and in Liang et al. [47], new data from the initial query reference such query to make the data lineage accessible. Scientific experiments exemplify an example of a data query/reuse: a dataset is, for example, used to instantiate an algorithm. In Honar et al. [33], the computation result will reference the dataset to keep track of this dependency. In Wittek et al. [53], a specific framework called ROCL (standing for Research Objects Chain Links) is developed to keep track of the dependencies between code, runtime environments, and data. The dependency tree is built using Merkle trees, as in Liang et al. [47].

Data Deprecation

The last lifecycle stage is data deprecation (corresponding to delete or update operations [7, 8, 34, 38–40, 47, 49, 54]. Smart contracts may support file versioning coupled with the update of the file status such as in Fernando et al. [8]. Additionally, several papers assign a status to notarized files (e.g., current version or deprecated version) [7, 33, 40, 42, 49]. It is coupled to a link to the update file version in [33, 40, 42, 49]. Data change approval decisions may be carried out using compliance to a voting threshold [49] or by requesting approval from the data owner [7]. Upon approval, the data dependency is specified in the new file versions. Proof of change is produced using the digital signatures of the data modifier, coupled to the change operation timestamp (Ramachandran et al. [49] and in Pradeep et al. [7]). One paper additionally checks whether a significant difference exists between two file versions to prevent data duplicates on-chain (Raman et al. [54]). Authors do so by computing the hash difference between the new file to be registered and archived files. If the difference is below a preset threshold, the file is considered unique and can be added to the blockchain ledger.

4.2.2 Data Quality

The notion of data quality is important when notarizing data on-chain to ensure experiments replications or audits [4]. Three categories of strategies have been noticed in the retrieved papers. The adoption of *provenance metadata standards* such as Prov in Bose et al. [38], or ProvOne in Cao et al. [55] helps select a set of data fields relevant for experiments traceability, which can foster the quality of the data lineage, and improve the reproducibility of the experiments [26, 31, 35, 43, 49]. Examples of such fields comprise runtime environments, activity states, and the notion of activity inputs and outputs. Neisse et al. [48] also record bulk events to ensure continuous recording even if data streams are null. One paper, Wen et al. [52], focuses on *data quality incentives*. It does so by assigning a quality status to each recorded data field. The mechanism leverages a reward system building on blockchain incentives to encourage high-quality status. A third strategy concerns the *certification or validation of notarized data* by dedicated authorities, e.g., a lab responsible or a certificating authority (Pradeep et al. [7] and Liang et al. [47]). Validation may consist of an approval function built into the smart contract. It may also consist of issuing a certificate to the data owner (Wittek et al. [53]) or a digital asset (Radha et al. [37]). Table 5 summarizes this list of scientific computation metadata notarized onchain.

Table 5 Scientific computation notarization—elements retrieved from the literature

Element notarized	References
Computational code/pipeline	[32, 38, 39, 53, 56]
Runtime environment (hardware/software)	[26, 53]
Experimental design/variables/steps	[26, 53]
Tests	[53]
Manuscript files	[53]
Results analysis	[26]
Execution time	[8]
Documentation (e.g., params and dependencies)	[39, 53]
Scientific workflow data	Cf Table 6

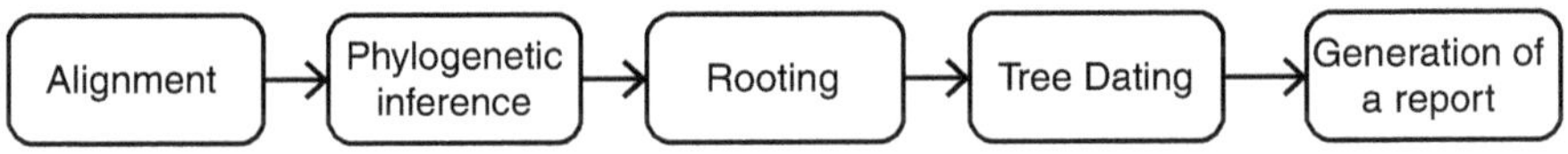

Fig. 5 Example of a directly follows graph depicting the ViReport workflow presented in Coelho et al. [26]

4.3 *Imperative Control-Flow Notarizations with Workflows*

Several research works include a workflow approach for data provenance traceability on-chain. These approaches explicitly trace the processes and the associated metadata (subprocesses, activities, role initiators and receivers, etc.). Workflow traceability helps standardize practices and improves data quality, as provenance traceability is included as part of the work to carry by responsible participants (Daidone et al. [45]).

Activities using previously archived data may also automate integrity checks through hash comparisons to verify if the data to be used corresponds to the version tracked on-chain, e.g., when using a previously notarized dataset for a new experiment, (Martinez et al. [32]).

The interest for workflows is specified but not implemented in [39, 40, 49]. Other works implicitly refer to workflows through the lens of provenance ontologies: TOVE in Kim et al. [31], ROCL in Wittek et al. [53], or ETL in Martinez [31, 32, 53]. Workflow specifications are, however, not implemented directly, which may hinder reproducibility.

Coelho [26] and Liang [47] introduce smart contract calls via an API to record data transactions in existing scientific workflow execution systems. However, the focus is on provenance management on-chain, and the workflow execution is not managed on-chain.

Finally, some approaches explicitly use workflow notation standards. Provenance traceability is introduced with directly follows graphs in several works [8, 34, 36, 38, 45]. BPMN is additionally used in one paper (Pradeep et al. [7]). Figure 5 illustrates one of the directly follows graphs executed on-chain in Coelho et al. [26].

Table 6 Workflow approaches: elements notarized

Element notarized	References
Subprocess id	[7]
Activity/activity id	[7, 8, 26, 32, 39, 43, 48, 49]
Role initiator	[5, 7, 8, 11, 33, 36, 39–43, 45–50, 52–54]
Affected roles/users	[7, 47]
Activity action or procedure	[38, 39, 47]
Data input/output	[8, 26, 32, 43, 49, 53]
Number of iterations	[26]

In these cases, the execution of the scientific workflows is managed directly on-chain via smart contracts, a strategy referred to as a Scientific Workflow Engine Smart Contract (SWE-SC) in Daidone et al. [45].

Table 6 summarizes the workflow elements notarized in smart contracts that we retrieved from the literature. Upon execution of each activity, the subprocess, the activity id, and the role initiator are specified. Optionally, for activities affecting data owned by another role or user, e.g., when sharing a document Liang et al. [47], affected roles or users can be notarized too. Activity actions refer to the list of actions accessible to activities that leverage notarized data coded into smart contracts, e.g., create, read, share, write, or delete an experiment file (c.f., Liang et al. [47] and Nandigam et al. [39]). In Bose et al. [38], these actions also leverage procedures such as methods, document templates, or techniques. The notion of data inputs, i.e., all the data elements that will be used to complete an action, and data outputs, i.e., all the data elements generated during the activity completion, are conceptualized and notarized too on the blockchain. Finally, Coelho et al. [26] notarize the iteration number of workflow segments to support scientific processes where activities or subprocesses can be executed several times.

4.4 *Security and Privacy Strategies*

Scientific workflows may comprise sensitive data. A set of approaches includes security and privacy policies to protect them. We summarize these approaches hereinafter through the lens of access control policies, privacy policies, and abstraction and encryption strategies.

4.4.1 Managing Actors and Digital Assets Access Control

Authentication and access control help, by the definition and enforcement of a set of policies, to prevent unwanted access to data elements stored on-chain or unwanted notarization behaviors.

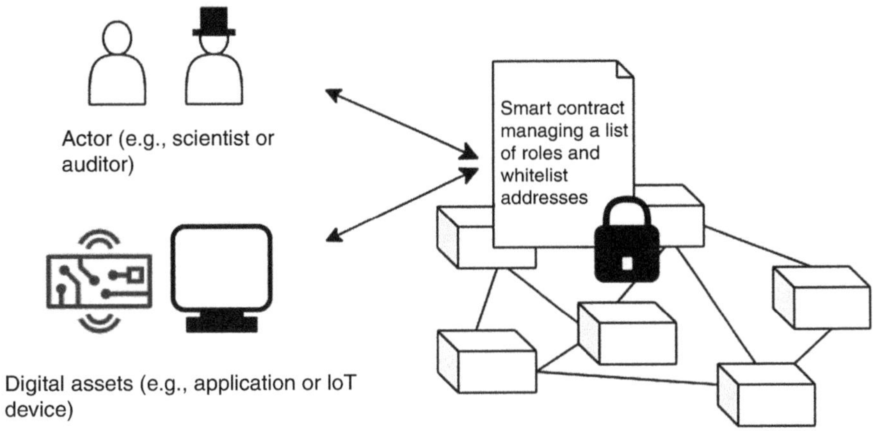

Fig. 6 Actors and digital assets of a scientific workflow must be authenticated in the blockchain before notarizing data of a scientific workflow

Access to blockchain logs can be limited to a set of predefined users at design time by choosing the blockchain permissions (c.f., Sect. 4.1, and permissioned ledgers [8, 26, 35, 48]). Action scopes can also be defined in the smart contract to allow notarization by the right roles.

Figure 6 schematizes the authentication of both actors and digital assets of a scientific workflow to the blockchain specified in several papers to a scientific workflow smart contract. Actors may refer to scientists or auditors [8, 37, 48, 49], while digital assets may concern digital applications such as machine learning training systems as in Radha et al. [37], or IoT devices for scientific workflows involving physical materials as in Honar et al. [33].

Users and digital assets are identified with a unique blockchain address. The scientific workflow smart contract assigns these addresses to the list of participants of the workflow. These addresses may be white-listed or linked to a role with a specific scope of actions. For example, (1) all workflow actors must be authenticated to be able to execute or access a workflow, notarize a data element, or access notarized data in Radha et al. [37] and in Sivagnanam et al. [40], (2) a sensor can only publish data (Honar et al. [33]), and (3) an auditor can only consult data [8, 37, 48, 49]. In Honar et al. [33], the gateway registration timestamp is notarized upon each data stream notarization request in the context of IoT device authentication.

Finally, it is to note auditors' authentication is not automatic; it is only specified in a subset of papers [8, 37, 48, 49]. Moreover, the read access queries of auditors may also be notarized on-chain, as suggested in Raman et al. [54] and Coelho et al. [26].

4.4.2 Smart-Contract-Based Sharing Policies for Anchored Data

Access control can also concern access to data traced using the anchoring technique (c.f., Sect. 4.1.2). Then, access control builds on the definition of sharing policies to ensure that off-chain data is transmitted securely (Daidone et al. [45]). Such sharing rights can be enforced with user groups as in Demichev et al. [35] and Coelho et al. [26], or more fine-grained access control, as in [5, 7, 34, 36, 48]. Neisse et al. [48] and Daidone et al. [45] propose to let users customize their privacy policies when notarizing data. The customization of Neisse et al. [48] takes the following format: the smart contract registers a set of policies, an event, a desired configuration, a condition, and an action. For example, the user defines the following policy: a file can be shared (event/condition) for a given purpose, a set of countries, or a group of users (configuration) if the query happens within a timeframe. According to these papers, the purpose of data collection and usage can be defined and optimized using purpose trees. Such policies, defined by data subjects, are then applied to data controllers or data providers.

4.4.3 Encrypting Sensitive Data

Encryption strategies are applied in several works to limit provenance data privacy leakage.

First, encryption layers can be added to the data and its metadata stored on-chain [5, 34, 47, 48, 52]. Metadata fields ciphered on-chain comprise the file location (c.f., Neisse et al. [48]) and the identity of data users (c.f., Ramachandran et al. [49] and Liang et al. [47]). Indeed, for Ramachandran et al. [49], the identity of file users should only be accessible to the file owner.

Additionally, data anchoring (or Merkle tree root anchoring) can be used to store data off-chain while preserving data integrity. It holds benefits concerning both storage efficiency and privacy considerations [7, 39, 47–49] (c.f., Sect. 4.1.2).

Finally, Tang et al. [5] focus on data provenance linkage privacy. The solution protects access from unauthorized blockchain peers to inheritance linkages between files, e.g., file transformations. They do so by separating the data elements (called the edges) from their linkages (called the vertices) in the ledger.

4.5 Auditing Track

The data provenance history is leveraged for audits. Auditors may be researchers trying to reproduce or build on the results. Auditors may also be certification entities, assessing the validity of the data after some time. Scientific workflow data history is registered on-chain. A set of retrieved papers propose auditing tools coupled with query efficiency strategies, as information retrieval is not straightforward in blockchain systems due to transaction recording.

4.5.1 Integrity Checking Tools

A set of papers propose *integrity checking* tools to retrieve an experiment or a group of data previously notarized (Sivagnanam et al. [40], Ramachandran et al. [49], and Liang et al. [47]). Anchored data can be searched and verified by computing the off-chain file's hash and retrieving or comparing the corresponding metadata from blockchain logs (c.f., Pradeep et al. [7] and Honar et al. [33]). Audits may also consist of estimating similarities between datasets, for example, by using the MinHash protocol in Chenli et al. [34], to detect data duplicates. The MinHash protocol is a type of locality sensitive hashing used to estimate the Jaccard index of two hashes (the Jaccard index being a measure of the degree of similarity between two sets). The higher the fraction of matching MinHash values, the closer their Jaccard similarity. It is to note that Fernando et al. [8] store unencrypted data purposefully to implement a Bloom filter on top of the blockchain for efficiency concerns. Each new data notarization generates a notarization log that is stored in a dedicated data structure named "logsBloom." To retrieve content assigned to a desired notarized hash efficiently, the data structure, which implements a Bloom filter, is queried.[1]

4.5.2 Backward Tracing

Audits may also consist of *backward tracing* queries. Backward tracing refers to identifying all precedence elements of a data element. Precedence elements comprise the original data and the transformations that lead to the data state audited [56] (illustrated in blue in Fig. 7). This auditing strategy may be used to understand why a stage is not reproducible and identify the corresponding deprecated workflow segments (c.f., Tang et al. [5], Chen et al. [11], or Chenli et al. [34]).

Querying data in the ledger can be time-consuming due to its size and complexity. A set of papers propose to enhance auditing queries efficiency in a backward tracing context.

The choice of blockchain architecture may impact the efficiency of queries. Private chains, smaller than public ones, hold fewer transactions. This choice is advocated in Neisse et al. [48] to ease information retrieval by design. Additionally, local auxiliary databases are leveraged to expedite blockchain queries in Fernando et al. [8] and in Costa et al. [44]. By fragmenting local storage in a distributed fashion, query processing becomes distributed, which enhances query efficiency (c.f., Sect. 4.1).

[1] The Bloom filter is a probabilistic data structure that efficiently tests the membership of an element in a set, providing a compact representation with potential false positives but no false negatives [57].

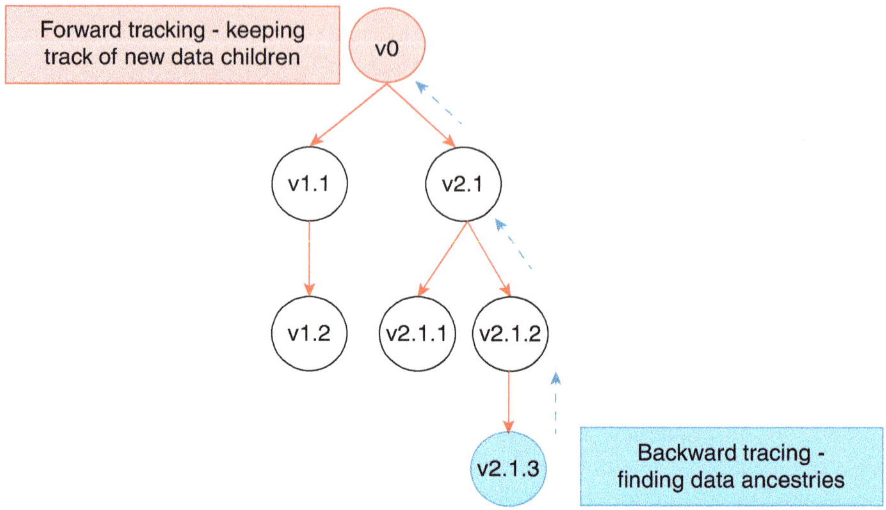

Fig. 7 Forward versus backward tracking

Moreover, two methods for efficient data indexing have been retrieved during this survey. *Ad-hoc index structures* are investigated in a blockchain context in Daidone et al. [45]. These index structures are temporary and are created when needed for a specific blockchain data query (i.e., for the specific data elements or attributes being queried). These indexes can be created in memory or stored in a separate database and discarded once the query is completed. Deterministic append only *skip list indexes* are proposed in Ruan et al. [50]. This method consists of building multiple levels of nodes, with the bottom level representing the original data and each level above it containing a subset of the nodes from the level below it. Each node in the skip list index includes a key and a pointer to the next node at the same level and pointers to higher-level nodes. This allows for faster searching by skipping over large sections of data.

4.5.3 Forward Tracking

Forward tracking refers to the identification of the data derivation history, i.e., all data transformations that are built on the data state audited (in red in Fig. 7). This auditing strategy may be used to detect all deprecated workflow experiments if the data state audited happens to be corrupted (be it the detection of full workflows or workflow segments leveraging the deprecated data element) [5, 8, 34, 47].

Enforcing dynamic forward tracking by design is challenging due to blockchain limitations according to Chenli et al. [34]. Indeed, a retrieved blockchain block holds no reference toward future blocks; hence, by design, it is not straightforward to automate the referencing of future data derivations during notarization. To circumvent this issue, Mendes et al. [56] and Stage et al. [58] declare future data relationships using workflow specifications. In these papers, an abstract workflow specification is leveraged to capture future data derivations before workflow instance execution, i.e., before the generation of execution logs.

5 Applications and Use Cases

The survey uncovers a range of potential scientific workflow management use cases for blockchain technology. Figure 8 shows the distribution of the surveyed papers based on their application fields.

Three *Biology and Medicine* use cases have been identified in the literature: (i) drugs development scenarios (phylogenetics, genomics, or bioinformatics) in Coelho papers [26, 43], (ii) a clinical drug trial in Ramachandran et al. [49], and (iii) DNA-related applications in Daidone et al. [45] and in Fernando et al. [8]. Four *Computer Science* use cases have been identified in the literature: (i) behavioral science deep learning analysis in Wittek et al. [53], (ii) collaborative distributed computing or experimental simulations in Demichev et al. [35] and Raman et al. [54], (iii) software provenance and experiments reproducibility in Bell et al. [42] and in Bose et al. [38], and (iv) IoT-based experiments in Martinez et al. [32] and in Honar et al. [33]. In the field of *Engineering*, three use cases for blockchain technology have been identified, namely marine traffic modelization and optimization (Möller et al. [36]), building information modeling (Pradeep et al. [7]), and food supply chains (Kim et al. [31]). In the field of Earth Science, *Geology* (Nandigam et al. [39]) and Biology (Costa et al. [44]) use cases have been identified for blockchain technology. Finally, a subset of surveyed papers focuses on scientific workflow management with blockchain without specific case studies.

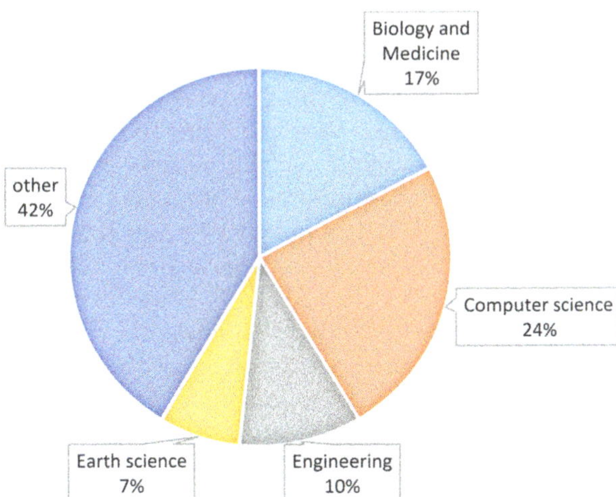

Fig. 8 Overview of the distribution of the surveyed papers based on their application fields

The following points were addressed in those papers: (i) scientific manuscripts and certification [37, 41, 46], (ii) scientific data objects sharing [11, 40, 51, 52], and (iii) scientific data provenance [5, 34, 47, 48, 50].

6 Research Challenges

The blockchain technology is an avenue worth exploring for trustworthy scientific workflow management. Nevertheless, this technology comes at the expense of added constraints on privacy, scalability, dynamicity, or tracing efficiency. In this section, we detail each one of these challenges.

6.1 Intellectual Property and Confidentiality of Scientific Experiments

Scientific experiments are considered intellectual property before being published in the literature [26]. Indeed, the traceability information generated by a scientific workflow may contain intellectual property belonging to multiple parties.

The provenance records [5, 45, 59], as well as sensitive data, held in research artifacts [7, 26, 37, 60], the size of record pointers [5, 59], or segments of participants interactions [61, 62] can be subject to legal issues related to intellectual property and ownership of the traceability information.

A need, therefore, arises for striking a balance between utility and privacy when releasing data provenance on-chain [59]. Moreover, using blockchain technology in scientific workflows involving personal or sensitive data may require implementing additional security measures to comply with privacy regulations [7, 26, 37, 60].

Efficient access control policies and privacy-preserving techniques related to data provenance can be leveraged to mitigate confidentiality concerns [49]. Research efforts have indeed been pointed toward the need for secure, fine-grained, and runtime access to provenance, extended to smart contracts [50, 52, 60]. The definition of user groups with specific rights has been proposed when releasing data provenance on-chain [35], as well as dedicated IoT device identity management solutions [33]. Abstraction trees hide provenance details in [59]. A coarse-grain view of processes or data hides more fine-grained and confidential information. Process views may also silo participants by providing only the information they need to carry on an assigned activity [59]. These approaches are yet to be implemented in blockchain-based scientific workflows.

Additionally, as user accesses may evolve through time—owners of scientific experiments may leave organizations as part of organizational changes, a requirement for delegating mechanisms preserving confidentiality when designing access control and privacy policies should be taken into account [4, 62]. Finally, for physical scientific workflows, ensuring the confidentiality of IoT data streams may involve specific research efforts, as low energy is available in devices to encrypt data [27].

6.2 Auditing Support

The execution of on-chain audits embodies specific challenges [4, 62].

The main limitation to the usability of blockchain-based scientific workflow auditing tools is the efficient tracing/tracking [5, 8, 11, 34, 44, 45, 50, 54, 59] of provenance information. Indeed, forward tracking remains inefficient, as smart contracts do not have references toward the future at runtime [34]. Smart contracts can only refer to blocks in the current blockchain designs. Another area for improvement to the usability of blockchain-based auditing tools may concern a need for dedicated blockchain data operators capturing provenance semantics, such as map, join, or union operators [50].

Challenges also embody investigating more auditing extensions such as the detection of cloning or plagiarism of documents previously registered on-chain by scientific workflow traceability smart contracts [4, 46, 48, 63]. Extensions to auditing tools worth investigating also comprise detecting abnormal operations from transaction logs or searching workflow segments from provenance data [11].

Finally, it is to note that the traceability of audits should also be reported in the blockchain for specific applications holding sensitive data [4]. The blockchain-based access control system design should include which auditors have access to which data items and to what degree and keep a record of the read accesses and the audits outcomes [4].

6.3 (Re)execution Flexibility

Another important factor in the design and execution of scientific workflows is flexibility. This includes the ability to adapt to changing requirements and conditions and to rerun parts of the workflow to validate and improve the accuracy and reliability of the results obtained from the experiment.

On the one hand, scientific workflows change during their lifecycle to adapt to changing requirements and conditions [61, 62]. For example, in federated data centers, the data usually follows a different path due to multiple copies of the data and the diversity of paths taken to ensure resilience [47]. Hence, scientific workflow management systems should reflect this temporal and dynamic flexibility [47, 61]. Additionally, scientific workflow provenance data may need to be updated [8]. Hence, it is necessary to invalidate the deprecated data references and the workflow computation results (e.g., invalidating all the research artifacts generated from the initial deprecated data) [8, 34]. A challenge arises concerning identifying the deprecated workflow segments and managing the on-chain records that are not modifiable on-chain [5, 60].

On the other hand, the ability to rerun experiments, or segments of experiments, is particularly important in scientific workflows, as it allows for the verification and validation of the results obtained from the experiment [9, 64]. This can help ensure the results' accuracy and reliability and identify and correct errors or inconsistencies in the experiment process [53]. Adaptive task scheduling is also evoked in [44] to foster execution flexibility. In this context, a duality between the flexibility of the scientific workflow management to support the trial and error cycles and the ability to reproduce workflow results, however, rises. This challenge is well-illustrated in [8], as "multiple derivation graphs may be computed for a specific scientific result due to the execution of a scientific workflow multiple times." Additionally, the ability to rerun experiments is tightly linked to the ability to support version control of the workflow data on-chain, which may be challenging for some use cases such as for the versioning of binary data [42], as well as off-chain. Strategies for persistent storage of off-chain data should, therefore, be defined to ensure the possibility to re-execute (or audit, c.f. Sect. 6.2) [4, 6, 7, 60, 62]. Moreover, managing the dependencies between data records (and their versions) in scientific workflows should be considered carefully. Data records may support transformations that make it difficult to deduce such dependencies by looking at the records only [4, 8, 63].

6.4 *Interoperability of Systems and Data Models/Data Understanding*

Another challenge is the lack of **standardization and interoperability** of both data used in scientific workflows and blockchain systems used for traceability.

Interoperability of data models between participants' views is necessary to foster the adoption of blockchain-based scientific workflow management systems [4, 61, 62]. Adopting such system requires a shared understanding of the data involved in the experiments [4]. This shared understanding may take the form of (i) conversions between ontologies and blockchain representations [31], (ii) standards for models, conversions, data, and data quality [31, 35, 37, 63], or (iii) formal specifications and axioms for cross-organizational verification [31].

Additionally, there are currently many different blockchain platforms and protocols available, and there is no consensus on which one is best suited for traceability in scientific workflows (c.f., Sect. 4.1). This can make it difficult for workflow practitioners to choose the right blockchain platform and integrate it into their workflow. For example, a scientific workflow may be designed using a specific blockchain platform. Still, other workflow systems or tools may not support the platform, making it difficult to share and reuse the workflow [44]. Such lack of interoperability may also be problematic for reusing data or experiment notarized with a given blockchain into a scientific workflow management system leveraging another blockchain system [8, 53].

6.5 Managing Data Complexity

Scientific workflows often involve (i) raw data [39, 40], (ii) derived products [36, 39, 40] (e.g., data sets being merged by a processing step or forks on the chain of data processing [36]), and (iii) research artifacts (e.g., algorithms [39, 40, 47, 59]). These data may involve large and complex data structures and algorithms, whose provenance should be stored at each step of the process in a meaningful and efficient way [53, 61]. Therefore, scientific workflows involving the management of massive (big) data call for blockchain scalability imperatives [33, 44, 54, 63]. Solutions coupling on-chain storage of data hashes with off-chain databases have been proposed in the literature [27, 33]. Some off-chain databases are distributed using, e.g., IPFS or Swarm, but more research is needed to comply with non-functional requirements such as performance and scalability [11, 35, 53, 59].

Some activities of scientific workflows may generate probabilistic, inconsistent [61], unstructured [6, 63], or heterogeneous data [27, 32, 56]. Additionally, lossy provenance links may exist during the workflow execution [27]. Hence, ensuring their traceability in a deterministic fashion through smart contracts calls for modularity and configurability at runtime and tools assessing the validity and quality of input data flows to be logged into the blockchain [6, 52]. This point refers to the blockchain oracle problem: blockchain oracles enable the storage of external data in a tamper-proof fashion in a blockchain system, but cannot ensure the validity of these external data points [35].

Overall, these challenges highlight the need for further research and development to overcome the barriers to using blockchain technology for traceability in experiments.

7 Comparison

This section compares the prototypes that leverage blockchain capabilities for scientific workflow management against the criteria detailed in the *challenges* section. Table 7 compares the aforementioned proofs-of-use. The framework columns summarize the works concerning the categorization proposed in Sect. 4.

The IP/confidentiality challenges are evaluated concerning access control mechanisms, delegation mechanisms, and privacy policy enforcement. *Access control* is frequently considered in the retrieved works. It is done first through the blockchain architecture's scope: public/permissioned blockchains, or multichains. It can also be fine-grained by combining off-chain data storage and smart contract-based access control patterns. Very few *delegation mechanisms* have been considered in the literature when data is stored off-chain. However, this criterion should be systematically considered for the wide and long-term adoption of blockchain-based scientific workflow management systems, as intellectual property and confidentiality concerns often appear in scientific workflows. *Privacy policy enforcement* has been studied

Table 7 Assessment of the surveyed papers with respect to the challenges listed in Sect. 6

	IP/confidentiality			Auditing support			(Re)execution flexibility			Data management		Systems interoperability	
Reference	Access control	Delegation mechanisms	Privacy policy enforcement/ obfuscation	Audit efficiency	Cloning/ plagiarism detection	Audit notarization	Data versioning management	Data/ workflow deprecation	Change/ flexibility support	Complex data handling	Data quality assessment	Provenance standard	Blockchain interoperability
[5]	x	–	x	–	–	–	x	x	–	–	–	–	–
[11]	–	–	–	–	–	–	–	x	–	–	–	–	–
[45]	x	x	x	x	–	–	–	–	x	–	–	–	–
[47]	–	–	x	–	–	–	x	x	–	–	–	–	x
[53]	–	–	–	–	–	–	–	–	–	–	–	x	x
[32]	–	–	–	–	–	–	–	–	–	–	x	x	–
[46]	–	–	–	–	–	–	–	–	–	–	–	–	–
[36]	x	–	–	–	–	–	–	–	–	–	–	–	–
[38]	–	–	–	–	–	–	x	–	–	–	–	x	–
[51]	–	–	–	–	–	–	–	–	–	–	–	–	–
[31]	–	–	–	–	–	–	–	–	–	–	–	x	–
[7]	x	–	x	–	–	–	x	–	–	–	–	–	x
[37]	x	–	x	–	–	–	–	–	–	–	–	–	–
[41]	–	–	–	–	–	–	–	–	–	–	–	–	–
[48]	x	–	x	x	–	–	–	–	x	–	–	–	x
[49]	x	–	x	–	–	–	x	–	–	–	–	x	–
[8]	x	–	–	x	–	–	x	x	–	–	–	–	–
[52]	–	–	–	–	–	–	–	–	–	–	–	–	x
[42]	x	–	–	–	–	–	x	–	–	–	–	–	–
[44]	–	–	–	x	–	–	–	–	–	–	–	–	–

[34]	x	–	x	–	x	–	x	x	–	–	–	–	–
[33]	x	–	–	–	–	–	x	–	–	–	x	–	–
[26]	x	x	–	–	–	x	–	–	–	–	–	x	–
[35]	x	x	x	–	–	–	–	–	–	–	–	x	–
[39]	–	–	–	–	–	–	x	–	–	–	–	–	–
[40]	x	–	–	–	–	–	x	–	–	–	–	–	–
[43]	–	–	–	–	–	–	–	–	–	–	–	x	–
[50]	–	–	–	x	–	–	–	–	–	–	–	–	–
[54]	–	–	–	–	x	x	x	–	–	–	–	–	–

more thoroughly in the literature. However, techniques vary, between on-chain and off-chain designs, data fingerprinting or Merkle root anchoring, off-chain storage types, and smart contract functions for data sharing. Hence, more research is necessary regarding developing modular designs for privacy policy enforcement at runtime.

The challenges related to auditing support are evaluated with respect to audit efficiency, cloning/plagiarism detection, and audit notarization. Blockchain *audit efficiency* is a common concern in several patterns. Different designs are proposed in the retrieved papers to address this challenge, be it through distributed off-chain storage, specialized block architectures, or data indexing strategies. More research is needed to evaluate the adequate architectures based on the scientific workflow use cases at stake. *Cloning/plagiarism detection* and *audit notarization* appear only in a few works. Trustworthy and optionally automatic cloning and plagiarism detection through blockchain archived data hashes appears as new audit patterns for blockchain-based scientific workflow notarization. More research is needed on the specifics of such patterns. Audit notarization appears as an interesting feature for data quality procedures: audits can then be certified and tracked on-chain in a trustworthy fashion.

Re-execution flexibility challenges are evaluated concerning versioning data management, data or workflow segment deprecation management, and change/flexibility support. *Data versioning* is considered in a significant subset of the surveyed papers. This pattern often involves using two metadata attached to the notarized element: a status and a link to the up-to-date data element if necessary. However, only a few papers do so by storing data on-chain or leveraging an off-chain distributed file system, which hinders trust in the system in the long run, as files may become deprecated. Alongside data versioning, some papers focus conjointly on *data or workflow segment deprecation* management, i.e., forwarding the detection of deprecated data to its lineage. This is often combined with data lineage and query efficiency strategies to ease this process. Finally, very few works, only two papers, investigate *flexibility support* for scientific workflows on-chain management. This is an avenue worth exploring as scientific methods may evolve over time.

The data management challenge is evaluated concerning massive data handling and data quality assessment. A few works investigate complex *data handling*, primarily through the lens of alternative architectures for massive data, e.g., using Merkle DAGs or Merkle root anchoring, to address scalability concerns. However, only works are lacking regarding probabilistic or heterogeneous data notarization and software environment notarization. A few works include *data quality* to their design scheme through verifiable credentials, provenance standards, or data quality status. However, the study of blockchain incentives for improved data quality notarization has been underlooked.

Finally, systems interoperability challenges are evaluated with respect to blockchain interoperability and the use of provenance standards. The use of *provenance standards* appears in several works, encouraging the interconnection between workflows and the reuse of qualitative archived data for new experiments. However, *blockchain interoperability* has not been investigated in the retrieved papers, which paves the way for exciting research efforts where scalable or privacy-preserving blockchain-based scientific workflow execution histories could be leveraged and connected.

8 Conclusion and Future Research

This chapter investigates the literature maturity concerning trustworthy scientific workflow notarization with blockchain. For this purpose, we conduct a survey to identify the proofs-of-use developed in the literature. We break down the study of current blockchain-based scientific workflow management research systems into five analytical perspectives: (1) blockchain architectures, (2) data archiving strategies (e.g., leveraging smart contracts for data, metadata notarization), (3) direct workflow management support, (4) security, and privacy strategies, and finally (5) audit strategies. We then determine the challenges linked to blockchain-based scientific workflows and compare retrieved work against these challenges to outline existing research gaps.

Five main challenges appear through the lens of these papers: Intellectual property and confidentiality of some or complete segments of scientific workflows when managed on-chain is challenging due to the transparent nature of blockchains. Blockchain-based scientific workflow management systems shall be adapted to forward tracking and backward tracing auditing, especially regarding indexing and off-chain storage strategies for efficient archived data queries. Re-execution and workflow notarization flexibility and workflow re-execution in a blockchain context have yet to be investigated thoroughly in the literature. Interoperability between blockchain systems and data frameworks is still in its infancy, hindering such systems' adoption. Research can also focus on developing new frameworks and standards for integrating and interoperability blockchain technology with existing scientific workflow management systems and tools. The notarization of data on-chain holds challenges related to the complexity of notarizing data: a blockchain-based scientific workflow management system should handle the large volumes of data and complex computational requirements of scientific workflows. These challenges highlight the need for further research and development in scientific workflow traceability with blockchains.

As research direction guidelines for the development of blockchain-backed scientific workflows, we foresee challenges related to several key areas: Notarization processes and formal verification have not been investigated extensively, and this presents a significant challenge for ensuring the security and reliability of scientific workflows. Integrating physical processes' notarization into blockchain workflows is another area that requires attention, as physical processes often have unique requirements that must be considered. Scientific workflow re-execution based on archived software versions and runtime environments is an important consideration for ensuring reproducibility and transparency. The design of blockchains for provenance traceability, including multichains, DAGs, and cross-blockchain interoperability, is an area that requires further investigation to develop effective and efficient blockchain-backed scientific workflows. Addressing these challenges will be critical to unlocking the full potential of blockchain technology in scientific research and development.

References

1. A. Barker, J.v. Hemert, Scientific workflow: a survey and research directions, in *International Conference on Parallel Processing and Applied Mathematics* (Springer, Berlin, 2007), pp. 746–753
2. E. Deelman, T. Peterka, I. Altintas, C.D. Carothers, K.K. van Dam, K. Moreland, M. Parashar, L. Ramakrishnan, M. Taufer, J. Vetter, The future of scientific workflows. Int. J. High Perform. Comput. Appl. **32**(1), 159–175 (2018)
3. S.B. Davidson, J. Freire, Provenance and scientific workflows: challenges and opportunities, in *Proceedings of the 2008 ACM SIGMOD International Conference on Management of Data* (2008), pp. 1345–1350
4. O.I. Abiodun, M. Alawida, A.E. Omolara, A. Alabdulatif, Data provenance for cloud forensic investigations, security, challenges, solutions and future perspectives: a survey. J. King Saud Univ. Comput. Inform. Sci. **34**(10), 10217–45 (2022)
5. W. Tang, C. Chenli, C. Ju, T. Jung, Trac2Chain: trackability and traceability of graph data in blockchain with linkage privacy, in *Proceedings of the 37th ACM/SIGAPP Symposium on Applied Computing* (2022), pp. 272–281
6. M. Giovanardi, T. Konstantinou, R. Pollo, T. Klein, Internet of things for building façade traceability: a theoretical framework to enable circular economy through life-cycle information flows. J. Cleaner Prod. **382**, 135261 (2023)
7. A.S.E. Pradeep, T.W. Yiu, Y. Zou, R. Amor, Blockchain-aided information exchange records for design liability control and improved security. Autom. Constr. **126**, 103667 (2021)
8. D. Fernando, S. Kulshrestha, J.D. Herath, N. Mahadik, Y. Ma, C. Bai, P. Yang, G. Yan, S. Lu, SciBlock: a blockchain-based tamper-proof non-repudiable storage for scientific workflow provenance, in *2019 IEEE 5th International Conference on Collaboration and Internet Computing (CIC)* (IEEE, Piscataway, 2019), pp. 81–90
9. R. Coelho, R. Braga, J.M.N. David, M. Dantas, V. Ströele, F. Campos, Blockchain for reliability in collaborative scientific workflows on cloud platforms, in *2020 IEEE Symposium on Computers and Communications (ISCC)* (IEEE, Piscataway, 2020), pp. 1–7
10. D. Karastoyanova, L. Stage, Towards collaborative and reproducible scientific experiments on blockchain, in *International Conference on Advanced Information Systems Engineering* (Springer, Berlin, 2018), pp. 144–149
11. W. Chen, X. Liang, J. Li, H. Qin, Y. Mu, J. Wang, Blockchain based provenance sharing of scientific workflows, in *2018 IEEE International Conference on Big Data (Big Data)* (IEEE, Piscataway, 2018), pp. 3814–3820
12. H. Wang et al., Blockchain challenges and opportunities: a survey. Int. J. Web Grid Serv. **14**(4), 352 (2018)
13. R. Hull, Blockchain: distributed event-based processing in a data-centric world, in *Proceedings of the 11th ACM International Conference on Distributed and Event-Based Systems* (2017), pp. 2–4
14. S. Leible, S. Schlager, M. Schubotz, B. Gipp, A review on blockchain technology and blockchain projects fostering open science, in *Frontiers in Blockchain* (2019), p. 16
15. R. Di Cosmo, S. Zacchiroli, Software heritage: why and how to preserve software source code, in *iPRES 2017-14th International Conference on Digital Preservation* (2017), pp. 1–10

16. A. Erri Pradeep, T. Yiu, R. Amor, Leveraging blockchain technology in a BIM workflow: a literature review, in *International Conference on Smart Infrastructure and Construction 2019 (ICSIC) Driving Data-Informed Decision-Making* (ICE Publishing, London, 2019), pp. 371–380
17. Y. Qu, M.P. Uddin, C. Gan, Y. Xiang, L. Gao, J. Yearwood, Blockchain-enabled federated learning: a survey. ACM Comput. Surv. **55**(4), 1–35 (2022)
18. A. Singh et al., Blockchain smart contracts formalization: approaches and challenges to address vulnerabilities. Comput. Secur. **88**, 101654 (2020)
19. E. Androulaki et al., Hyperledger fabric: a distributed operating system for permissioned blockchains, in *Proceedings of the Thirteenth EuroSys Conference*, EuroSys '18, New York (2018)
20. G. Wood et al., Ethereum: a secure decentralised generalised transaction ledger. Ethereum Project Yellow Pap. **151**(2014), 1–32 (2014)
21. N. Szabo, Formalizing and securing relationships on public networks. First Monday **2**(9) (1997). https://doi.org/10.5210/fm.v2i9.548
22. K. Christidis, M. Devetsikiotis, Blockchains and smart contracts for the internet of things. IEEE Access **4**, 2292–2303 (2016)
23. J. Mendling et al., Blockchains for business process management – challenges and opportunities. ACM Trans. Manage. Inform. Syst. **9**(1), 1–16 (2018)
24. K. Wüst, A. Gervais, Do you need a blockchain?, in *2018 Crypto Valley Conference on Blockchain Technology (CVCBT)* (2018), pp. 45–54
25. M.J. Sembay, D.D.J. de Macedo, M.L. Dutra, A proposed approach for provenance data gathering. Mob. Netw. Appl. **26**(1), 304–318 (2021)
26. R. Coelho, R. Braga, J.M.N. David, V. Stroele, F. Campos, M. Dantas, A blockchain-based architecture for trust in collaborative scientific experimentation. J. Grid Comput. **20**(4), 1–31 (2022)
27. S. Suhail, R. Hussain, M. Abdellatif, S.R. Pandey, A. Khan, C.S. Hong, Provenance-enabled packet path tracing in the RPL-based internet of things. Comput. Netw. **173**, 107189 (2020)
28. I. Altintas, Lifecycle of scientific workflows and their provenance: a usage perspective, in *2008 IEEE Congress on Services-Part I* (IEEE, Piscataway, 2008), pp. 474–475
29. J. Webster, R. Watson, Analyzing the past to prepare for the future: writing a literature review. MIS Quart. **26**, xiii–xxiii (2002)
30. H. Zhang et al., Identifying relevant studies in software engineering. Inform. Softw. Technol. **53**(6), 625–637 (2011)
31. H.M. Kim, M. Laskowski, Toward an ontology-driven blockchain design for supply-chain provenance. Intell. Syst. Account. Finance Manage. **25**(1), 18–27 (2018)
32. C. Martinez-Rendon, J. González-Compeán, D.D. Sánchez-Gallegos, J. Carretero, CD/CV: Blockchain-based schemes for continuous verifiability and traceability of IoT data for edge–fog–cloud. Inform. Proces. Manage. **60**(1), 103155 (2023)
33. H. Honar Pajooh, M.A. Rashid, F. Alam, S. Demidenko, IoT big data provenance scheme using blockchain on Hadoop ecosystem. J. Big Data **8**, 1–26 (2021)
34. C. Chenli, W. Tang, F. Gomulka, T. Jung, ProvNet: networked bi-directional blockchain for data sharing with verifiable provenance. J. Parallel Distrib. Comput. **166**, 32–44 (2022)
35. A. Demichev, A. Kryukov, N. Prikhod'ko, Business process engineering for data storing and processing in a collaborative distributed environment based on provenance metadata, smart contracts and blockchain technology. J. Grid Comput. **19**, 1–30 (2021)
36. J. Möller, S. Fröschle, A. Hahn, Permissioned blockchain for data provenance in scientific data management, in *Innovation Through Information Systems: Volume III: A Collection of Latest Research on Management Issues* (Springer, Berlin, 2021), pp. 22–38
37. S.K. Radha, I. Taylor, J. Nabrzyski, I. Barclay, Verifiable badging system for scientific data reproducibility. Blockchain Res. Appl. **2**(2), 100015 (2021)
38. R.J.C. Bose, K.K. Phokela, V. Kaulgud, S. Podder, Blinker: a blockchain-enabled framework for software provenance, in *2019 26th Asia-Pacific Software Engineering Conference (APSEC)* (IEEE, Piscataway, 2019), pp. 1–8

39. V. Nandigam, K. Lin, M. Shantharam, S. Sakai, S. Sivagnanam, Research workflows – towards reproducible science via detailed provenance tracking in open science chain, in *PEARC '20: Practice and Experience in Advanced Research Computing, Portland, July 27–31, 2020*, ed. by G.A. Jacobs, C.A. Stewart (ACM, New York, 2020), pp. 484–486
40. S. Sivagnanam, V. Nandigam, K. Lin, Introducing the open science chain: protecting integrity and provenance of research data, in *Proceedings of the Practice and Experience in Advanced Research Computing on Rise of the Machines (Learning), PEARC 2019, Chicago, July 28–August 01, 2019*, ed. by T.R. Furlani (ACM, New York, 2019), pp. 18:1–18:5
41. M.R. Hoffman, L.-D. Ibáñez, H. Fryer, E. Simperl, Smart papers: dynamic publications on the blockchain, in *Proceedings of the Semantic Web: 15th International Conference, ESWC 2018, Heraklion, June 3–7, 2018* (Springer, Berlin, 2018), pp. 304–318
42. J. Bell, T.D. LaToza, F. Baldmitsi, A. Stavrou, Advancing open science with version control and blockchains, in *2017 IEEE/ACM 12th International Workshop on Software Engineering for Science (SE4Science)* (IEEE, Piscataway, 2017), pp. 13–14
43. R. Coelho, R.M.M. Braga, J.M.N. David, M.A.R. Dantas, V. Ströele, F. Campos, Integrating blockchain for data sharing and collaboration support in scientific ecosystem platform, in *54th Hawaii International Conference on System Sciences, HICSS 2021, Kauai, January 5, 2021* (ScholarSpace, 2021), pp. 1–10. https://scholarspace.manoa.hawaii.edu/items/ca9cfa55-df1c-46b4-a883-9ce13f1919c3
44. F. Costa, D. De Oliveira, M. Mattoso, Towards an adaptive and distributed architecture for managing workflow provenance data, in *2014 IEEE 10th International Conference on e-Science*, vol. 2 (IEEE, Piscataway, 2014), pp. 79–82
45. F. Daidone, B. Carminati, E. Ferrari, A blockchain-based framework in support of privacy preferences enforcement for scientific workflows, in *2022 IEEE International Conference on Web Services (ICWS)* (IEEE, Piscataway, 2022), pp. 428–437
46. B. Gipp, C. Breitinger, N. Meuschke, J. Beel, CryptSubmit: introducing securely timestamped manuscript submission and peer review feedback using the blockchain, in *2017 ACM/IEEE Joint Conference on Digital Libraries (JCDL)* (IEEE, Piscataway, 2017), pp. 1–4
47. X. Liang, S. Shetty, D. Tosh, C. Kamhoua, K. Kwiat, L. Njilla, ProvChain: a blockchain-based data provenance architecture in cloud environment with enhanced privacy and availability, in *2017 17th IEEE/ACM International Symposium on Cluster, Cloud and Grid Computing (CCGRID)* (IEEE, Piscataway, 2017), pp. 468–477
48. R. Neisse, G. Steri, I. Nai-Fovino, A blockchain-based approach for data accountability and provenance tracking, in *Proceedings of the 12th International Conference on Availability, Reliability and Security* (2017), pp. 1–10
49. A. Ramachandran, M. Kantarcioglu, Smartprovenance: a distributed, blockchain based dataprovenance system, in *Proceedings of the Eighth ACM Conference on Data and Application Security and Privacy, CODASPY 2018, Tempe, March 19–21, 2018*, ed. by Z. Zhao, G. Ahn, R. Krishnan, G. Ghinita (ACM, New York, 2018), pp. 35–42
50. P. Ruan, G. Chen, T.T.A. Dinh, Q. Lin, B.C. Ooi, M. Zhang, Fine-grained, secure and efficient data provenance on blockchain systems. Proc. VLDB Endow. **12**(9), 975–988 (2019)
51. P. Wang, W. Cui, J. Li, A framework of data sharing system with decentralized network, in *Big Scientific Data Management: First International Conference, BigSDM 2018, Beijing, November 30–December 1, 2018, Revised Selected Papers 1* (Springer, Berlin, 2019), pp. 255–262
52. L. Wen, L. Zhang, J. Li, Application of blockchain technology in data management: advantages and solutions, in *Big Scientific Data Management: First International Conference, BigSDM 2018, Beijing, November 30–December 1, 2018, Revised Selected Papers 1* (Springer, Berlin, 2019), pp. 239–254
53. K. Wittek, N. Wittek, J. Lawton, I. Dohndorf, A. Weinert, A. Ionita, A blockchain-based approach to provenance and reproducibility in research workflows, in *2021 IEEE International Conference on Blockchain and Cryptocurrency (ICBC)* (IEEE, Piscataway, 2021), pp. 1–6

54. R.K. Raman, R. Vaculin, M. Hind, S.L. Remy, E.K. Pissadaki, N.K. Bore, R. Daneshvar, B. Srivastava, K.R. Varshney, A scalable blockchain approach for trusted computation and verifiable simulation in multi-party collaborations, in *2019 IEEE International Conference on Blockchain and Cryptocurrency (ICBC)* (IEEE, Piscataway, 2019), pp. 277–284
55. Y. Cao, C. Jones, V. Cuevas-Vicenttín, M.B. Jones, B. Ludäscher, T. McPhillips, P. Missier, C. Schwalm, P. Slaughter, D. Vieglais et al., ProvONE: extending PROV to support the dataONE scientific community, in *PROV: Three Years Later* (2016)
56. Y. Mendes, R. Braga, V. Ströele, D. de Oliveira, Polyflow: a SOA for analyzing workflow heterogeneous provenance data in distributed environments, in *Proceedings of the XV Brazilian Symposium on Information Systems* (2019), pp. 1–8
57. S. Tarkoma, C.E. Rothenberg, E. Lagerspetz, Theory and practice of bloom filters for distributed systems. IEEE Commun. Surv. Tutorials **14**(1), 131–155 (2011)
58. L. Stage, D. Karastoyanova, Provenance holder: bringing provenance, reproducibility and trust to flexible scientific workflows and choreographies, in *Business Process Management Workshops: BPM 2019 International Workshops, Vienna, September 1–6, 2019, Revised Selected Papers 17* (Springer, Berlin, 2019), pp. 664–675
59. D. Deutch, A. Frankenthal, A. Gilad, Y. Moskovitch, On optimizing the trade-off between privacy and utility in data provenance, in *Proceedings of the 2021 International Conference on Management of Data* (2021), pp. 379–391
60. T. Eifert, U. Schilling, H.-J. Bauer, F. Krämer, A. Lopez, Infrastructure for research data management as a cross-university project, in *Human Interface and the Management of Information: Supporting Learning, Decision-Making and Collaboration: 19th International Conference, HCI International 2017, Vancouver, July 9–14, 2017, Proceedings, Part II 19* (Springer, Berlin, 2017), pp. 493–502
61. S. Abiteboul, M. Arenas, P. Barceló, M. Bienvenu, D. Calvanese, C. David, R. Hull, E. Hüllermeier, B. Kimelfeld, L. Libkin et al., Research directions for principles of data management (abridged). ACM SIGMOD Rec. **45**(4), 5–17 (2017)
62. D.D. Sánchez-Gallegos, D. Di Luccio, S. Kosta, J. Gonzalez-Compean, R. Montella, An efficient pattern-based approach for workflow supporting large-scale science: the DagOnStar experience. Fut. Gen. Comput. Syst. **122**, 187–203 (2021)
63. K. Devaki, L. Leena Jenifer, A study on challenges in data security during data transformation, in *Computer Networks, Big Data and IoT: Proceedings of ICCBI 2021* (Springer, Berlin, 2022), pp. 49–66
64. A. Weiß, V. Andrikopoulos, M. Hahn, D. Karastoyanova, Model-as-you-go for choreographies: rewinding and repeating scientific choreographies. IEEE Trans. Serv. Comput. **13**(5), 901–914 (2017)

Part III
NFT Applications

Tokenization of Fine Arts: Revolutionizing the Fine Arts Industry with Blockchain

Teesta Bhandare and Mohamed Ali Kandi

Abstract This chapter analyzes the scope of possibility for tokenizing proof of ownership, evidence of valuation, and quality of alternative investment classes that are tangible in nature. The focus of this treatise is specifically on on-boarding fine arts on the blockchain and the utilization of the blockchain to improve security associated with these assets. Through the assessment of Art Garde, a blockchain based cataloging tool and marketplace for fine arts, this chapter analyzes a commercial use case for blockchain that merges the benefits of digitization of physical assets. While there are numerous benefits to the tokenization of physical assets on the blockchain, there are certain problems associated with the same that are addressed by Art Garde. These problems stem primarily from the need for creating a link between the physical asset and its digital twin to ensure that the integrity and identity of either asset is not compromised. Art Garde establishes a forward-looking approach and model that can be used to secure any high value physical assets from luxury goods and watches to real estate.

Keywords Blockchain · Art · Investment · Provenance · Digital twin · NFT

1 Introduction

1.1 Bringing Art to the Future

The global art market is currently valued at $67.8 billion [1] in terms of the value of art sales that took place in 2023. Art has long been considered a valuable alternative

---

T. Bhandare (✉)
Art Garde, New Delhi, India
e-mail: teesta@teestabhandare.com

M. A. Kandi
Art Garde, New Delhi, India

IRIT, Université Paul Sabatier, Toulouse, France
e-mail: ali@artgarde.io

N. El Madhoun et al. (eds.), *Blockchain and Smart-Contract Technologies for Innovative Applications*, https://doi.org/10.1007/978-3-031-50028-2_6

investment class, which is a class of investment that is alternative to traditional equity markets, bonds, or cash [2]. Last year, Knight Frank tracked ten luxury classes, and art outperformed each of these asset classes, which include cars, wine, and jewelry [3]. The sheer value and volume of transactions makes the art industry one that is ripe to benefit from technological advancements that secure not only the transactions but also the multiple stakeholders, as discussed below, that are reliant on the currently opaque workings of the industry.

This chapter addresses how through the utilization of blockchain technology, stronger systems of accountability, transparency while maintaining privacy and accurate record keeping are implemented. This utilization of blockchain creates a more robust ecosystem for various parties involved. Art Garde is a unique blockchain based cataloging system that has identified the key weaknesses in the art market, which are largely caused by heavy reliance on intermediaries, lack of information sharing, and inefficient documentation methods.

Since 1997, developments within the art world have been tracked and forecasted almost as carefully as those with property in the United Kingdom [1]. Since then art has been gaining traction amongst more than just those individuals who fall within the category of ultra-high net worth individuals as a lucrative investment choice. However, a problem that has plagued the art market for years is the opacity and unregulated nature of the market.

1.2 The Obstacle of Opaqueness

The value of art is determined by inter alia the size, year of creation, condition, provenance, name, and esteem of the artist. Individual works are usually accompanied with documentation that corroborates this information. Furthermore, the final price conveyed to a buyer does not account for the various hidden costs associated with a work, such as taxation, transportation, commissions, and insurance. These costs result in higher transaction costs for art than most other investment classes. For instance, auction houses charge anywhere between 12% to 30% "hammer price" on sold lots. Hammer price is a commission charged on the winning bid, over and above the bid price [4]. It should be noted that not everyone who buys art buys it exclusively as an investment. Unlike traditional asset classes, there is an aesthetic and emotional value that is also associated with art. There is always the buyer who buys for the simple pleasure of admiring a work. However, regardless of the motivation to buy art, the process of art valuation, the hidden and visible costs remain the same. Additionally, the extent of the lack of regulation and organization of the market leaves various stakeholders unprotected, ranging from artists to dealers and collectors. Furthermore, the barriers to sharing information often prevent the overall value of a work or collection.

In an interview in April 2022, Hena Kapadia, Director of Tarq, a gallery in Mumbai, stated that one of the most lacking tools when transacting with a third party is the inability to follow the lifespan of a work. She purports that a published work is more valuable than an unpublished or uncatalogued one. This refers to a work that has been published in a third-party publication or catalog. Evidence of this publication goes to the heart of the matter of authenticity of an artwork as well as the overall value whether, cultural, financial, or aesthetic, of that artwork or artist. It denotes that the publishing party is corroborating the authentic nature or strength of the work. Furthermore, the identity of the author of the catalog or publication further adds to the strength of the argument in favor of authenticity and overall value. A catalog published by a world reputed international museum like the Metropolitan Museum of Art carries more credibility than if a private individual publishes a catalog. Kapadia claims that once she sells an artwork, she loses contact with the work so if any subsequent information about the work or artist becomes available to her after the sale, she is unable to update the literature associated with the same. The current owner of the work is the party that is most inconvenienced as should they wish to sell the work; they may lose out on realizing its true value by not having the most up to date information about the artwork. It is often private collectors, individuals and smaller institutions who are disadvantaged by this unequal access to information. The lack of information sharing is detrimental to the various players in the art market.

Data is essential for the survival of supply chains [5]. Art is an industry that relies heavily on provenance, on supply chains. Understanding the source of a work or a collection at large speaks to the authenticity of the work. Information about the levels before or after a certain amount facilitate decision making throughout the chain. This data-driven decision making and process of evaluation is true for the art industry. Data related to certain assets and the "flowing of that information to those who need it within a network gives individuals and organizations visibility." [6] Visibility here does not mean just making information visible to individuals or organizations, but rather these parties using the information as a basis to make certain decisions. For example, specifically in the art industry going back to the scenario of discovering whether an artwork has been published in a catalog or not, in such a case if an individual knew an artwork was published, they would trust its authenticity more and make an informed decision to pay a certain price for that work. Whereas, if a work is not published yet an expensive price point is associated with that work, the buying party may be apprehensive to pay the price. This party has no independent unbiased third-party validation of authenticity, less visibility of information, and is ultimately taking a riskier decision.

The almost stone walling of information sharing and inability to trace assets after they leave a certain level in the supply chain negatively impacts the larger art ecosystem, as well as individual assets and members of the ecosystem. There is an urgent need for mechanisms that ensure better connectivity and communication between the different parties involved throughout the lifespan of a work.

2 Understanding Blockchain

Blockchain is a decentralized and secure storage technology. In simple terms, it is a secure virtual ledger, made secure by the fact that it is not controlled or managed by any one authority, in the case of public blockchains (discussed below). It is named after its structure, which consists of a chain of blocks. Each block stores transactions and the cryptographic hash of the previous block (Fig. 1). Transactions serve as the building blocks of the blockchain. The security of blockchain lies in the linking of blocks through their hashes. Modifying a block requires altering the hash of all subsequent blocks in the chain, essentially overwriting their contents [7].

2.1 Blockchain Architectures

Blockchain architectures can be classified into three categories [8]: public, consortium, and private. By understanding the distinctions among public, private, and consortium blockchains, organizations can choose the most suitable architecture based on their specific requirements, security considerations, and collaborative needs.

2.1.1 Public Blockchain

A public blockchain is open to anyone who wants to join and participate in managing its data. It offers a high level of immutability since data stored in the blockchain cannot be modified. Examples of public blockchains include Bitcoin [7] and Ethereum [9]. However, the decentralized nature and open accessibility of public blockchains can result in resource-intensive data management.

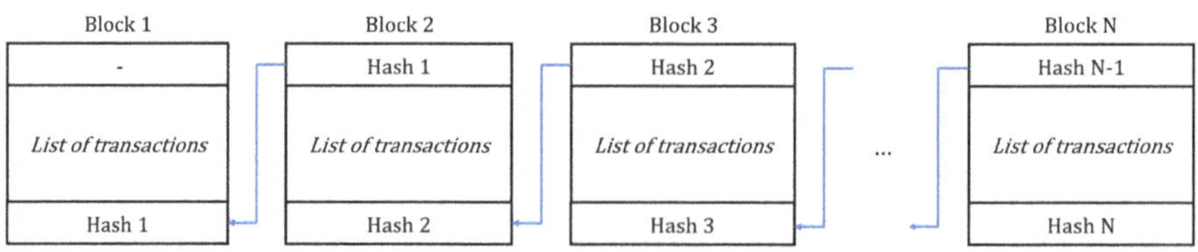

Fig. 1 Blockchain structure

2.1.2 Private Blockchain

A private blockchain is accessible only to authorized users within a specific organization or entity. It provides a more efficient architecture due to a limited number of participants. However, it may be less secure compared to public blockchains. Private blockchains, such as Ripple [10] and Tendermint [11], are often utilized by organizations for internal purposes or specific collaborations.

2.1.3 Consortium Blockchain

A consortium blockchain is accessible and managed by authorized users from multiple organizations. It offers a balance between the privacy and control of a private blockchain and the accessibility and collaboration of a public blockchain. Consortium blockchains, like Quorum [12] and Hyperledger [13], are particularly useful for industries and sectors where multiple organizations need to share and verify data while maintaining certain levels of control and confidentiality.

2.2 *Blockchain Features*

Blockchain applications are on the rise due to the multitude of features that the technology offers, making it valuable to various industries. The below sections outline the key features of blockchain [14–16].

2.2.1 Distribution

Every participant in the network possesses a copy of the ledger. This public ledger contains information regarding network participants and transactions. The distribution of computational power across multiple computers contributes to improved outcomes and there are no single points of failure.

2.2.2 Decentralization

Apart from distributing computation and storage, blockchain decentralizes control, eliminating the reliance on a central authority to make decisions. This enhances trust among network participants. Also, by removing the need for a central authority, the blockchain system becomes more resilient, secure, and less prone to manipulation.

2.2.3 Immutability

The permanence of data stored in the blockchain is a key attribute that contributes to enhancing trust and ensuring data integrity. Once information is recorded on the blockchain, it becomes immutable and cannot be altered or deleted. This feature provides a high level of assurance, as it prevents unauthorized modifications and maintains the integrity of the data. Consequently, parties involved can have confidence in the accuracy and reliability of the information stored on the blockchain.

2.2.4 Traceability

Blockchain enables tracking the origin of each transaction, preventing tampering and facilitating quick identification of issues.

2.2.5 Pseudonymity

Blockchain uses generated addresses to provide a level of anonymity for users. Participants can engage in blockchain activities without revealing personal information.

2.2.6 Transparency

The content of the blockchain is accessible to all, and transactions are recorded in multiple locations. This transparency enhances security and makes the blockchain resistant to hacking attempts.

2.2.7 Security

Blockchain employs cryptography, including digital signatures, to ensure the confidentiality, integrity, and authenticity of transactions. By leveraging these features, blockchain technology offers innovative solutions in various industries and continues to expand its range of applications.

2.3 *Blockchain Consensus*

Consensus algorithms play a crucial role in ensuring that participants agree on adding new blocks to the blockchain. These algorithms can be classified into two categories [17]: leader-based mechanisms and Byzantine fault tolerance-based algorithms.

2.3.1 Leader-Based Mechanisms

In this class of algorithms, participants compete to elect a leader who will validate the next block. Proof of Work (PoW), used in cryptocurrencies like Bitcoin [7], is an example of a leader-based mechanism. Miners solve complex cryptographic puzzles through a process called mining, with the first miner to solve the puzzle becoming the leader and forging the next block. Other participants verify the solution and add the block to their copies of the blockchain. PoW requires significant computational power and energy consumption. Proof of Stake (PoS) [18] is another leader-based mechanism that is more energy-efficient. Validators are chosen based on various criteria, such as their economic stake in the cryptocurrency, without the need for solving complex problems. Ethereum officially switched to a Proof of Stake (PoS) consensus mechanism in 2022.

2.3.2 Byzantine Fault Tolerance-Based Algorithms

These algorithms focus on communication. The Practical Byzantine Fault Tolerance Algorithm (PBFT) [19] is an example. In PBFT, a participant (proposer) is elected in a round-robin fashion. The proposer broadcasts its block to all other participants, who become validators. Each validator checks the received block, broadcasts a response (positive or negative), and awaits responses from others. When a validator receives a sufficient number of positive responses (usually two-thirds of the participants), it broadcasts a commit and waits for responses from others. When a validator receives a sufficient number of commits, it adds the block to the blockchain.

These consensus algorithms ensure that all participants agree on the order and content of blocks added to the blockchain, enabling trust, security, and the decentralized nature of the technology.

2.4 Blockchain and Smart Contracts

Smart contracts represent a highly promising application of blockchain technology. These autonomous and immutable computer programs are typically stored on a blockchain and can be executed automatically by participants to facilitate contract settlements between organizations or individuals [20]. The advantages of smart contracts are discussed in the following sections.

2.4.1 Security

Due to the fact that they are stored on a blockchain, smart contracts leverage the inherent security features of the technology, ensuring that the terms of the contract cannot be modified or tampered with, by malicious actors. This means that businesses can rest assured that their contracts are secure from external threats.

2.4.2 Automated Execution

Unlike traditional contracts, smart contracts automate the execution process, reducing the risk of contract violations as they are self-executable based on predefined conditions.

2.4.3 Cost Reduction

Smart contracts eliminate the need for intermediaries such as notaries and lawyers, resulting in reduced costs associated with contract preparation, monitoring, and signing.

2.4.4 Application of Smart Contracts

Smart contracts find application in various scenarios. In cryptocurrency, they facilitate automatic coin exchanges between users based on predefined conditions, eliminating the need for third-party involvement. They are also used in travel insurance to automatically compensate passengers when flights are delayed or canceled, streamlining the process without requiring form filling or manual processing. Additionally, smart contracts enable the sale of personal items by automatically transferring ownership to the buyer once the payment is received by the seller. This eliminates the need for intermediaries and mitigates the risk of fraud (Fig. 2).

(a) Normal contract.

(b) Smart contract.

Fig. 2 Normal contract vs. smart contract

2.5 *Non-fungible Tokens*

Non-fungible tokens or NTFs are cryptographic digital assets which are stored in a blockchain. As their name suggests, non-fungible means that each NFT is unique and cannot be replaced. The unique identification and authenticity of NFTs are guaranteed by powerful cryptographic functions. Leveraging blockchain technology, NFTs facilitate seamless transfer of ownership and verification. Each transfer of ownership for an NFT is transparently (and usually publicly) recorded and can be traced back to the wallet address from which it originated. By verifying that the wallet address aligns with the original artist's wallet, one can confirm the authenticity of the NFT [21].

2.5.1 NFT Minting

In the world of cryptocurrency, specifically in the realm of NFTs, the term "mint" refers to the act of creating an NFT and permanently recording it on the blockchain (Fig. 3). This step holds immense significance because once minted, the NFT is inscribed in the tamper-proof and publicly accessible ledger, ensuring its existence for an indefinite period (or at least, as long as the blockchain remains operational). By being recorded on the blockchain, the NFT's ownership history, metadata, and transactional details are securely and publicly stored. This ensures that the existence and ownership of the NFT can be independently verified by anyone at any time [23].

Note that the inherent nature of the blockchain makes it unsuitable for storage purposes. Instead, Inter Planetary File System is usually used for storing significant volumes of data. IPFS is a distributed file system specifically designed to enhance the speed, security, and openness of the web. By addressing the issues with HTTP, IPFS strives to establish a more adaptable and dynamic Internet environment. For more details about IPFS, please refer to the work presented in [24].

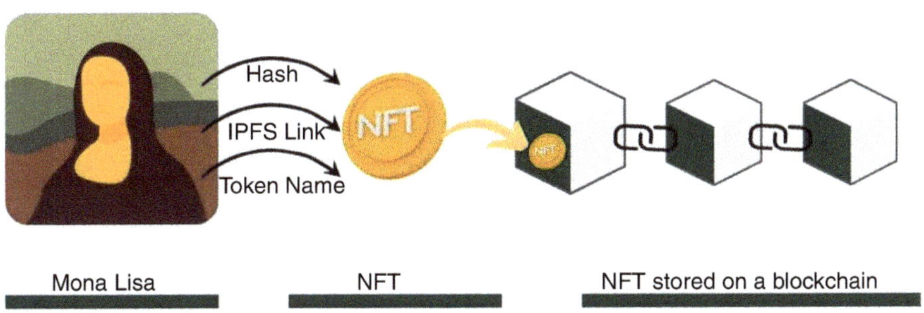

Fig. 3 NFT minting [22]

Marketplace platforms play a vital role in the minting process. These platforms provide the necessary tools and infrastructure for creators to mint their NFTs easily and efficiently. Examples of popular marketplace platforms for NFT minting include OpenSea, Rarible, SuperRare, and others. These platforms not only facilitate the minting process but also serve as marketplaces where NFTs can be bought, sold, and traded among collectors [25].

2.5.2 NFT Uses

NFTs encompass various digital formats such as photos, videos, audio files, and more. Examples of NFTs include artwork, comic books, sports collectibles, trading cards, games, and numerous other items. The value of an NFT is determined by market dynamics of supply and demand, allowing them to be bought and sold like physical assets. One notable advantage of NFTs is their ability to represent real-world objects, such as real estate and fine art. This allows for more efficient transactions, as ownership can be transferred digitally through tokenization. By utilizing NFTs to represent physical items, the risk of fraud can potentially be reduced, as the authenticity and provenance of the asset can be easily traced through the blockchain [26].

2.5.3 NFT and Metaverse

The emergence of the Metaverse has heightened the relevance and impact of NFTs [27]. Indeed, despite being in the early stages, the Metaverse are real societies. Every society, whether virtual or real, needs a functioning economy. In a Metaverse, the economy is decentralized and essentially depends on the authentication of digital assets. The blockchain guarantees the transparency and stability of an open and decentralized economy, governed by the law of supply and demand. NFTs are therefore pivotal in the Metaverse economy, authenticating individuals' possessions, as most objects in the Metaverse are basically NFTs.

3 Blockchain Enters the Fine Art Industry

3.1 Digital Art and the Protection of Intellectual Property Through Blockchain

One of the early applications of blockchain was to empower artists; however, for too long this application limited itself to digital artworks and assets, exclusively [6]. Blockchain technology was used and continues to be used to trace provenance and ownership of intellectual property of digital artworks, so relevant parties i.e., intellectual property (IP) owners received adequate attribution and where applicable compensation [6]. With works available on the Internet, the creator of the work often does not know who all have access to the work. Google is the most visited website in the world. It is viewed 89.3 billion times in a month [28]. Google and search engines like it provide users with access to images of all kinds of creations, a simple copy-paste feature allows users of these search engines to use and disseminate information without considering IP ramifications of these actions.

Of the different types of IP, the primary one concerning artistic works is copyright. The Berne Convention for the Protection of Literary and Artistic Works, also known as the Berne Convention was the first international concerted effort to standardize IP protection, most specifically copyright protection. It was initially signed in 1886 by 10 European countries, but today 180 countries have ratified the convention. The tenets Berne Convention lay the foundation for national copyright legislation of various countries, including India and the United Kingdom and stipulate a minimum standard of protection [29]. The implementation of the Berne Convention through national legislation demonstrates that globally there is an acceptance in favor of protecting the artist and author. However, 1886 was well before the era of the Internet. It did not foresee the ease with which platforms like Google would allow users to access works created by others. The problem lies in that the creator does not know when and how their works are accessed. "In the digital domain the visibility of who owns what and who is using what is obscure." [6]. Blockchain provides a solution to that obscurity.

By uploading content on the blockchain creators of this work can trace who was accessing what and for what purpose. By tracking every use, creators were able to access the royalties owed to them. Most national legislations empower the creator of a work with economic rights, which include payment of royalties when another party uses the copyright protected work [29]. These blockchain based platforms for digital art created a robust marketplace for digital artwork. Artworks that did not need to exist in the physical world.

3.2 *Digital Art NFTs*

In 2008 Bitcoin became the first cryptocurrency to be deployed on blockchain. Bitcoin and other cryptocurrencies are just one application that can be used via blockchain. The future of these cryptocurrencies and the debate about their long-term viability is not within the scope of this chapter. The sole focus of this chapter is understanding non-cryptocurrency based applications of blockchain within the art industry. Part of the debate around the viability of cryptocurrencies stems from its nature as a highly speculative investment. Issues pertaining to how to categorize it from a regulatory perspective is another concern. Some jurisdictions view them as assets, while others classify them in the same category as fiat currencies. Digital art in the form of non-fungible tokens or NFTs share many of the same uncertainties and are, therefore, treated with the same apprehension. Those in the fine arts industry view NFTs as digital art, they see them as a market that has nothing to do with fine arts. This perception is true as each market is dealing in two different products. NFTs in the digital art sense typically refer to branded tokens, where a large part of the value of the token is derived from the excitement created around it. For instance, CryptoKitties was a brand of these NFTs that would "drop" on NFT platforms. The hype created around this particular brand resulted in steady price increase. Furthermore, the price of these tokens is not monitored in fiat currency, instead it is directly linked to the value of the cryptocurrency on whose blockchain it is minted on. As mentioned earlier, each blockchain uses gas fee of a cryptocurrency associated with that blockchain. CryptoKitties were minted on the Ethereum blockchain; therefore, their value was directly linked to the value of Ether. So 1 Ether may be worth 1 USD, but depending on its demand it may be worth USD 1,825.60 [30], which is what it is at the time of writing this chapter. In 2017 the most expensive CryptoKitty token, named Founder Cat #40, sold for an estimated USD 1,064,022 [31]. However, the crypto crash of 2022, witnessed a decline in the value of cryptocurrencies, as well as other crypto assets like these tokens. Post the crash, the average price of CryptoKitties was between USD 40 and USD 50 [32].

Art Garde overcomes the volatility of NFTs by providing a use case whereby their value is not derived from cryptocurrencies or from the volatility of the fact that they are assets that simply do not exist in the physical realm. Each token on Art Garde is backed by an existing tangible physical asset whose worth is determined by factors including market forces and other traditional markers of fine arts valuation. Further, due to problems of creating an actual link between physical products and their digital counterparts, the efficiency of these platforms currently does not apply to lend itself to moveable physical assets. Art Garde is the first of its kind platform that seeks to overcome this disconnect and bring the benefits of blockchain to the world of fine arts and protect traditional as well as new mediums i.e., oil, acrylic on canvas, sculptures as well as videos, digital graphics, and so on.

3.3 Tokenization Challenges

To create an NFT for a physical artwork, there are mainly two methods [22]. The first, and easiest, is to take a photograph of the artwork and mint the image as an NFT. However, it is important to note that the NFT will represent the photo, not the physical artwork itself. As a result, the value of the photo may not equal the value of the physical art. The second method is to generate a Certificate of Authenticity for the physical art and mint the certificate as an NFT. Although this second method is more promising, the NFT and the physical art are still distinct entities. It is, therefore, still possible for an owner to sell the NFT and the physical piece of art separately, which would result in a potential separation of ownership.

For these reasons, the majority of existing marketplaces focus solely on the sale of digital artwork and do not support the sale of physical pieces. Conversely, platforms that specialize in physical assets typically do not facilitate direct sales.

Instead, they tokenize the data associated with physical objects to offer additional services like traceability, cataloguing, and authentication. To the best of our knowledge, and following our study of the existing, Art Garde is the first solution to secure any high value physical fine art assets that actually provides a link ensuring that each token is directly corresponding to only one artwork. To address the above-mentioned issue, Art Garde plans to use the Trace Crystal Technology, which aims to link digital assets with their physical counterparts using fiducial markers with patterns defined by cholesteric spherical reflectors (CSRs), selective retroreflectors produced using liquid crystal self-assembly [33].

4 Art Garde: A Secure Solution to Bring Fine Arts to the Future

4.1 Understanding Art Garde

At its very base level Art Garde is a blockchain based digital locker for documentation associated with fine arts, both for individual pieces and collections at large. Art Garde targets all types of users within the art market, from individual collectors, art enthusiasts to larger institutions and commercial players such as galleries. Each user is empowered to create a token of individual physical artworks by uploading images and relevant documentation. They can then trade these tokens to new owners of the work, while conducting the actual sales of the physical asset either offline or online. Art Garde brings the ease of doing transactions that are seen in NFT marketplaces for digital art to a physical fine art. Additionally, it accounts for the specific problems of traceability of an artwork throughout its lifespan by allowing parties who have been associated with a work at any stage to follow that work. If a gallerist wishes to update publishing information about a work that they have sold, they can simply search for the work on the platform, provided the sale took place on it or that the token was transferred via the platform. They can then request access from the current owner to update and add any new relevant information to the token, provide the current owner permits them.

Art Garde provides two features, the first is a cataloging platform. The second is a marketplace where these tokens are traded. A key factor for the value of the tokens is that they do not have any value of their own. While the transfer of ownership of tokens involves the minting of gas fee, the actual value of the token is exclusively the same as the value of the physical artwork. The gas fee is merely the transaction cost.

The user base extends itself to all key players in the fine arts industry, but it is vital to note that these players are almost always not crypto natives. They are used to the opacity of the fine arts industry, an industry which is largely still based offline. Therefore, there are certain issues that Art Garde has faced in terms of market adoption and practical implementation.

4.2 Obstacles in Implementation

4.2.1 Apprehension from the Fine Arts Market

As mentioned earlier in this chapter, stakeholders of the fine arts industry have been complacent about technological advancements in the industry, or the lack thereof. They view NFTs and blockchain as tools that are not relevant to them. Explaining blockchain use cases that are not derivative of cryptocurrency trading and having users test the product has been one of the biggest hurdles in the market launch of Art Garde.

4.2.2 Apprehensions Regarding Cryptocurrencies

Stemming from a combination of social stigma, current affair and tepid, if not outright opposing government views, many including those in the fine arts industry are nervous to engage with cryptocurrencies. For this reason, Art Garde accepts payments exclusively in fiat currencies. Additionally, all routing of token transfers is done through the administrator, which ensures that users are not liable for gas fees. The cataloging feature is offered as a subscription-based model for users to limit interaction with cryptocurrencies. Additionally, since the value of the tokens has no relation to the value of any cryptocurrency and payments are accepted exclusively in fiat currencies, the user experience has virtually no crypto interaction but is able to harness the benefits of blockchain.

The marketplace also accepts payment only in fiat currencies. Transfer of ownership is done through the admin, thereby once again preventing the user experience from being confused due to their perception of cryptocurrency. In terms of market adoption, Art Garde has noted that where the interaction with cryptocurrency can be limited, if not completely removed, users are more receptive to engaging with the platform.

4.2.3 Creating a Secure Link

The first issue regarding the effective and actually secure implementation of Art Garde is ensuring that each token is linked only to one unique artwork. Even if an artwork is part of an edition of prints i.e., it is not a unique work and there are multiple works that look exactly the same, the paperwork for each artwork still varies. For instance there may be a difference in the condition between two prints of the same edition, or the chain of provenance is different. The latter refers to the chain of ownership changes a work has faced since its inception, e.g., from the artist to a gallery to a third-party buyer and so on.

A secure link between the two assets ensures that a user is not creating multiple tokens for one work, and then selling a token for which they do not have the corresponding physical asset.

One way to create this link is through QR codes, which have gained popularity in utilization since 2019 with the advent of COVID-19. However, these codes are problematic because while they are inexpensive to produce, they are easy to copy. Additionally, they cannot be placed on the body of the work. They can only be placed as stickers on the frame of the work. The stickers themselves can be removed and applied on different works, or the frame can be changed and put on a different work. While QR codes provide a link between physical and digital, it is not a very secure one. Another alternative is radio frequency based ID codes or RFID codes. However, they face problems of durability, inability to be placed on the body of art works so easy to tamper and are visible to the naked eye, therefore, easy to clone. Art Garde proposes to create this secure link through CSR crystals, which outperforms existing technology in terms of inability to be cloned, its ability to be hidden, placed on the work itself, durability, low production cost and quick authentication. This technology is explained in detail in the section below.

4.3 Understanding CSR Crystals

Conventional methods of identification such as QR codes have a significant vulnerability: they can be easily replicated. A team from the University of Luxembourg has developed a novel approach to combat counterfeiting in identity cards, works of art, and industrial products. This innovative method harnesses the intricate light patterns generated by liquid crystals, which are commonly encountered in LCD screens [34].

Liquid crystal combines properties of liquids and crystals. It flows as a liquid and exhibits macroscopic anisotropic properties as if it was a crystal. Cholesteric Liquid Crystals shells, which have micron-sized diameters, can be densely packed in a small surface area to obtain a tag (CSR Tag). The unique structural properties of these shells enable them to reflect light and generate intricate and unpredictable colorful patterns.

These patterns are computationally challenging to predict, making them suitable for security applications, especially in object authentication. The pattern generated by illuminating the CSR tag with light is captured and stored as a reference in a database. To authenticate the object, the end-user simply needs to capture a picture of the tag using their smartphone and compare it with the original image stored in the database. This comparison process enables easy and reliable object authentication (Fig. 4).

A CSR tag has the following properties:

4.3.1 Uniqueness

Dynamic patterns formed by random arrangements of smart reflectors are more individualized than fingerprints.

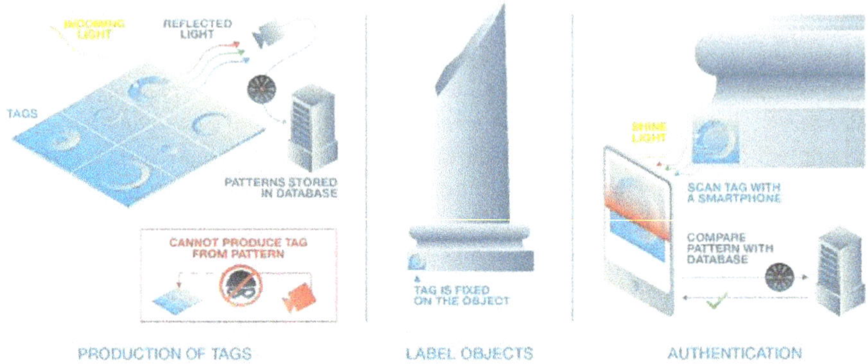

Fig. 4 CSR tags [34]

4.3.2 Unclonable

Even the creators cannot replicate or make a copy of the CSR tag.

4.3.3 Tamper-Proof

Adapted to the specific use case, the tag is designed to break if an attempt is made to transfer it without authorization.

4.3.4 Versatility

The CSR tag can be attached to objects of any shape or integrated as a coating. It offers the option to be visible or invisible, depending on the requirements.

4.3.5 Optical Readout

Standard optics found in mobile phone cameras and 3D cinema goggles can be used to read and interpret the CSR tag.

For more detail about the CSR technology, please refer to the works presented in [33, 35–42].

Thanks to these properties, CSR tags allow us to establish a connection between an NFT and its physical counterpart, effectively bridging the two. Therefore, when an owner intends to sell the NFT or the physical artwork, both must be acquired together. This requirement ensures that the true owner of the NFT associated with the artwork can be easily identified by simply scanning the CSR tag on the physical object.

4.4 Understanding How Art Garde Works

4.4.1 NFT Minting

The process begins with the owner of the artwork submitting a service request to Art Garde. The request entails authentication of the artwork. Upon receiving the request, Art Garde proceeds to provide the owner with a certificate of authentication, ensuring the artwork's legitimacy. Additionally, Art Garde supplies the owner with a CSR tag, which is affixed to the artwork for enhanced security.

Before providing the owner with the CSR tag and its associated certificate, both are securely stored in the Art Garde database. This storage serves as a crucial backup measure in case the physical tag becomes damaged or inaccessible. By securely storing this information, Art Garde ensures that even if the CSR tag is compromised or rendered unreadable, the ownership and authenticity of the artwork can still be verified. This database serves as a reliable source of reference to validate the artwork's identity, providing an additional layer of protection and assurance for both the owner and potential buyers of the NFT.

Upon the owner receiving the certificate of authentication and the artwork being tagged with the CSR tag, the process continues by leveraging the owner's crypto wallet to mint the NFT. This step entails using one or multiple images of the artwork, along with the certificate of authentication. The integration of these elements ensures the NFT's distinctiveness and authenticity, establishing a verifiable digital representation of the artwork (Fig. 5).

4.4.2 Ownership Transfer

Once a buyer has selected and decided to purchase the NFT through the Art Garde platform, the transaction is initiated. The ownership transfer of the NFT is automatically facilitated by the blockchain technology underpinning the platform. This ensures a secure and immutable transfer of ownership from the seller to the buyer, with clear proof recorded on the blockchain.

Furthermore, the buyer and the seller have the option to mutually decide whether the physical artwork will be sent. In some cases, it may not be necessary to physically transfer the artwork. However, it is important to note that even if the physical art is not sent, the true owner can be verified by performing a quick scan of the CSR tag attached to the artwork. This scan provides a reliable way to ascertain the rightful owner based on the unique pattern associated with the tag, as further explained in the subsequent section (Fig. 6).

4.4.3 Ownership Verification

Anyone with a phone can easily scan the CSR tag on the artwork using the Art Garde application. Once scanned, the application queries the database to provide necessary

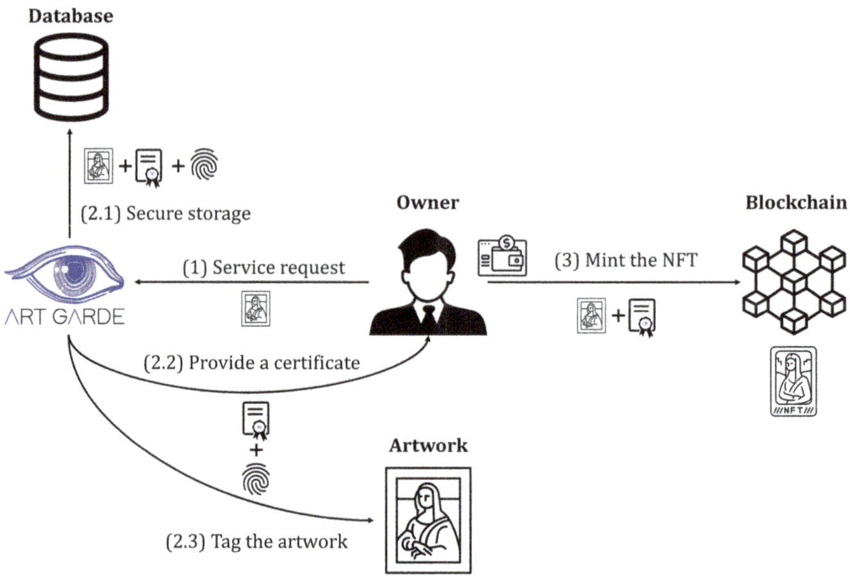

Fig. 5 NFT minting

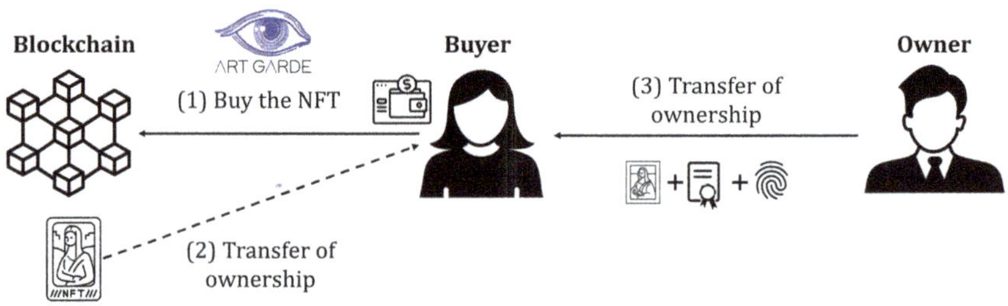

Fig. 6 Ownership transfer

information about the certificate associated with the tag. This information can then be utilized to search for the corresponding NFT in the blockchain and retrieve essential details, including the identity of the owner. The Art Garde application serves as a convenient tool for accessing and obtaining the pertinent information stored within the blockchain, ensuring a seamless and transparent experience for users (Fig. 5).

It is important to note that due to the uniqueness and security provided by the CSR tag, the Art Garde application establishes a powerful link between the physical and digital arts. The intricate patterns generated by the CSR tag, along with its unclonable nature, ensure that each artwork possesses a distinct identity in both the physical and digital domains. Art Garde creates a connection between the physical object and its corresponding NFT in the blockchain. This enhances the authenticity,

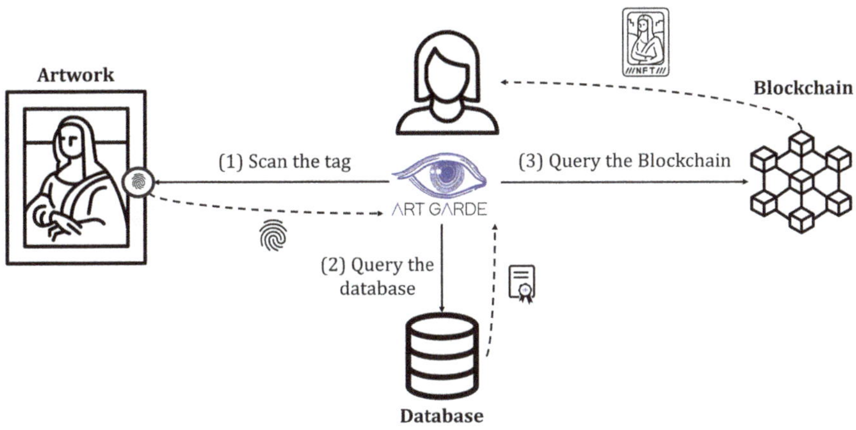

Fig. 7 Ownership verification

provenance, and overall value of the artwork, creating a seamless integration of the art world with the digital ecosystem (Fig. 7).

4.4.4 Tag Replacement

In the event that the CSR tag gets damaged, the owner should approach Art Garde for assistance. The owner will need to provide proof of their identity, demonstrate ownership of the NFT associated with the artwork and of the authenticity of the work. Once the provided information is validated, Art Garde facilitates the process of issuing a new tag to the owner. Additionally, Art Garde updates its database to establish a linkage between the certificate and the new tag, ensuring the continuity of the authentication and ownership records. This seamless procedure guarantees that the owner maintains the integrity and security of their artwork, further reinforcing the trust and reliability of the Art Garde platform (Fig. 8).

5 Conclusion

The fine arts industry suffers from various drawbacks due to opacity, haphazard and fragmented documentation and a heavy reliance on intermediaries, which all result in the reliance on blind trust. The infusion of blockchain technology into a largely stagnant, yet extremely financially lucrative industry like the fine arts, promises to disrupt the industry for the better. Art Garde is a unique solution that overcomes the barriers of adoptions and creates a secure link between physical art and its digital twins. It is a truly viable platform that seeks to revolutionize the market, without discriminating against any stakeholder. Art Garde is currently in the process of

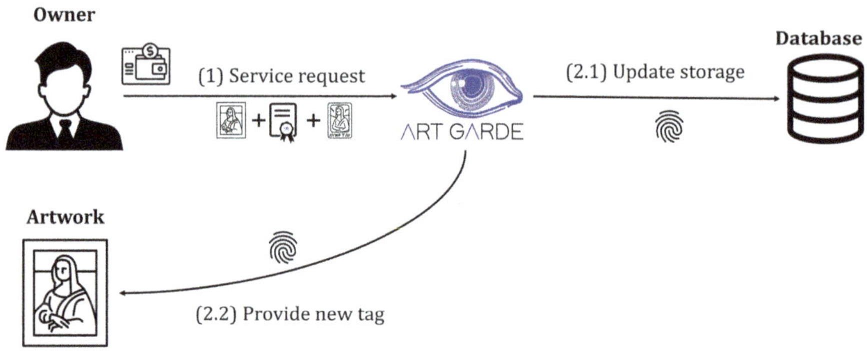

Fig. 8 Tag replacement

testing its product amongst various art industry insiders. It plans to launch for the entire market in November 2023.

References

1. R. Mamarbachi, M. Day, G. Favato, Art as an alternative investment asset. Available at SSRN 1112630 (2020)
2. N. Chladek, 7 Types of Alternative Investments Everyone Should Know (2023). https://online.hbs.edu/blog/post/types-of-alternative-investments
3. K. Frank et al., The Wealth Report (2023)
4. C. Spaenjers, S. Lovo, Art Markets: What You Should Know Before Investing In Art (2019). https://www.hec.edu/en/knowledge/articles/art-markets-what-you-should-know-investing-art
5. M. Attaran, RFID: an enabler of supply chain operations. Supply Chain Manage. Int. J. **12**, 249–257 (2007)
6. M. McConaghy, G. McMullen, G. Parry, T. McConaghy, D. Holtzman, Visibility and digital art: Blockchain as an ownership layer on the Internet. Strategic Change **26**, 461–470 (2017)
7. S. Nakamoto, Bitcoin: a peer-to-peer electronic cash system. Decentralized Bus. Rev. 21260 (2008)
8. L.S. Sankar, M. Sindhu, M. Sethumadhavan, Survey of consensus protocols on blockchain applications in *2017 4th International Conference on Advanced Computing and Communication Systems (ICACCS)* (2017), pp. 1–5
9. G. Wood et al., Ethereum: a secure decentralised generalised transaction ledger. Ethereum Project Yellow Paper **151**, 1–32 (2014)
10. F. Armknecht, G.O. Karame, A. Mandal, F. Youssef, E. Zenner, Ripple: overview and outlook in *International Conference on Trust and Trust-Worthy Computing* (2015), pp. 163–180
11. J. Kwon, Tendermint: consensus without mining. Draft v. 0.6, fall **1**(11), 1–11 (2014)
12. A. Baliga, I. Subhod, P. Kamat, S. Chatterjee, Performance evaluation of the quorum blockchain platform. Preprint. arXiv:1809.03421 (2018)
13. C. Cachin et al., Architecture of the hyperledger blockchain fabric, in *Workshop on Distributed Cryptocurrencies and Consensus Ledgers*, vol. 310 (2016), p. 4
14. M. Ma, G. Shi, F. Li, Privacy-oriented blockchain-based distributed key management architecture for hierarchical access control in the IoT scenario. IEEE Access **7**, 34045–34059 (2019)

15. W. Viriyasitavat, D. Hoonsopon, Blockchain characteristics and consensus in modern business processes. J. Ind. Inf. Integr. **13**, 32–39 (2019)
16. M.A. Kandi et al., A decentralized blockchain-based key management protocol for heterogeneous and dynamic IoT devices. Comput. Commun. **191**, 11–25 (2022)
17. I. Bashir, *Mastering Blockchain* (Packt Publishing Ltd., Birmingham, 2017)
18. P. Vasin, Blackcoin's proof-of-stake protocol v2. vol. 71 (2014). https://blackcoin.co/blackcoin-pos-protocol-v2-whitepaper.Pdf
19. M. Castro, B. Liskov et al., Practical Byzantine fault tolerance in *OSDI*, vol. 99 (1999), pp. 173–186
20. K. Christidis, M. Devetsikiotis, Blockchains and smart contracts for the internet of things. IEEE Access **4**, 2292–2303 (2016)
21. Q. Wang, R. Li, Q. Wang, S. Chen, Non-fungible token (NFT): overview, evaluation, opportunities and challenges. Preprint. arXiv:2105.07447 (2021)
22. H. Investments, Can You Make NFTs From Physical Art? 5 Things To Know (2023). https://hanginginvestments.com/make-nfts-physical-art
23. K.A. Houser, J.T. Holden, Navigating the non-fungible token. Utah L. Rev. 891 (2022)
24. J. Benet, IPFS-content addressed, versioned, P2P file system. Preprint. arXiv:1407.3561 (2014)
25. K.B. Wilson, A. Karg, H. Ghaderi, Prospecting non-fungible tokens in the digital economy: stakeholders and ecosystem, risk and opportunity. Business Horizons **65**, 657–670 (2022)
26. S. Bhujel, Y. Rahulamathavan, A survey: security, transparency, and scalability issues of NFT's and its marketplaces. Sensors **22**, 8833 (2022)
27. R. Scheiding, Designing the future? The metaverse, NFTs, & the future as defined by unity users. Games and Culture 15554120221139218 (2022)
28. M. Mohsin, 10 Google Search Statistics You Need to Know in 2023 (2023). https://www.oberlo.com/blog/google-search-statistics
29. Act, P. Berne Convention for the Protection of Literary and Artistic Works. Director (1979)
30. Coinbase. Convert Ethereum to US Dollar (2023). https://www.coinbase.com/converter/eth/usd
31. STREET, N. Top 10 Most Expensive CryptoKitties Ever Sold in 2021 (2021). https://www.nftsstreet.com/top-10-most-expensive-cryptokitties-ever-sold-in-2021
32. M.S. Smith, The Spectacular Collapse of CryptoKitties, the First Big Blockchain Game (2022). https://spectrum.ieee.org/cryptokitties
33. M.P. Arenas, H. Demirci, G. Lenzini, An analysis of cholesteric spherical reflector identifiers for object authenticity verification. Mach. Learn. Knowl. Extract. **4**, 222–239 (2022)
34. Fund, L.N.R. Kaleidoscopic microbeads to fight counterfeiting (2023). https://www.fnr.lu/research-with-impact-fnr-highlight/kaleidoscopic-microbeads-to-fight-counterfeiting
35. H. Agha et al., Unclonable human-invisible machine vision markers leveraging the omnidirectional chiral Bragg diffraction of cholesteric spherical reflectors. Light Sci. Appl. **11**, 309 (2022)
36. M. Schwartz et al., Linking physical objects to their digital twins via fiducial markers designed for invisibility to humans. Multifunct. Mat. **4**(2), 1–19 (2021)
37. Y. Geng, R. Kizhakidathazhath, J.P. Lagerwall, Encoding hidden information onto surfaces using polymerized cholesteric spherical reflectors. Adv. Funct. Mat. **31**, 2100399 (2021)
38. H. Agha, Y.-S. Zhang, Y. Geng, J.P.F. Lagerwall, Pixelating structural color with cholesteric spherical reflectors. Adv. Photon. Res. **4**(4), 2200363 (2023)
39. M. Schwartz et al., Cholesteric liquid crystal shells as enabling material for information-rich design and architecture. Adv. Mat. **30**, 1707382 (2018)
40. Y. Geng, J. Noh, I. Drevensek-Olenik, R. Rupp, J. Lagerwall, Elucidating the fine details of cholesteric liquid crystal shell reflection patterns. Liq. Cryst. **44**, 1948–1959 (2017)
41. Y. Geng et al., Through the spherical looking-glass: asymmetry enables multicolored internal reflection in cholesteric liquid crystal shells. Adv. Opt. Mat. **6**, 1700923 (2018)
42. Y. Geng et al., High-fidelity spherical cholesteric liquid crystal Bragg reflectors generating unclonable patterns for secure authentication. Sci. Rep. **6**, 1–9 (2016)

VeraciTIX: An NFT-Based Event Ticketing System

Ryan Ward, Komei Jo, Naipeng Dong, Bill Owens, and Marius Portmann

Abstract Among the disruptive technologies that have emerged since the inception of the Internet, blockchain stands out more than most. It serves as the foundation for all cryptocurrencies, digital wallets, and smart contracts. The problems with event ticketing systems, such as fraud, counterfeiting, and lack of transparency, can potentially be addressed through the implementation of blockchain technology. This chapter outlines the potential benefits of using a decentralized platform for event ticketing, including increased security, transparency, accountability, and customer confidence. The advent of Layer-2 blockchains has made it more feasible to achieve scalability for large systems without compromising on security. In this chapter, a proof-of-concept decentralized event ticketing system is designed and implemented, with a focus on shielding the user from all complex blockchain interactions. A requirements analysis was conducted, comparing the prerequisites of a traditional event ticketing system, and how they can be applied to a decentralized event ticketing system. Several experiments were performed on the system, which aims to quantize the transaction times and fees when a user interacts with the system. An evaluation was conducted on the results and contrasted with the requirements analysis of traditional systems.

Keywords Blockchain · Cryptocurrency · Non-fungible token (NFT) · Mobile event ticketing · Ticket fraud

---

R. Ward · K. Jo · N. Dong (✉) · M. Portmann
School of Electrical Engineering and Computer Science, The University of Queensland, St Lucia, Australia
e-mail: ryan.ward@uqconnect.edu.au; komei.jo@uqconnect.edu.au; n.dong@uq.edu.au; marius@eecs.uq.edu.au

B. Owens
Veracity Business Solutions, Woolloongabba, Australia
e-mail: billo@veracity.com.au

N. El Madhoun et al. (eds.), *Blockchain and Smart-Contract Technologies for Innovative Applications*, https://doi.org/10.1007/978-3-031-50028-2_7

1 Introduction

The concept of worldwide networking is central to the technological revolution and has resulted in the digitization (the conversion of "physical" information to a digital format) of many objects, documents, images, and sounds. This digitization, paired with the increased use of mobile devices in the last decade, has resulted in the ability to convert tickets to a digital medium, rather than the traditional paper form [1]. However, the increased use of mobile and electronic tickets has attracted scammers and fraudulent actors, due to social media and ticket advertising platforms giving a wide and easy reach to potential victims. It is estimated that in 2022, multiple millions of dollars have been lost to ticket fraud [2], and with demand for event tickets doubling in the same year, the scams and fraudulent ticket sales are certain to increase [3]. The current traditional ticketing systems enable fraudulent transactions through many factors—allowing the use of paper tickets, having a weak association between the ticket and its owner, and having easily replicated barcodes or QR codes.

The domain of event ticketing is starting to undergo a significant transformation, with the introduction of blockchain technology to act as a decentralized and trust-less platform on which ticketing systems can operate. Being distributed and immutable, blockchain technology enables secure and transparent transactions between event organizers and attendees, without the need for intermediaries [4]. These attributes can provide a system that presents a solution to many long-standing industry problems—fraud, scamming, counterfeiting, and scalping. This chapter proposes a system design and implementation, VeraciTIX, that is aimed at providing a traditional user experience whilst using the technological advantages of a blockchain.

The VeraciTIX system aims to prioritize the user experience and interaction, by ensuring that any complex blockchain interaction is handled "behind the scenes" by backend servers and interfaces. In a contemporary blockchain-based system, a user has to configure a cryptocurrency wallet and manage their private key and relevant funds independently. Instead, this key and wallet management can be handled by a service provider, with the user only needing traditional account details (i.e., an email/username and password) to use the system.

This chapter serves as an important contribution to the literature on blockchain-based event ticketing systems. By providing a comprehensive overview of the current state of the industry and the potential benefits and challenges of using blockchain technology, this chapter lays the groundwork for future research and development in this area.

Organization The remaining part of this chapter is organized as follows: Sect. 2 provides essential concepts of blockchain and NFT. Section 3 discusses existing solutions and their limitations and presents the requirements that this work aims to meet. Then we propose the VeraciTIX solution and describe the design and implementation details in Sect. 4. The solution and implementation are validated by experiments in Sect. 5 and evaluated on functionality, security, and costs in Sect. 6. Finally, this chapter is concluded in Sect. 7.

2 Background

2.1 Blockchain

Blockchains are a decentralized, distributed ledgers that record transactions across a network of computer nodes in a verifiable and permanent way [5]. The concept of a blockchain and its first application was introduced in the Bitcoin whitepaper released in 2008 by an author with the pseudonym Satoshi Nakamoto [6]. Bitcoin is a decentralized digital currency that uses the blockchain to record and validate digital financial transactions. Since then, blockchain has been applied to a variety of other industries and applications including real estate, digital identity, voting systems, supply chain management, and healthcare [7].

The name "blockchain" originates from the use of data "blocks" linked by cryptographic "chain." Each block generally contains three types of information including the data which can represent some assets, its hash, and the hash of the previous block [8]. The cryptographic hash of a block is its unique identifier which will always depend on its contents. The use of the previous block's hash enables the immutability of the blockchain. The infeasibility of altering the recorded transactions stored in blocks is also why blockchains are considered trust-less.

Blockchain is decentralized because information is distributed across the entire network, instead of it being managed by a single entity [9]. Most well-known blockchains, including Bitcoin, Ethereum, and Polygon, are public blockchains, meaning anyone can participate and view the state of the network. Blockchains rely on consensus algorithms, where nodes verify the validity and determine the order of transactions [10]. These mechanisms eliminate the need for a central authority or a trusted third party and play a significant role in maintaining the security and efficiency of the blockchain.

2.2 Smart Contracts and Non-Fungible Tokens (NFTs)

Smart contracts are self-executing programmable digital contracts that are stored and run on the blockchain [11]. They eliminate the need for intermediaries as they automatically and securely execute agreements when specific conditions are met. The agreements and state of smart contracts are always transparent and tamper-proof as they are stored on the blockchain. Generally, a smart contract cannot be altered once it has been deployed on the chain. Because they are based on code, they can be programmed to perform various functions such as transferring funds, managing digital assets, or tracking real-world data.

Smart contracts are popularly used on the Ethereum blockchain and can be interacted with by sending a transaction through a remote procedure call (RPC) [12]. Contracts are accessible by sending a transaction to its 20 byte address, similar to how a transaction is sent to transfer funds between two users in the network. The code within the smart contracts are executed on the miners' computers, and all miners in the network are required to update and verify the state after the contract call has finished executing [12]. The Ethereum community has created a standard, named the "Ethereum Request for Comments" (ERC) which defines standards for smart contracts to enable interoperability between different decentralized applications [13]. ERC-20 and ERC-721 are the two most common standards, with the former being a standard for a fungible token, which is a token that can be used for currency, voting rights, staking, and more. The latter is a standard for a non-fungible token (NFT), which is used for representing uniqueness (no ERC-721 token is alike, each one has a unique identifier). Throughout this project, the use of NFTs for event ticketing systems will be examined.

3 Related Works

The authors of [14] and [15] analyzed modern ticketing systems for events, highlighted flaws and possible improvements using a blockchain-based system. They described how current event ticketing systems are susceptible to fraud, invalid tickets and scalping, and highlighted the improvements a blockchain-based system would provide.

The authors also developed a prototype implementation that addressed the concerns of two main stakeholders, the event organizers and the attendees. It included functionality to handle ticketing for both the primary and secondary market. The primary market in event ticketing refers to the initial sale of tickets by the event organizer or authorized ticketing company. The secondary market consists of the resale of tickets initially purchased from the primary market. The prototype introduced various techniques to improve the accessibility of the secondary market to ticket buyers and sellers. This prototype is a valuable example of preliminary work for the use of NFTs in event ticketing. However, the design lacks features for integrating to a traditional ticketing system. The design objectives focused mainly on utilizing the benefits of the blockchain which accompanies with added usage complexity for stakeholders.

The paper [16] proposed a smart contract for event ticketing on the Hyperledger Fabric blockchain. It includes basic functionality such as ticket enrolment, ticket resale, and ticket invalidation. Each event ticket is represented as a tuple which contains a unique identifier, customer's public key, event organizer's public key, a Boolean for ticket validity, and ticket age.

The smart contract employs digital signatures and timestamps for validating ticket and event ownership. This covers important security considerations including double re-selling and ticket theft from posted images.

However, the practicality of the designed system is limited. All stakeholders including the ticket seller, customers, and the event organizer are required to directly call to the contract and know how to use the digital signatures. The resale functionality assumes the ticket re-seller already knows the new ticket buyer.

Guaranteed Entrance (GET) Protocol [17], formerly known as GUTS, is one of the leading NFT ticketing solutions in the market. GET Protocol offers an API solution, named Digital Twin, and a full stack ticketing solution known as White-Label. For all ticketing processes, GET Protocol uses its native ERC-20 $GET token on both Ethereum and Polygon blockchains. Digital Twin allows existing ticketing infrastructures to mimic on-chain a number of states and interactions to verify access rights of tickets. This product is not a replacement of traditional ticketing systems but provides additional visibility over ticket sales and the possibility to convert tickets into digital collectables [17]. The White-Label solution is a full stack NFT ticketing solution that allows integrators to control the entire ticketing lifecycle including the secondary market.

NFT-TiX [18] is another popular Layer 1 Ethereum-based NFT ticketing solution. The platform is multichain-capable meaning it can integrate with other EVM-compatible blockchains. However, it requires event organizers and attendees to connect and use their cryptocurrency wallets through Metamask or Walletconnect to sign transactions. For general customers, this ticket purchasing and redeeming process are complex and confusing if they are not familiar with NFTs and cryptocurrency wallets.

SeatLab [19] is an NFT ticketing project in development with key features being continuously rolled out. The main product is a full stack ticketing solution with an easy-to-use dashboard design for ticket sellers. It is built on the Gen-3 NEAR protocol, which is a Proof-of-Stake blockchain with a sharding architecture. The main appeal of this product for event attendees is their use of NFTs in their own marketplace and mobile ticket wallet. Similarly to the previous platform, the complexity of using cryptocurrency wallets and marketplaces can discourage inexperienced consumers.

In summary, there is a clear gap in the traditional ticketing market that has the potential to be disrupted through the use of a blockchain-based system. Traditional systems lack customer trust, and service and transaction fees can quickly escalate to become a significant cost. The mentioned systems all provide capable solutions to the gap, however, many of them require previous cryptocurrency knowledge and for the customer to consistently interact with the given blockchain. VeraciTIX solves this issue by abstracting all blockchain and cryptocurrency wallet transactions away from the users, by designing the system to behave like a traditional ticketing system, yet leverage the security of the underlying blockchain.

3.1 Requirements Analysis

Established ticketing vendors are aware of the security shortcomings of existing online ticketing services so they take steps to reduce fraud, theft, and scalping by introducing control measures in their sales processes such as withholding ticket distribution until a few days before an event or limiting the number of tickets sold to a single buyer. In our burgeoning e-commerce world, where trust and convenience matter, these measures are clumsy and are unlikely to mitigate the risk of fraud or theft, but they will almost certainly damage the customer experience where customers do not know for sure if their tickets will be released on time before an event or will be valid on entry. In order for the VeraciTIX system to surpass and replace traditional ticketing platforms, several functions and features need to be implemented. These functional requirements provide a basis to characterize the system both during development and testing.

1. Secure Ticket Issuance—The system must provide a secure and tamper-proof way of issuing event tickets at the point of purchase. The tickets should be digitally signed and recorded securely, ensuring that they cannot be duplicated or forged.
2. Ticket Transfer Functionality—The system should allow users to transfer their tickets securely and transparently to other users. This will prevent ticket scalping and enable the legitimate exchange of tickets between users.
3. Event Capacity Management—The system should be able to manage the capacity of the event venue and prevent overbooking. This can be achieved by recording the number of tickets sold on the blockchain and updating the availability of tickets in real-time.
4. Payment Integration—The system should integrate with payment gateways to enable seamless and secure payments for the tickets. The payment information should be stored safely and securely to ensure no information is compromised. This integration will come in the form of cryptocurrency on and off ramps, which provide seamless fund conversion.
5. User Privacy and Security—The system must ensure the anonymity and security of user data, including personal information and payment details. Sensitive user data should never be stored on the blockchain, and instead should be held in a secure database. The user shall be hidden behind a pseudo-anonymous address, making all public transactions untraceable to a specific user.
6. Analytics and Reporting—The system should provide real-time analytics and reporting on ticket sales, revenue, and attendance. This will enable event organizers to make informed decisions about event planning and marketing.
7. Scalability and Performance—The system should be able to handle large volumes of transactions and scale up or down as per demand.
8. Efficiency—The system should also be able to perform at least as quickly and efficiently as existing online ticketing transactions to meet user expectations and ensure a smooth ticketing experience.

4 VeraciTIX Ticketing Solution

The VeraciTIX system is described in the following subsections, starting with the key components, which describe and elaborate on all the necessary features the system provides. The architecture describes how these separate components work together to produce a rounded event ticketing system, supporting primary sales, secondary market sales, and administrator functionality for the event organizers. Section 4.3 describes the various use cases of the system through a step-by-step analysis of how data travels through different parts of the VeraciTIX system.

4.1 Key Components

4.1.1 Custodial Wallets

There are two main types of wallets that can be used to interact with the blockchain—custodial and non-custodial (also referred to as self-custodial). A wallet contains a public key and a private key—the public key is used by external programs to send or verify cryptographic transactions. The private key is linked to the public key and allows the creation of cryptographic transactions as well as proof of ownership of the wallet. Whilst identical in high level functionality, the two wallets differ when it comes to the management of the private keys associated with the wallet. A custodial wallet is defined as a wallet in which a third party holds the private key, meaning that the third party has full control over the wallet and all funds contained. A non-custodial wallet requires the owner of the wallet to manage its private keys; hence, the user has more control over their funds at the cost of being responsible for the private keys.

4.1.2 On/Off Ramping

In order to interact with the blockchain for the creation and transfer of tickets, a form of currency that is compatible with the blockchain must be used. This form of currency is different to traditional (fiat) currency and must be purchased from an exchange, or in this case, an on-ramp. An on ramp allows a user to purchase cryptocurrency with fiat currency using a credit/debit card, and upon a successful transaction, the cryptocurrency is sent to the user's wallet. In the context of program flow, when it is time for the user to purchase a ticket, they will be prompted for their credit card details. There are some possible exchange rate risks when transferring fiat currency to cryptocurrency; however, there are two possible ways to mitigate the associated risk—keep the time between currency transfer to a minimum, or use an existing "stablecoin" cryptocurrency, which never deviates in value.

4.1.3 Rolling QR Codes

It is quite standard that cellular data does not work at events, due to the large concentration of mobile devices. There is not enough bandwidth to accommodate the large number of devices attempting to communicate with the cell towers. Hence,

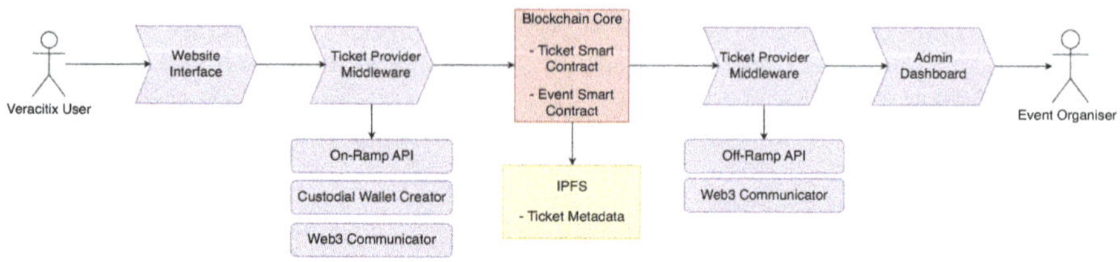

Fig. 1 System architecture

an "offline" method needs to be used to store tickets, as it can be assumed that the users will not be able to access their tickets from the cloud. In order to maintain security and fraud prevention, the mobile app will create a QR code representation of the ticket based off the ticket ID, the private key, and a "salt" (pseudo-random value) that changes with time.

4.2 Architecture

Throughout development, there were multiple design decisions that needed to be made, to allow for further development and operation success. The single most important objective was a minimal user interface, with the goal being to purchase a ticket in as few actions and input from the user as possible (i.e., being able to purchase a ticket within five "clicks" of a mouse).

Figure 1 is the system architecture diagram, comprising of the blockchain core, the ticket provider middleware, and the connecting interfaces (web interface for users and an admin dashboard for event organizers).

The blockchain core contains the smart contract logic and, therefore, constantly maintains the state of the system. All state changes must be executed and verified on the blockchain to be valid and accepted. The core ensures all transactions are passed through the smart contract, which imposes the rule that a ticket cannot exist outside the blockchain core, which provides a high level of fraud and counterfeit mitigation. The smart contracts are based on the ERC-721 standard and operate on the chosen Polygon blockchain for its efficiency and low cost. Polygon is a Layer-2 blockchain, meaning that it sits atop an existing blockchain, and bundles transactions together to be verified by the underlying blockchain, which in turn lowers transaction fees.

The user/admin will only interact with the web interface, with the bulk of all processing being handled in the "middleware" communication interface, which creates transactions, provides authentication measures, and communicates to the blockchain. The is achieved by a standard front-end and server to handle all user interactions and send the generated transactions to the blockchain. The front-end ticketing interface will handle all user interactions, including the viewing, purchasing, and checking-in of all tickets. Attached to the user front-end will also be an event organizer interface, in which event organizers can activate ticket sales, collect ticket earnings, and alter information about the event when required. This admin interface will still communicate with the same smart contract but with elevated privileges, allowing the event organizer to execute functions to which the general user would not have access.

The front-end interface is linked to a user-specific cryptocurrency wallet that is instantiated upon user account creation. Tatum, a cryptocurrency wallet management service, is used to generate public private key pairs on the Polygon blockchain [20], which get immediately stored in a private database. The user is completely abstracted from this process and will only ever use a traditional username and password combination to log into their account, rather than having to memorize a standard twenty word passphrase and other security measures that current cryptocurrency wallets use. Custodial wallets and cryptocurrency on-off ramps are used in conjunction to optimize the user experience. The ramping component was handled using the third party product Ramp [21], which allowed for users to enter their preferred payment details when buying a ticket, with the fiat currency being transferred to cryptocurrency for use with the ticket smart contract.

4.3 Use Cases

The following use cases describe how user data is passed between the various components of the system—the web interface, database, server, and blockchain. The use cases handle the primary functionality sectors of the system, being user account creation, primary ticket purchasing, secondary market on-selling, on-ramping and off-ramping, and the various administrator functions available to the event organizers.

4.3.1 User Account Creation

Figure 2 shows the process flow of the front-end which is identical to any other traditional ticketing system. The user will arrive at the website's landing page, where they can choose to purchase a ticket if they already have an account or sign up if they do not (see Step 1 and 2). The sign-up process records basic details (i.e., name, email, and username) and will create a link to a blockchain wallet upon a successful sign-up. The user will not have access to this wallet; however, it will be used for all interactions with the smart contracts. Table 1 provides detailed communication information for this use case.

4.3.2 Ticket Purchase

Figure 3 and Table 2 show the most common transaction handled by the system—a ticket purchase. It is a straightforward process, with the front-end website being

Fig. 2 User account creation

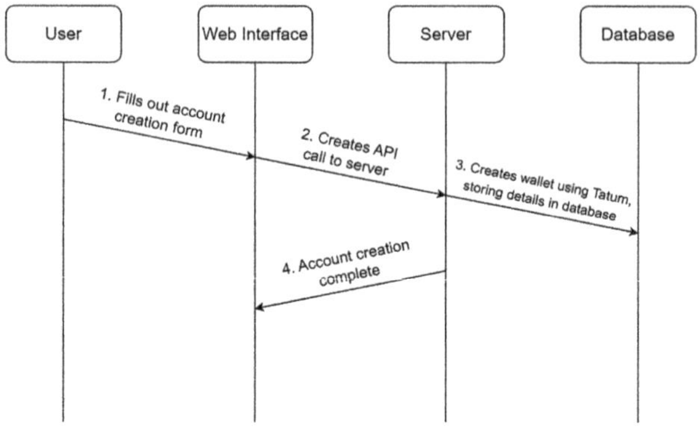

Table 1 Information sent in user account creation

Operation step	Information sent
1	User ID, email, requested username, passwords
2	User ID, email, requested username, passwords
3	Password and wallet key pairs
4	Account creation acknowledgement

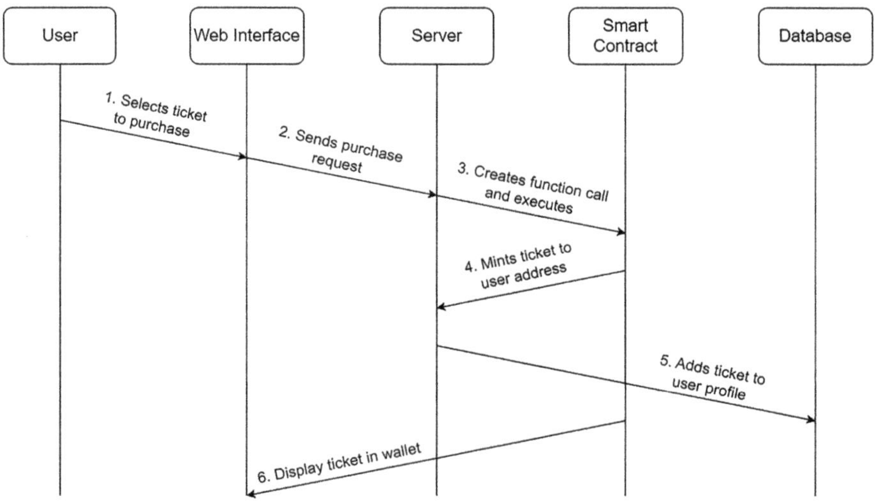

Fig. 3 Ticket purchase

used as an interface for the smart contract, to handle new user sign ups and to display the user's tickets. A server is attached to the front-end via a REST API, which is responsible for the previously stated wallet creation, as well as calls to the blockchain to allow for ticket purchasing, selling, and the various admin controls. The server operates using the Alchemy SDK [22] for blockchain communication and uses the NextJS framework to handle API calls between the server and client.

Table 2 Information sent in ticket purchase

Operation step	Information sent
1	Ticket selection
2	Ticket selection, user ID
3	Signed transaction to generate NFT, public key
4	Acknowledged transaction
5	Transaction completion acknowledgement
6	Transaction completion acknowledgement

This ensures that there are never any private keys sent to the browser, as they could potentially be intercepted by a malicious actor during any smart contract call.

When the user selects a ticket to purchase, if they do not have sufficient cryptocurrency funds in their account from a previous purchase (which will be the case for many users), an on-ramping widget will provide a way of transferring fiat currency from a traditional form of payment to cryptocurrency, which is stored in the user's blockchain wallet and is used for current and future ticket purchases and secondary market sales. Figure 3 demonstrates this process, with Steps 1 through 3 containing the ticket purchase transaction details, and Step 4 through 6 describing the result of the transaction. This on-ramp widget is provided by Ramp [21], a Web3 developer toolkit that handles Know Your Customer (KYC), regulation and compliance and fiat payment methods, allowing users to buy cryptocurrency within the VeraciTIX application. Currently, the application uses a pop-up widget with a simplified interface, purposely omitting the receiving address for a streamlined user experience. Instead, the receiving wallet is the address that is currently in use within the state of purchase, which will be used to execute the ticket purchase smart contract when the cryptocurrency funds are converted.

4.3.3 Secondary Market Transfer

Figure 4 and Table 3 show the process a secondary market transfer between two users. The two actors, Alice and Bob, wish to engage in a transaction, with Alice listing her ticket for sale, and Bob purchasing it with fiat currency. Alice initiates the transaction by first signing into the web interface, where she can view all their currently owned tickets. Alice will then select a ticket to sell by interacting with the web interface, which will then prepare a smart contract transaction to send to the server. The server will receive said transaction and validate it, to ensure that the ticket does indeed exist and belong to the user who wishes to sell it. Once the transaction has been executed, the state of the ticket will change to "on sale." Bob can browse all "on sale" tickets in a secondary market interface on the web application and can opt to purchase one or multiple tickets. Similar to the selling process, the buying process also generates a smart contract call, to be sent to the server to ensure Bob has sufficient funds, and that the ticket is still on sale, as the

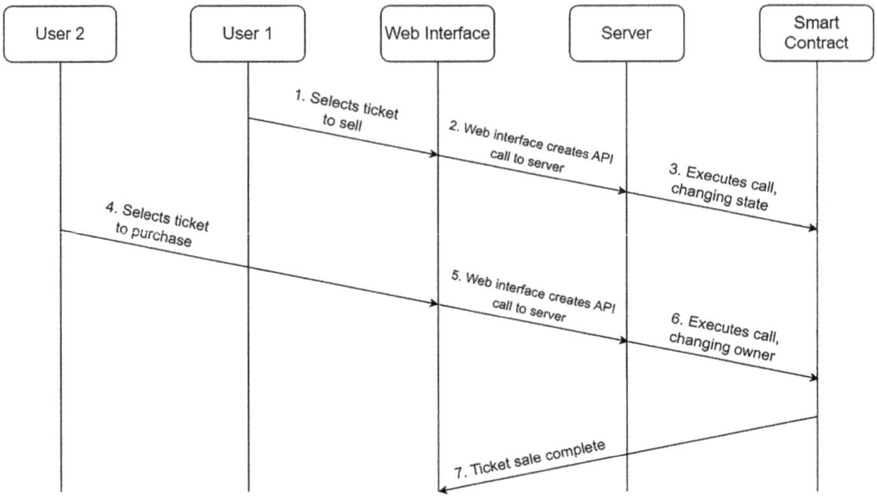

Fig. 4 Secondary market transfer

Table 3 Information sent in secondary market transfer

Operation step	Information sent
1	Ticket ID
2	Ticket ID, user ID
3	Signed transaction to sell
4	User ID, Ticket ID
5	User ID, Ticket ID
6	Transaction to buy
7	Transaction completion acknowledgement

state could have changed in between the time it takes for the web interface function call to be executed on the blockchain.

4.3.4 Administrator Functions

In order for both users and admins to collect revenue from ticket sales, the functionality to securely release funds in the form of fiat currency was implemented. By using a cryptocurrency off-ramp, the system will automatically convert the cryptocurrency stored in their wallet (from secondary market transfers) to fiat currency whenever the funds are withdrawn. In Fig. 5 and Table 4, there are two processes shown—the event organizer starting and stopping ticket sales, and the event organizer collecting funds from the event smart contract. The first transaction involves signing in to the web interface with administrator credentials, which allows access to high privilege function calls, including changing the state of ticket sales,

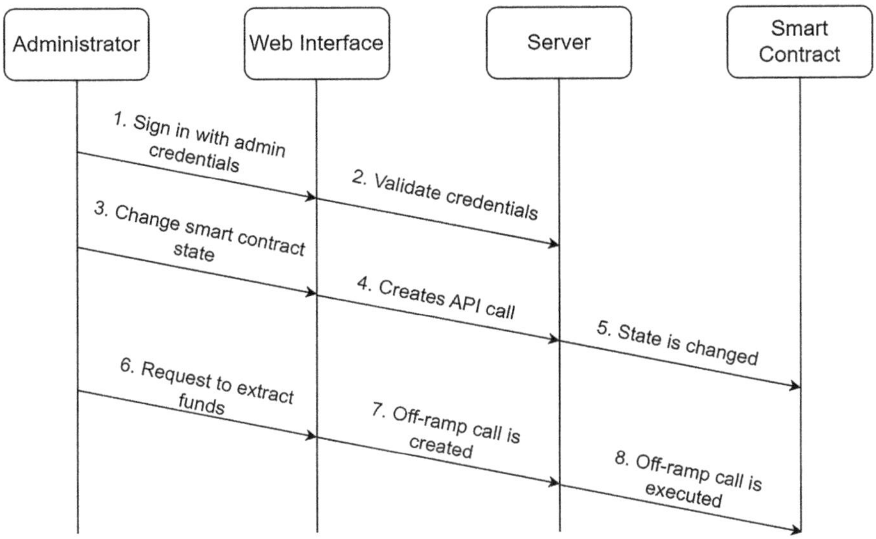

Fig. 5 Administrator functions

Table 4 Information sent in administrator functions

Operation step	Information sent
1	Administrator login details
2	Administrator login details
3	Transaction data
4	Signed blockchain transaction
5	Transaction RPC
6	Transaction data
7	Signed transaction data
8	Transaction RPC

changing the number of total and available tickets, and the most critical function of all—withdrawing funds from the smart contract.

The second transaction shown in Fig. 6 and Table 5 is a simple fund release, in which the admin account in this instance is the owner of the wallet that is being interacted with. The user will send a claim request to collect their funds, and a contract call will be created and executed, which then gets sent to the respective wallet. However, at this point in time, the currency is not in a fiat domain and needs to be "off-ramped," which is shown in Fig. 6. After the funds have been transferred to the user's wallet, the off-ramp API is called to transfer all funds to the user-specified fiat account, then the process is complete.

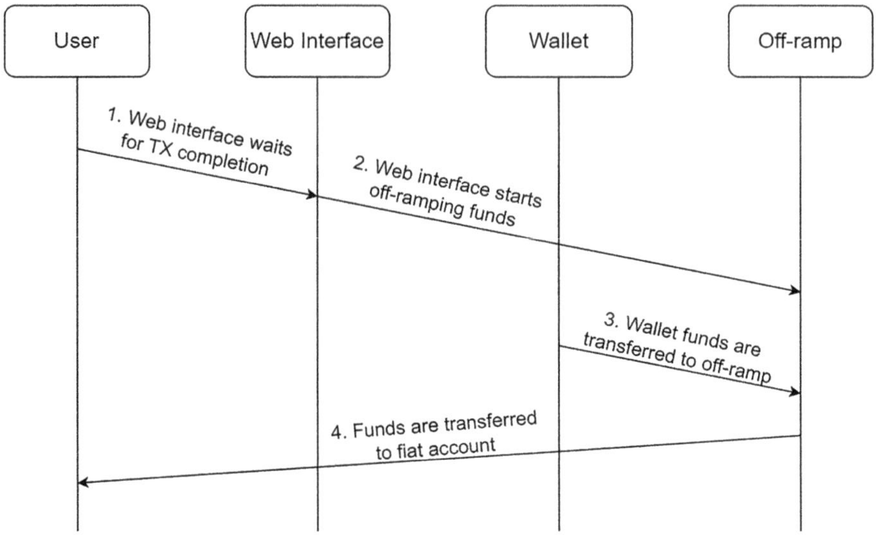

Fig. 6 Off-ramp transaction

Table 5 Information sent in off-ramp transaction

Operation step	Information sent
1	User and wallet details
2	Wallet transaction request
3	Off-ramp transaction details
4	Transaction acknowledgement

4.4 Implementation

In order to develop a secure and operational prototype, multiple third party tools were used. This practice is extremely common in developing decentralized applications, as there are multiple prerequisites that need to be satisfied before an application can communicate with a given blockchain. These include having a public and private key that is recognized and valid on the particular blockchain (also known as a wallet or address), as well as having a provider that can make smart contract calls to the selected network, and finally an on/off ramp tool that can convert fiat currency into cryptocurrency for use on the selected blockchain. Tatum is a blockchain development framework and provides many different tools that can be used to build decentralized applications, with one being the Tatum Key Management System (KMS) [20], which can create, store, and manage wallets on a selected blockchain. It provides a JavaScript SDK that allows keys to be created server side then stored securely in the MongoDB database, such that no critical information is sent to or stored in the browser.

After the necessary keys have been created, they may be used in conjunction with a blockchain provider to send transactions to the network. The Alchemy SDK

[22] is used for preparing and sending transactions to the blockchain, extending the Ethers.js library—a popular resource for interacting with Ethereum-based blockchains. All Alchemy API calls are sent from the server, as the caller's keys are required to sign and execute the transaction, which are stored on the database and only sent to the server when required. Before sending certain transactions, such as a ticket purchase, the library is used to check the balance of the address, as well as any other parameters that would cause the transaction to fail. Due to the blockchain interaction is abstracted away from the user, this checking step is crucial to the user experience, as any "runtime" error (i.e., any error that gets thrown during the execution of the smart contract call) will consume gas and increase the time taken to purchase a ticket.

The platform was integrated with the Polygon blockchain, which is used to keep a trust-less (i.e., not having to rely on the integrity of a third party actor) record of all ticket sales and transfers. The smart contract on the blockchain maintains the total "state" of the VeraciTIX system, with each smart contract call (e.g., a purchase ticket call) updating the current state. When the designated event starts, the current state is captured and used for ticket check-in. This state is public and immutable, two crucial factors for creating a transparent and trust-less system on which a decentralized event ticketing system can operate. A new smart contract will be created for each event, allowing the event organizer to have an administrative role within the smart contract logic. This will allow them to start and stop ticket sales, set date and time information for events, and withdraw the funds that are held within the smart contract from primary ticket sales. The public blockchain design can be a "double-edged sword," in that because all data is publicly verifiable, any user can see the state of the smart contract. This means that all users will see the total funds kept in the smart contract, as well as all transactions that go through it. This is why all transactions are abstracted away from users, and the only sensitive data that is sent to the blockchain is a request with addresses and keys that have no public link to any users. The link is instead contained within a private database maintained by the VeraciTIX system; therefore, the state and integrity of the system itself is maintained, with the privacy and security of a traditional ticketing system.

5 Experiments

5.1 Methodology

Experimentation on the developed prototype was conducted to investigate the scalability of using blockchain technology for event ticketing. The focus was to evaluate the blockchain component of the prototype to understand the integrability of blockchain usage with traditional ticketing systems.

Tests were conducted to measure the system's ability to handle large numbers of ticket transactions simultaneously. The experimental setup involved the development of the Python script and the deployment of the smart contract. There were two test domains—the Polygon testnet, accessed through the blockchain service providers Alchemy and Infura [23], and a local blockchain, using the Hardhat [24] development environment.

5.1.1 Testnet Experimentation

Similarly with the proposed system, tests were conducted on the Polygon Mumbai test network, allowing large volumes of transactions to be sent without incurring any real-world costs. The peak usage of a ticketing system would generally occur during the initial release of event tickets, and this was simulated by the experimental script which asynchronously calls the blockchain service providers with specified numbers of ticket purchase requests (batches).

The scalability of the system was determined by measuring the transaction time of each ticket purchase. Transaction time is the duration it takes for a transaction to be processed and confirmed in the blockchain after being submitted to the network. This was measured by subtracting the timestamp recorded in the block the transaction was included in by the time the script recorded when the transaction was sent.

5.1.2 Local Experimentation

In order to test the scalability and functionality of the smart contract without the limitations of a testnet blockchain, the system was also evaluated on a local node. The local blockchain was forked from the Polygon Mumbai blockchain with the Hardhat [24] development environment and the same Python script and smart contract was tested against this local chain.

5.2 Results

Figures 7 and 8 show that regardless of the API service, the time it takes for a transaction to be confirmed on the block is constant. The approximate 1 second difference between Alchemy and Infura can be attributed to varying network traffic. These tests were conducted on the Polygon Mumbai testnet and do not necessarily reflect transaction times that would be seen on the Polygon mainnet. Other factors that would affect transaction times on the mainnet include network size, network traffic, and gas price. Moreover, these results show that the smart contract developed includes scalable functions and does not negatively impact the performance and usability of the system.

To understand whether the smart contract or API services caused bottlenecks in the system, the experimentation was also conducted on a local chain. Transaction

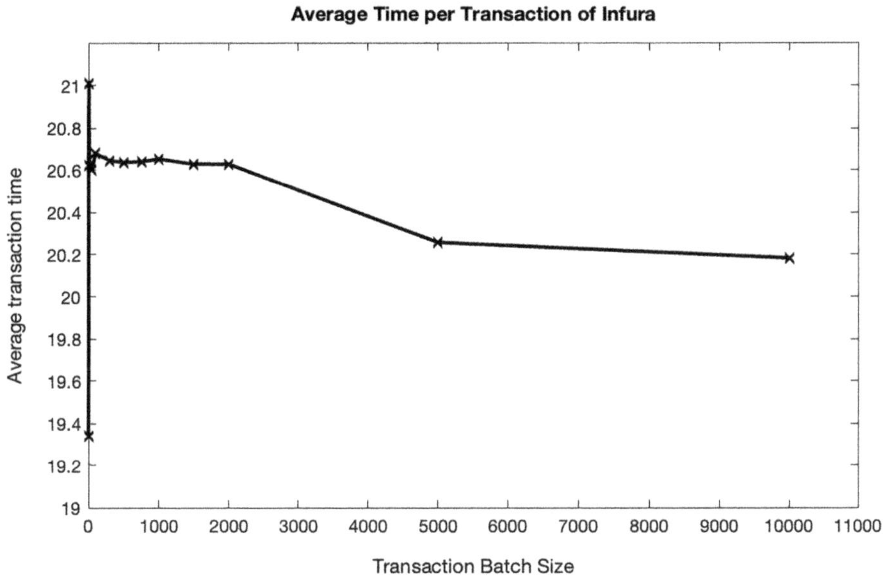

Fig. 7 Average transaction time of varying batch sizes with Infura API

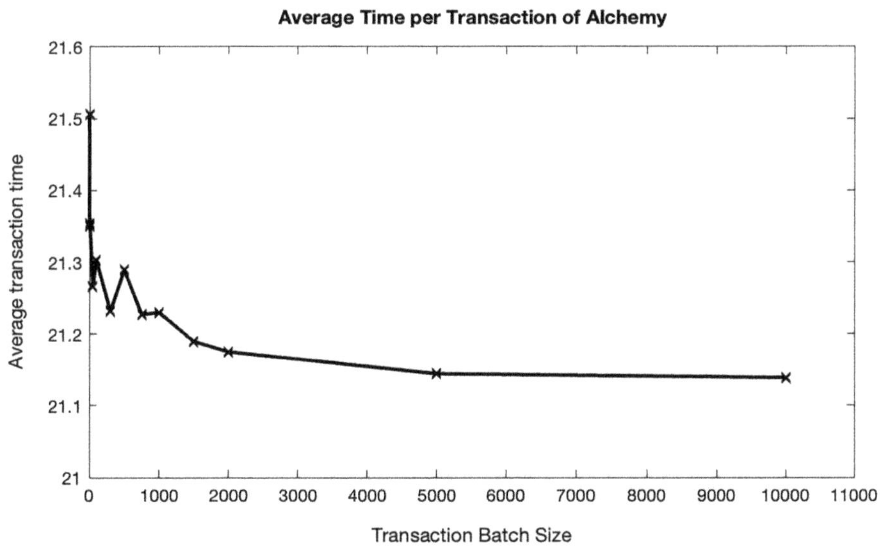

Fig. 8 Average transaction time of varying batch sizes with Alchemy API

times were approximately 1 second as seen in Fig. 9. This reflects the default configuration of the Hardhat node where each transaction is mined in a new block and the block time is set at one second. This data provides further evidence that transaction time is reflective of the blockchain. The proposed system can

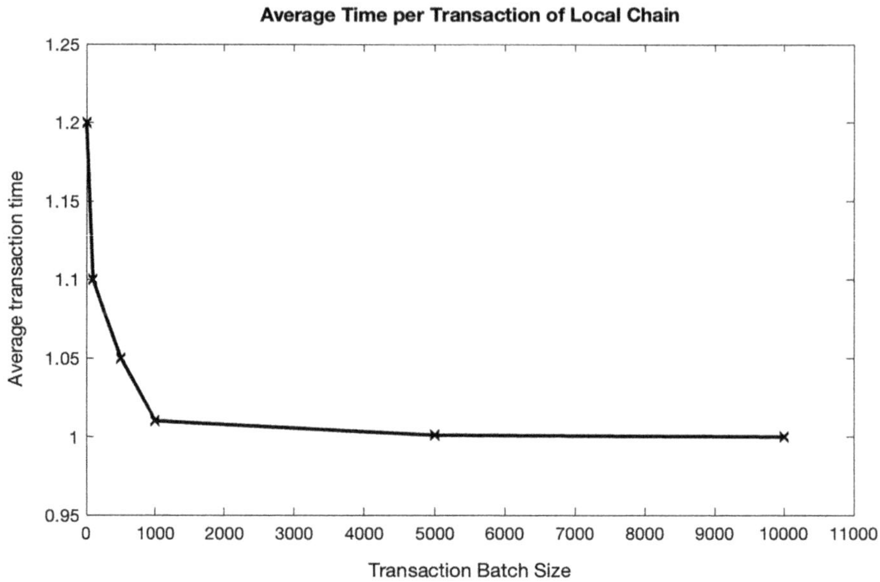

Fig. 9 Average local transaction time

achieve its desired scalability with the scalability solutions on the Polygon network. Furthermore, the VeraciTIX provider would operate and maintain their own Polygon node in a production environment which would allow all transactions to be routed through the hosted node without the limitations of a third party API provider.

6 Evaluation

Ticketing and related management is a critical component of the events management industry, as it allows organizers to manage their attendees, control access to events, and generate revenue. Traditional event ticketing systems have several shortcomings, including lack of transparency, inefficiencies, high transaction costs, and counterfeiting issues. However, blockchain technology can address these issues by providing a secure and transparent event ticketing system.

The system was designed to satisfy the User Requirements described in Sect. 3.1. The system has been shown to provide payment integration and event capacity management, as described in the use cases of Sect. 4.3. The reporting and analytics are provided to the event organizer by the various third party tools including MongoDB, Alchemy, and Tatum. The efficiency, scalability, and performance have been demonstrated in Sect. 5. The requirements of secure ticket issuance and ticket transfer functionality are discussed in the following Sect. 6.1. In addition, the Sect. 6.1 further discusses scalability and other important factors including cost, lifespan of the system. The requirement of user privacy and security is systematically analyzed in Sect. 6.2. Finally, Sect. 6.3 presents a quantitative cost analysis.

6.1 Functionality

The ticket purchasing process is one of the most important functionalities of the VeraciTIX system. With blockchain technology, ticket purchasing would be streamlined and made more efficient, removing third party ticket providers and routing the funds directly between users and event organizers. This in turn would eliminate other intermediaries and reduce transaction fees, making it easier for consumers to purchase tickets. Smart contracts are used to automate the ticketing process, which reduces the risk of fraud by making the system trust-less via a public state. The smart contract logic is programmed to verify the authenticity of tickets and ensure that they are not duplicated or resold at inflated prices.

One of the biggest challenges facing event ticketing systems is scalability, both in terms of speed and storage. Traditional ticketing systems can become overwhelmed when there is a high demand for tickets, leading to slow transaction times and a poor user experience. This, however, is mitigated by fast servers which can scale up and down elastically based on current system traffic. Blockchain-based ticketing systems, however, do not have the option to dynamically scale and instead will bottleneck at the blockchain mempool, where transactions sit and wait to be executed. This bottleneck can be mitigated by using a Layer-2 solution, Polygon.

With blockchain technology, ticket sales can be managed on a decentralized network, which can handle large volumes of transactions without slowing down. Additionally, blockchain-based ticketing systems can use distributed ledger technology to store and verify ticket data, which can help to prevent ticket fraud and ensure that tickets are not duplicated or resold at inflated prices, increasing customer confidence.

Another important factor to consider when evaluating event ticketing systems is cost. Traditional ticketing systems can be expensive, with high transaction fees and hidden costs that can quickly add up. Blockchain-based ticketing systems would help to reduce costs by eliminating intermediaries and automating the ticketing process. All costs are also transparent, as each transaction is public and contains all fee and cost data. With VeraciTIX, smart contracts managing ticket transactions, transaction fees are reduced and the need for intermediary third parties are eliminated. Additionally, blockchain-based ticketing systems can provide transparency and accountability, which can help to reduce the risk of fraud and minimize the costs associated with ticket disputes.

The lifespan of an event ticketing system is an important consideration. Traditional ticketing systems can become outdated quickly, requiring costly upgrades and maintenance. Blockchain-based ticketing systems can provide a long-term solution to event ticketing, with a secure and scalable platform that can adapt to changing industry needs. Blockchain technology is constantly evolving, and new innovations are being developed that can help to improve the functionality of blockchain-based ticketing systems. As the technology continues to mature, blockchain-based ticketing systems are likely to become more efficient, cost-effective, and scalable, providing a long-term solution to event ticketing.

When fully functional, a blockchain-based event ticketing system can provide a secure, efficient, and transparent platform for ticket sales. With blockchain technology, ticket purchasing can be streamlined, scalability can be improved, costs can be reduced, and the lifespan of the ticketing system can be extended. As the event industry continues to evolve, blockchain-based ticketing systems are likely to become an increasingly important component of event management, providing a long-term solution to the challenges facing traditional event ticketing systems.

6.2 *Security*

Whilst blockchain technology has been operational for many years, applications of smart contract are still immature. With the rapid development of decentralized applications, bugs and security flaws are found in both the smart contract language, Solidity, as well as the underlying blockchains [25]. The security of a blockchain-based system largely depends on the security of its smart contract code, as the security of the underlying blockchains are out of control of the developers. In the VeraciTIX system, a smart contract is used to handle all ticket transactions, and is replicated for each event, hence, the security risks fall in the correctness of the smart contract. The security of VeraciTIX was analyzed from two perspectives: the application design with respect to a network attacker, and the smart contract implementation against common vulnerabilities.

6.2.1 Application Design Analysis

VeraciTIX is an integrated system that contains multiple components. Therefore, it is essential to analyze the entire system as a whole. However, directly verifying the entire system implementation is impractical as it is a known open challenge. Hence, we analyze the security of the entire system on the design level.

There have been many security practices published with respect to web security [26] and database security [27], and we do not duplicate them here. Instead, we focus on the security of their integration, i.e., their communications. With this in mind, we made the following assumptions:

1. The web interface (normally a web browser) is secure. Comprehensive lists of vulnerabilities exist for web browsers, but its security is out scope for this work.
2. Similarly, it is assumed that the server and database, as well as their communication is safe, as they are typically hosted in an isolated and protected intranet.

3. All messages sent to the blockchain/smart contracts are public information, since the messages/transactions can be seen by all nodes on the blockchain, potentially including an attacker node in a public network.

With the given assumptions, the threat model of this analysis is the attackers to the communication. The strongest attacker is considered as the adversary that controls the entire network. Known as the Doleve-Yao [28], this attacker is able to read all messages sent through to the internet, inject their own messages, and block or redirect messages. If a system of interest is secure against this type of adversary that implies that it is secure against any weaker adversaries.

Analysis of User Account Creation This system ensures that the user information is secure from the network attackers in the Users Account Creation process (refer to Fig. 2). This is achieved by adopting the standard TLS (Transport Layer Security) secure communication channels, which ensures that the communicated information over the Internet (i.e., the message in Step 2 containing user information like name, email, and address, and the message in Step 4—an acknowledgement in Fig. 2) are authenticated and are not known to the network attacker. Note that Step 1 is local on the client side and Step 3 is a local action on the server side, both of which are secure by assumption. The exception is that wallets are created with external calls. However, this would not leak the account information, as the creating wallet does not involve using the user account information.

Analysis of Ticket Purchase The system ensures that the relayed information is secure in the Ticket Purchase process. By assumption, Steps 1 and 5 in Fig. 3 are internal communications and are assumed to be secure. The communication between the web interface and the server (message 2 in Fig. 3) is protected by the TLS protocol, which ensures that any personally identifiable information are not leaked to the attacker. While messages 3, 4, and 6 are public messages, these messages only contain information public keys, which are not sensitive.

Analysis of Secondary Market Transfer The secondary market involves the resale of tickets initially purchased from the primary market. Similarly, Steps 1 and 4 in Fig. 4 are secure by assumption, and messages 2 and 5 are protected by TLS for authentication and integrate. The messages 3, 6, and 7 are public information. However, no sensitive information is involved.

Analysis of Administrator Functions Steps 1 and 2 in the process (Fig. 5) contain sensitive administrator authentication data but are protected by TLS. Steps 3 to 8 are all public messages, with Steps 4 and 7 being signed blockchain transactions. If a malicious actor had the private key of the administrator account, they would have full control the smart contract; therefore, it is only Steps 1 and 2 that require security measures.

Off-Ramp Transaction Steps 1 and 2 in the process (Fig. 6) would similarly have TLS encryption as the request gets sent over the internet via the Off-ramp API. Steps 3 and 4 are public messages, and the transfer of funds occur in the off-ramp domain, which is assumed to be secure.

6.2.2 Smart Contract Vulnerability Analysis

Several smart contract analysis tools have been developed to detect vulnerabilities of a given smart contract implementation. If a contract is not written correctly, it can lead to increased gas costs, failed transactions, and in the worst case, loss of funds due to exploited vulnerabilities. Two such verification check tools are Slither and Mythril that are widely adopted in smart contract auditing and bug bounty platforms for vulnerability detection. Slither uses static analysis techniques while Mythril is a representative tool of dynamic analysis. Following an empirical study of the tools, the combination of Slither and Mythril would detect the most unique vulnerabilities, including the following 6 known types defined as follows [29]:

1. Access Control: Insecure visibility setting of contract values can give attacker access to a contract's private information or logic.
2. Arithmetic: Integer overflows and underflows.
3. Front Running: Users specify higher fees to have their transactions mined more quickly.
4. Re-entrancy: External contract calls are allowed to make new calls to the calling contract before the initial execution is completed, which may result in a contract behaving in an unexpected way.
5. Time Manipulation: If a contract logic is dependent on block timestamps, a malicious miner could intentionally manipulate when a block is mined for their own benefit.
6. Unchecked Low Calls: Low-level blockchain functions calls fail and is not checked, which is a common root cause of many vulnerabilities.

Analysis Results Mythril did not find any vulnerabilities. Slither reported a vulnerability of re-entrancy of a function in our smart contract implementation and provided suggestion on changing variables into a constant to save gas, which have been addressed. Slither also reported false positives—identifying the standard ERC function calls as public functions.

After identifying and fixing issues, the analysis results from Slither and Mythril suggest that the smart contracts are secure with respect to the mentioned vulnerabilities.

6.3 Costs

For any blockchain system, transaction fees are an important consideration for both event organizers and end users. High transaction fees may limit the adoption of blockchain technology and if customers incur these fees, they may be discouraged from participating in the system.

The gas costs for deploying the smart contract, purchasing (minting) a ticket, transferring a ticket, and redeeming a ticket were calculated. Seen in Table 6, the MATIC and USD cost for each operation was calculated with gas price of 130 Gwei.

Table 6 VeraciTIX blockchain transaction fees

Operation	Units of gas	MATIC	USD
Contract deployment	2,179,575	0.283	$0.34
Ticket purchase	485,957	0.063	$0.076
Ticket transfer	29,804	0.004	$0.005

Table 7 VeraciTIX ramp transaction fees

Fee type	USD
Credit card purchase fee	$2.76
Network fee	$1.00

In addition to smart contract execution fees, there are also fees associated with using the on-ramp to purchase tickets. Ramp is the third party tool used to convert fiat currency to cryptocurrency and has fees associated with its use. All of these fees are charged on a "per transaction" basis, meaning that there is no preliminary fee for event organizers. The table of costs shows the on-ramp fees per ticket purchase, assuming that the user has no cryptocurrency in their account, and needs to convert the full ticket price from fiat to perform the purchase. The costs in Table 7 were also calculated using a ticket price of $100 USD. The credit card purchase fee is set by Ramp and is not dependent on the size or value of the purchase; however, the network fee is dynamic and can change based on the network traffic and value of the cryptocurrency. Despite this, it is safe to assume that it will stay at approximately 1% of the purchase value.

7 Conclusion

The rapid growth of digital assets and mobile computing has provided an ideal environment for an adaptable, user-centered online ticketing model. However, with a single entity controlling the state of the system, the ticket provider has complete control over any customer's ticket, which reduces consumer confidence and freedom. Whilst the current solutions provide the capability of a decentralized ticketing system, their primary targets are users with previous blockchain knowledge and experience. VeraciTIX abstracts this by using a conglomerate of existing tools to produce a completely decentralized and secure system that has the front-facing functionality of a traditional ticketing interface. Instead of having to use native cryptocurrency tokens in a personal wallet, the user can simply login with their chosen details and interact with the VeraciTIX system through a standard web interface.

The sale of tickets are not managed by a third party ticket provider, instead rely on the event organizers to interact with the VeraciTIX system to start and stop ticket sales, and collect proceeds after the event has ended. This restores control of revenue and sales back to the event organizer. In order to prepare the system for a production environment, extensive testing must be performed, and security measures must be implemented to ensure no sensitive data can be retrieved by malicious actors. The system would be required to undertake a UI redesign before being released, for an optimal user experience. This chapter proposes a new system that has the potential to disrupt the current traditional event ticketing model by utilizing the inherent security of the blockchain—providing a more secure and robust ticketing infrastructure to end users which in turn increases consumer trust, confidence, and experience.

References

1. R. Couto, J. Leal, P.M. Costa, T. Galvão, Exploring ticketing approaches using mobile technologies: QR codes, NFC and BLE, in *2015 IEEE 18th International Conference on Intelligent Transportation Systems* (2015), pp. 7–12
2. ActionFraud. Ticket to nowhere: don't let ticket fraudsters take off with your cash (2022). https://www.actionfraud.police.uk/alert/ticket-to-nowhere-dont-let-ticket-fraudsters-take-off-with-your-cash
3. Vikki Davies. Ticket fraud reports double since the start of 2022 (2022). https://cybermagazine.com/articles/ticket-fraud-reports-doble-since-the-start-of-2022
4. S.-C. Cha, W.-C. Peng, T.-Y. Hsu, C.-L. Chang, S.-W. Li, A blockchain-based privacy preserving ticketing service, in *2018 IEEE 7th Global Conference on Consumer Electronics (GCCE)* (2018), pp. 585–587
5. M. Iansiti, K.R. Lakhani, et al. The truth about blockchain. Harvard Bus. Rev. **95**(1), 118–127 (2017)
6. S. Nakamoto, Bitcoin: A peer-to-peer electronic cash system (2008). http://www.bitcoin.org/bitcoin.pdf
7. U. Bodkhe, S. Tanwar, K. Parekh, P. Khanpara, S. Tyagi, N. Kumar, M. Alazab, Blockchain for industry 4.0: a comprehensive review. IEEE Access **8**, 79764–79800 (2020)
8. R. Zhang, R. Xue, L. Liu, Security and privacy on blockchain. ACM Comput. Surv. **52**(3), 1–34 (2019)
9. K. Wüst, A. Gervais, Do you need a blockchain?, in *2018 Crypto Valley Conference on Blockchain Technology (CVCBT)* (IEEE, Piscataway, 2018), pp. 45–54
10. D. Mingxiao, M. Xiaofeng, Z. Zhe, W. Xiangwei, C. Qijun, A review on consensus algorithm of blockchain, in *2017 IEEE International Conference on Systems, Man, and Cybernetics (SMC)* (2017), pp. 2567–2572
11. K.E.C. Levy, Book-smart, not street-smart: blockchain-based smart contracts and the social workings of law. Engaging Sci. Technol. Soc. **3**, 1–15 (2017)
12. W. Metcalfe et al., Ethereum, smart contracts, dApps. Blockchain and Crypt Currency **77**, 77–93 (2020)
13. P. Zheng, Z. Zheng, J. Wu, H.-N. Dai, Xblock-eth: extracting and exploring blockchain data from Ethereum. IEEE Open J. Comput. Soc. **1**, 95–106 (2020)
14. F. Regner, A. Schweizer, N. Urbach, Utilizing non-fungible tokens for an event ticketing system, in *Blockchains and the Token Economy* (Springer, Berlin, 2022), pp. 315–343
15. F. Regner, N. Urbach, A. Schweizer, NFTs in practice–non-fungible tokens as core component of a blockchain-based event ticketing application, in 40th International Conference on Information Systems (ICIS) (2019)
16. B. Tackmann, Secure event tickets on a blockchain, in *Data Privacy Management, Cryptocurrencies and Blockchain Technology* (Springer, Berlin, 2017), pp. 437–444
17. GET Protocol. https://www.get-protocol.io/
18. NFT-TiX. https://nft-tix.com/
19. Seatlab NFT. https://www.seatlabnft.com/
20. Tatum Key Management System (2023). https://tatum.io

21. Ramp (2023). https://ramp.network/
22. Alchemy API (2023). https://alchemy.com
23. Infura API (2023). https://infura.io
24. Hardhat Ethereum Development Environment (2023). https://hardhat.org
25. H. Chen, M. Pendleton, L. Njilla, S. Xu, A survey on Ethereum systems security: vulnerabilities, attacks, and defenses. ACM Comput. Surv. **53**(3), 1–43 (2020)
26. X. Li, Y. Xue, A survey on web application security. Nashville, TN USA **25**(5), 1–14 (2011)
27. I. Basharat, F. Azam, A.W. Muzaffar, Database security and encryption: a survey study. Int. J. Comput. Appl. **47**(12), 28–34 (2012)
28. D. Dolev, A. Yao, On the security of public key protocols. IEEE Trans. Inform. Theory **29**(2), 198–208 (1983)
29. T. Durieux, J.F. Ferreira, R. Abreu, P. Cruz, Empirical review of automated analysis tools on 47,587 Ethereum smart contracts, in *ICSE '20: 42nd International Conference on Software Engineering*, ed. by G. Rothermel, D.-H. Bae (ACM, New York, 2020), pp. 530–541

Part IV
Ontologies and Telecommunications Applications in Blockchain Technology

Auto-generation of Blockchain-Based Distributed Applications Using Ontologies

Muhammad Uzair Qureshi, Damien Graux, Fabrizio Orlandi, and Declan O'Sullivan

Abstract Blockchain has already promoted itself to solving business issues within every major domain, from supply chain to financial institutions to the healthcare industry. This is marked as the transition to Blockchain 2.0. However, this mass migration of industries cannot yet be a reality due to the limitations in standards and expertise of smart contracts within the various domains and the concern of the legal validity of smart contracts. Therefore, it is necessary to standardize concepts of smart contracts within blockchain frameworks in relation to legal agreements and to provide a direct mapping of agreements to code. This would allow for standardization and reuse of smart contracts across domains and make them legally enforceable. We target the R3 Corda blockchain framework and propose a novel Ontology, CordaO, that can be used to model Corda Smart Contracts (CorDapps). We also develop a tool, CordaOntoG, that auto-generates the relevant state, contract, and flow code in Java that can be deployed and run on a Corda network. The ontology and code generator are then evaluated with elementary domain-specific agreements such as clinical trial patient registration, car rental, and invoices.

Keywords Blockchain · Ontology · Corda · Semantic Web · Smart Contract · Code Generation

1 Introduction

Blockchain over the years has aimed to provide a secure distributed ledger for all parties to transact on. With the advent of this disruptive technology, businesses have seen the potential in increased productivity for their business processes with other parties in their network. Namely, with permissioned blockchain frameworks such as R3 Corda, IBM Hyperledger, and Quorum. Moreover, "smart contracts" on the blockchain increase the autonomy of the network by self-executing recurrent flows

M. U. Qureshi · D. Graux (✉) · F. Orlandi · D. O'Sullivan
Trinity College Dublin, Dublin, Ireland
e-mail: mqureshi@tcd.ie; grauxd@tcd.ie

N. El Madhoun et al. (eds.), *Blockchain and Smart-Contract Technologies for Innovative Applications*, https://doi.org/10.1007/978-3-031-50028-2_8

and validating them against predefined business rules. This technology helps us move closer to the vision of autonomous industries.

Blockchain has already promoted itself to solving business issues within every major domain from supply chain to financial institutions to the healthcare industry. This is marked as the transition to Blockchain 2.0. However, this mass migration of industries cannot be a reality yet due to the limitations in standards and expertise of smart contracts within the various domains and the ambiguity of legally binding smart contracts.

Therefore, it is necessary to standardize concepts of smart contracts within blockchain frameworks in relation to legal agreements and to provide a direct mapping to code that can be deployed and run on the network and be enforceable in court. This standardization can be related to a semantic layer between paper contracts and the smart contracts.

Within this chapter, we explore the landscape of blockchain frameworks and their respective representation of smart contracts that model business processes. We also explore their approach to increasing legal validity of the code that runs on their network. Furthermore, the chapter explores literature on approaches taken to provide semantics of blockchain and frameworks to generate code for blockchain using Ontologies. We also provide an introduction to ontologies, code generation tools, and key concepts of a Corda distributed application that are used. Finally, we develop an ontology to represent key concepts and their relationships in Corda Distributed Applications and provide details into the implementation of the tool that auto-generates code from triple statements. The ontology and tool are then evaluated with elementary domain-specific agreements such as patient registration, car rental and invoice issuance, and settlement.

Key Results—We were able to road map the process of building a semantic medium between smart contract code and natural-language legal agreements. The amalgamation of the fields of blockchain, law, and semantic web pose a new paradigm for the future of autonomous industries. There is still much research to be done with collaboration from all fields to move toward standards for code as law and law as code. Moreover, a better understanding of smart contracts in different blockchain frameworks has been gained for purposes of abstraction to semantic layers. Domain-specific knowledge of the smart contract platform is necessary to be able to build new code generation tools. Hopefully in future a hybrid ontology could morph agreements into any choice of smart contract.

Contributions—We have developed a novel and extensible Ontology that can model Corda smart contracts and business agreements as a knowledge graph, alongside a code generator that can query the agreement representation to generate smart contracts that run on the R3 Corda platform.

The rest of the chapter is organized as follows. Section 2 provides background to an overview of current block chain frameworks and their how they view smart contracts. In Sect. 3, we discuss the literature review, and we lay out the work of the relevant authors in the field of smart contract templates and legal smart contracts and provide critique, if any. Section 4 presents the Methodology; we lay out the development process of the Ontology and the Auto-generation Tool. Section 5 present three case studies using the program. Finally, Sect. 6 briefly provides future work, challenges, and final remarks.

2 Background

2.1 Blockchain Frameworks

The term "blockchain" originated from the Bitcoin protocol and network published by the anonymous Satoshi Nakamoto in 2008 [23]. Since then, blockchain technology has been morphed and integrated into several other platforms and applications. Blockchain now generally refers to the varying nascent distributed ledger technologies that evolved from Bitcoin. Therefore, distributed ledgers and blockchain may be used interchangeably throughout the paper.

2.1.1 Bitcoin

Digital artifacts are a staple of our technological revolution. However, they can be easily copied, presenting a concern in the representation of valuable assets such as money. Typically, third parties, such as banks, are the trusted ledger keeps and prevent "double spending" [11]. "Double spending" is the risk that an asset can be split twice. In banks, this is avoided by lengthy reconciliation and settlement processes between different banks' ledgers. When two parties transact, such as a customer and a shop, we involve a trusted third party for the settlement process of our assets. This is before bitcoin presented a unique consensus protocol to solve this double spending problem.

At a high level, bitcoin replaces a single ledger keeper with a network of ledger keepers. Every node on the network holds a copy of the ledger and helps in reaching consensus about proposed transactions that update the ledger state. Bitcoin demonstrated that mutually distrusting parties can reach a consensus on a transaction based on a simple protocol. However, bitcoins use-case limits the use of consensus on shared data to simply moving funds, but the possibilities are numerous.

Proof-of-Work In Bitcoins blockchain, the block of transactions are appended using hashes of previous blocks as inputs of subsequent blocks. These transactions can contain several inputs and outputs.

To append a valid block to the chain, Bitcoin uses a proof-of-work algorithm that also helps in transaction ordering. A valid block is a well-ordered set of transactions, contains a hash of the previous block, and contains a "nonce," which is produced from proof-of-work. Nodes have to compete against each other to add these transactions to the ledger and get rewarded. Proof-of-work is necessary in Bitcoins platform as it thwarts various forms of attack. This process implies a considerable commitment of computing resources to gain incentive (Bitcoin reward) and acts as a substitute for accountability.

2.1.2 Ethereum

In 2014, Vitalik Buterin published a paper titled "A Next Generation Smart Contract & Decentralised Application Platform" [5], which birthed the second wave of blockchain technology and the underlying Ethereum blockchain protocol, the public network was running by 2015. Beyond a distributed ledger like Bitcoin, Ethereum proposed a new paradigm of a distributed computation platform. This allowed for distributed code execution implemented on the Ethereum Virtual Machine (EVM). The EVM is Turing-complete and enables the deployment of smart contracts via transactions.

Gas Ethereum introduced "gas" as a fee for computation, with every block having a gas limit. This limits the computation executed per block. The "gas" price is then converted into Ether spent, used to provide incentives to the proof-of-work protocol.

The original intent of Blockchain was to provide Public Networks and still remains its most powerful use. It provides ease of accessibility, as you only require the client software and an Internet connection (Bitcoin and Ethereum networks). There exists no hierarchy, since all nodes are equal. There will always be a basis for incentive based on the consensus algorithms to prevent attacks.

2.1.3 Issues with Public Blockchains

Bitcoin and Ethereum were deployed as Public Blockchains. Their aim is to provide a single global network for all members of society. It is a compelling vision and accounts for the collaboration of great minds to advance distributed ledger technology. However, the vision is not fit for enterprise needs. These public blockchains use proof-of-work, and studies have presented proof-of-work's issues with scalability along with its vulnerability. Rational nodes can collude to obtain a revenue share larger than their fair share [10]. Second, as all data are shared with all parties and in their ledgers, it poses privacy concerns for enterprises with valuable data as it exposes a global attacker model. It is also to be noted that malicious actors will always find ways to exploit the network for their needs. There have been several exploitation's targeting Ethereum smart contracts [2]. A famous exploitation being of a prominent smart contract in Ethereum, which left the exploiter with 60M dollars. It was hard to flag it as illegal as there was no non-code specification of what the program was meant to do; however, the funds were revoked after a harshly debated revision of the blockchain. Therefore, this vision of removing third parties completely seems utopian as of yet and constantly requires a third party, such as the maintainers of the blockchain, to resolve disputes as they see fit.

2.2 *Permissioned Blockchains*

Due to the powerful consensus capable by blockchain on shared data, soon, several enterprise-targeting blockchain platforms were created. These enterprise blockchains focused on building collaborative networks between parties with pre-built trust and known identities. These platforms are referred to as permissioned blockchains as they require permissions to join the network to maintain privacy and security to standard within the distributed ledger. The main permissioned distributed ledgers are Quorum, Hyperledger Fabric, and Corda.

2.2.1 Quorum

Quorum [22] is a fork of the Ethereum blockchain platform backed by JP Morgan and Microsoft. It is an essentially private Ethereum network for enterprise with flexible choices in consensus algorithms and privacy options.

2.2.2 Hyperledger Fabric

With blockchain evolving over the years, the Hyperledger foundation was set up in late 2015. It is an open-source consortium for developing business applications hosted by the Linux foundation. Hyperledger Fabric [1] is a permissioned distributed ledger platform that allows modularity for different industry use cases. Fabric also supports private transactions and confidential contracts.

2.2.3 Corda

In early 2016, R3 announced its working on a distributed ledger, Corda [17]. Corda focused from the beginning on heavily regulated industries, namely, the financial institution. Corda was designed and developed to record and automate legal agreements between identifiable parties. Its explicit purpose is to record, manage, synchronize, and enforce business agreements between trading partners. It takes a unique approach to data distribution and transaction semantics while emphasizing features of distributed ledgers attractive to enterprises. Namely, reliable execution of contracts in an automatable and legally enforceable fashion. Corda Distributed Applications (CorDapps) are applications that run on the Corda Blockchain. CorDapps closely model business processes with legal contracts in mind. It keeps close to legal contracts by implementing collections of approvals and signatures between parties involved for every transaction.

2.3 Smart Contracts

Ethereum was first to introduce smart contracts on their distributed ledger.[1] A smart contract in the EVM is an autonomous agent consisting of an internal account that encodes agreement logic. These contracts execute deterministic code, with the correct execution being part of the transaction validation process. Smart contracts in Ethereum are written in Solidity, a purpose-built high-level programming language.

Quorum, being a fork of Ethereum, supports both public and private contracts. Specific private contracts can only be executed by the specified nodes in the network.

Hyperledger Fabric uses the term "Chaincode" for their smart contracts, and they are used to model business logic in their applications. Chaincode is able to be written in Node/Go/Java.

In Corda, smart contracts can include legal prose, defining what a contract is meant to do. A @LegalProseReference annotation is provided, which provides the URL/URI to a specification document. It is up to parties in the network to add legal weight to the code that automates this process. This is necessary for heavily regulated industries. This form of smart contract is referred to as a Ricardian Contract [15] and requires signatures from parties involved to verify proposed transaction. Smart contracts in Corda can be developed in any programming language that targets the JVM (Kotlin and Java).

The following figure from "Comparison of Ethereum, Hyperledger Fabric and Corda" [34] provides a summary of the characteristics of the different Blockchain/Distributed Ledger platforms (Fig. 1).

By exploring the blockchain platforms, we explore which is best suited for modelling legal agreements as smart contracts. We conclude that R3 Corda provides the best basis for the roadmap of Legal Smart Contracts, due to its target toward regulated industries and foundation of combining legal prose and smart contract code.

2.4 Semantic Web

2.4.1 Semantic Web

Sir Time Berners-Lee proposed a distributed information management system that interlinked hypertext documents, called the World Wide Web. This provided a revolutionary platform for all to collaborate, learn, earn create, and dream. However, rather than solely being a Web of Documents, Sir Tim Berners-Lee hoped for a Web of Data. *"I have a dream for the Web [in which computers] become capable of*

[1] Smart contracts have been used in many applications, such as online game, see, e.g., [14, 31] for studies powered by Semantic Web technologies.

Characteristic	Ethereum	Hyperledger Fabric	R3 Corda
Description of platform	– Generic blockchain platform	– Modular blockchain platform	– Specialized distributed ledger platform for financial industry
Governance	– Ethereum developers	– Linux Foundation	– R3
Mode of operation	– Permissionless, public or private[4]	– Permissioned, private	– Permissioned, private
Consensus	– Mining based on proof-of-work (PoW) – Ledger level	– Broad understanding of consensus that allows multiple approaches – Transaction level	– Specific understanding of consensus (i.e., notary nodes) – Transaction level
Smart contracts	– Smart contract code (e.g., Solidity)	– Smart contract code (e.g., Go, Java)	– Smart contract code (e.g., Kotlin, Java) – Smart legal contract (legal prose)
Currency	– Ether – Tokens via smart contract	– None – Currency and tokens via chaincode	– None

Fig. 1 Comparison of Ethereum, Hyperledger Fabric, and Corda [34]

analyzing all the data on the Web – the content, links, and transactions between people and computers. A 'Semantic Web', which makes this possible, has yet to emerge, but when it does, the day-to-day mechanisms of trade, bureaucracy and our daily lives will be handled by machines talking to machines. The 'intelligent agents' people have touted for ages will finally materialize." [3] To make Web of Data a reality, it requires a standard format, reachable and manageable by Semantic Web tools. The Semantic Web is thus an extension of the Web and builds upon it to provide standards and tools for semantic interoperability on the Web. Semantic Web aims to:

1. Provide standards to promote common data formats for heterogeneous data integration, whereas the Web focused on interchange of human-oriented documents.
2. Define modelling languages and protocols to allow for interlinking of data, foundationally, RDF.

2.4.2 Semantic Web Architecture

The following components form the building blocks for the Semantic Web:

– URIs—providing unique identification of resources.
– RDF (Resource Description Framework)—abstract model to express knowledge by combining simple statements about resources.

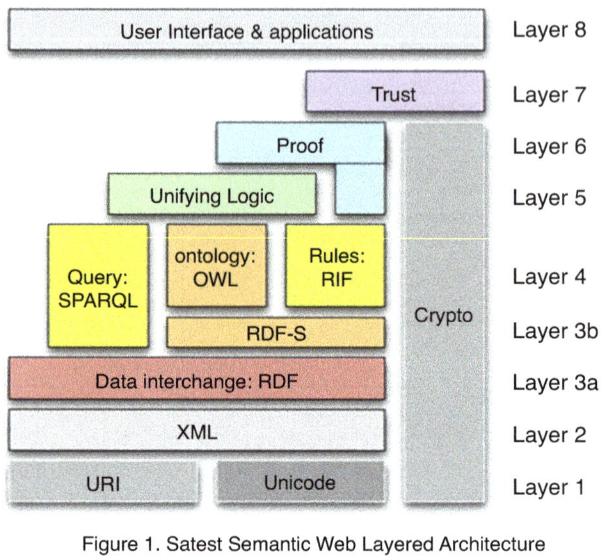

Fig. 2 Semantic Web architecture. Source: [9]

- SPARQL—query language and a set of related standards for querying RDF graphs.
- Ontologies—describe formal, shared conceptualizations of a domain of interest [16]. Ontologies establish common semantics between applications on the Semantic Web.
- RDF Schema—Simple language for defining RDF Ontologies that allow developers to define classes, properties, and relations.
- Web Ontology Language—ontology language built on top of RDFs and extends it with more complex constructs, such as restrictions and cardinality.

Finally, a logic layer consists of rules that enable inference and define semantics of RDF data and ontologies (Fig. 2).

2.4.3 RDF

RDF is the standard format for representing labelled graph data on the Semantic Web. RDF allows expression of knowledge with triple simple statements consisting of a subject (thing), predicate (property), and object (value).

"In RDF, a document makes assertions that particular things (people, Web pages or whatever) have properties (such as 'is a sister of,' 'is the author of') with certain values (another person, another Web page). This structure turns out to be a natural way to describe the vast majority of the data processed by machines." [4]

Fig. 3 Sample Turtle Syntax

```
@base <http://example.org/> .
@prefix rdf: <http://www.w3.org/1999/02/22-rdf-syntax-ns#> .
@prefix rdfs: <http://www.w3.org/2000/01/rdf-schema#> .
@prefix foaf: <http://xmlns.com/foaf/0.1/> .
@prefix rel: <http://www.perceive.net/schemas/relationship/> .

:green-goblin rel:enemyOf :spiderman ;
    a foaf:Person ;
    foaf:name "Green Goblin" .

:spiderman rel:enemyOf :green-goblin ;
    a foaf:Person ;
    foaf:name "Spiderman", "Человек-паук"@ru .
```

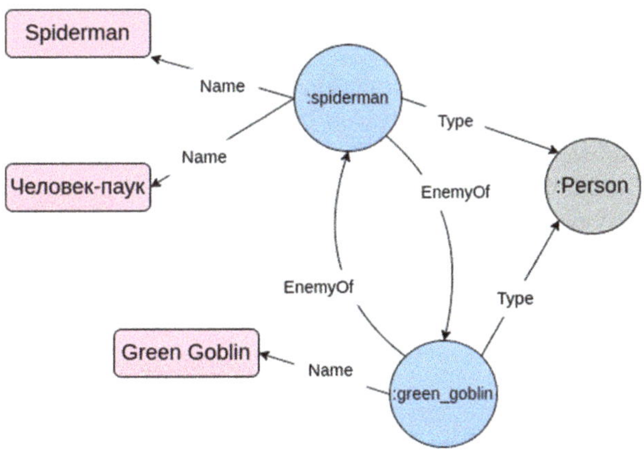

Fig. 4 Visualization of Sample Spiderman Turtle Statements from Fig. 3

The use of URIs allows for merging of statements allowing a combination of knowledge graphs to present collective information. This makes RDF useful for integrating information from multiple sources, be it from the Web or enterprise information systems.

In order to exchange RDF information, RDF can be serialized with N-triples syntax or Turtle syntax. We use the Turtle syntax, since it is compact and in natural text form.

The following example represents statements of the Spiderman universe in Turtle Syntax. The statements can be read as: "Green Goblin" is a Person and he is the enemy of spiderman. "Spiderman" is a Person and he is the enemy of Green Goblin (Figs. 3 and 4).

Vocabularies Vocabularies define the concepts and relationships used to describe and represent an area of concern. In practice, RDF is written in combination with vocabularies and ontologies. Main Vocabularies are RDF, RDFS, and Web Ontology Language (OWL). The URI prior to ":" prefix defines the vocabulary of the resource.

2.4.4 SPARQL

SPARQL is the query language standardized by the W3C for RDF. SPARQL can express queries across data sources, based on matching required and optional RDF graph patterns, the syntax is similar to SQL, but it queries graphs rather than relational databases. Although SPARQL is "data-oriented" in that it is only able to query information in models, providing no inference, many tools such as Jena [20] and Stardog provide the impression of certain triples existing with use of OWL Reasoning (Fig. 5).

2.4.5 Ontologies

Decisions we make (intelligent systems) are based on knowledge we understand the context of the situation/environment. Humans require common understanding on basic concepts to understand each other, that is, we require the same semantics. Similarly, for computers to exchange information and reason about it, they need a syntactic representation that all software agents can process and have a shared conceptualization of, within a domain. Semantic web formalizes knowledge in a way that improves decision-making and can form basis for autonomous reasoning.

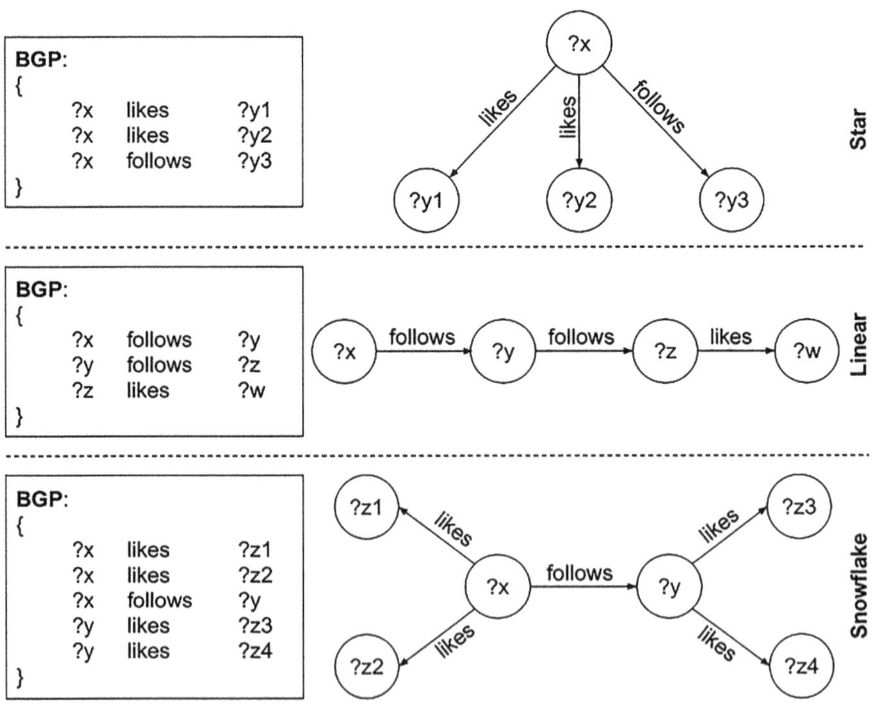

Fig. 5 SPARQL queries match graph patterns in dataset from [28]

Such shared conceptualizations are called Ontologies, and they are widely used in Computer Science fields such as Information Management and Natural-Language Processing.

"An ontology is a formal, explicit specification of a shared conceptualization. Conceptualization refers to an abstract model of some phenomenon in the world by having identified the relevant concepts of that phenomenon. Explicit means that the type of concepts used, and the constraints on their use are explicitly defined. Formal refers to the fact that the ontology should be machine-readable. Shared reflects the notion that an ontology captures consensual knowledge, that is, it is not private of some individual, but accepted by a group." [30]

Ontologies are modelled as a set of concepts (classes) and a set of properties that link the concepts. Additional axioms can be defined to further constrain and formalize the ontology.

Classes or concepts represent general qualities and properties of a group of individuals. Instances of classes can be defined using `rdf:Class` as their type and further defined with instances of `rdf:subClassOf` and more.

Properties represent relationships between subject and object. The `rdf:domain` and `rdfs:range` define the Objects that pertain to the property's subject and object instances. There are several vocabularies and terms that can be used to define concepts in Ontologies.

The following figure represents a simple Ontology and Instances using the Ontology to define a University (Fig. 6).

Thus, an Ontology can be defined being a:

- Knowledge model defining a set of concepts and relationships between those concepts within a specific domain
- Model that supports automated reasoning and inference of data using logical rules
- Provides knowledge sharing and reuse among people and software agents

Web Ontology Language (OWL) OWL provides an extended vocabulary for describing ontologies with classes, properties, and more specific relations between

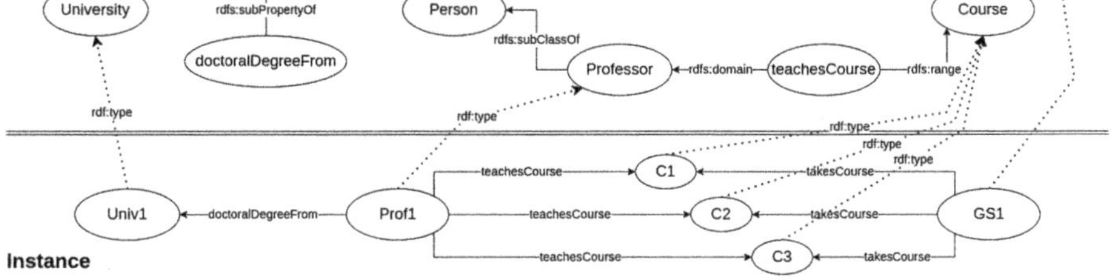

Fig. 6 Simple University Ontology. Source: [25]

them (e.g., range, domain, disjoint classes, etc.). OWL facilitates greater machine interpretability of Web Content.

2.5 JavaParser

Code Generation is a powerful tool for programmers who write repetitive boilerplate code for applications and services based on a specification, and these can be company code practices or a design methodology. For example, Data Access Object Classes from database schema files.

Source code in many programming languages can be represented as a tree. As with most trees, they have a starting point, and then branches form independently from one another as methods, statements, conditions, expressions, and so on. From a single class declaration, we can reach many nodes, representing fields, methods, and constructors of the class. An important distinction is that humans and compilers can understand that a specific variable reference or method call relates to another part of the source code, a syntax tree does not, but can be with use of another library, JavaSymbolSolver. To model the Java language concepts, Javaparser has tried to remain true to the official grammar specification[2] for the naming of classes.

The following is an example extracted from JavaParser [29].

```
package com.github.javaparser;

import java.time.LocalDateTime;

public class TimePrinter {
    public static void main(String args[]){
        System.out.print(LocalDateTime.now());
    }
}
```

Figure 7 demonstrates the three children of the source code rooting from the Compilation Unit, the PackageDecleration, ImportDeclerations, and ClassDecleration. The method declaration can be expanded further to show its children in Fig. 8. Now we see the name, return type, and parameters for the method with a block statement that can again be further expanded to show its children (Fig. 9).

This shows how complex these trees can get from this simple example. These abstract syntax trees are crucial for allowing automated generation of code as the recurrent nodes can be manipulated and replaced with another node or another subtree.

JavaParser [29] is a library that allows for interaction with Java Source Code as a Java Object Representation, and this Object Representation is referred to as an Abstract Syntax Tree. The library provides functionalities such as:

- Representation of Abstract Syntax Tree of code.
- Manipulate underlying structure of source code.
- Build own Code Generating Software.

[2] https://docs.oracle.com/javase/specs/jls/se7/html/jls-18.html.

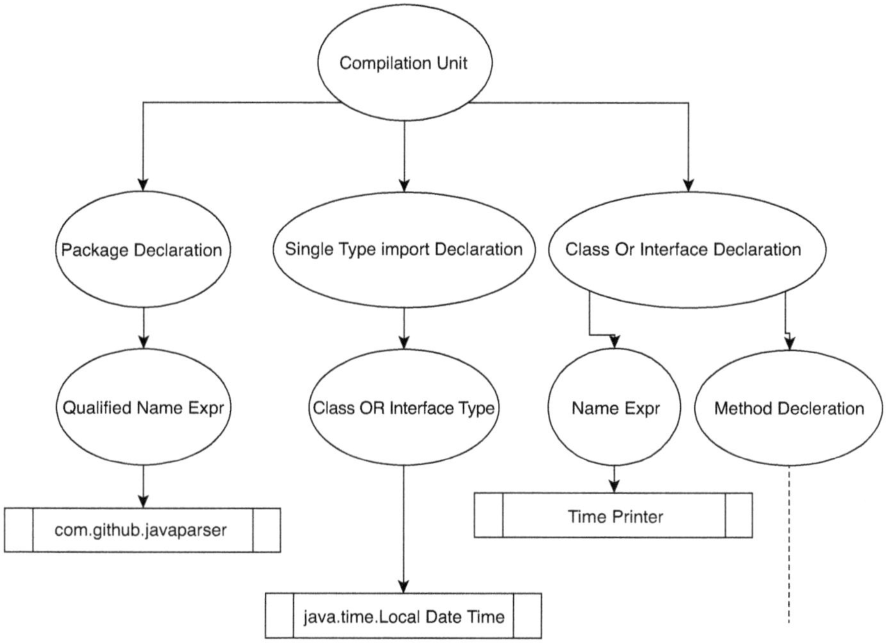

Fig. 7 Compilation Unit

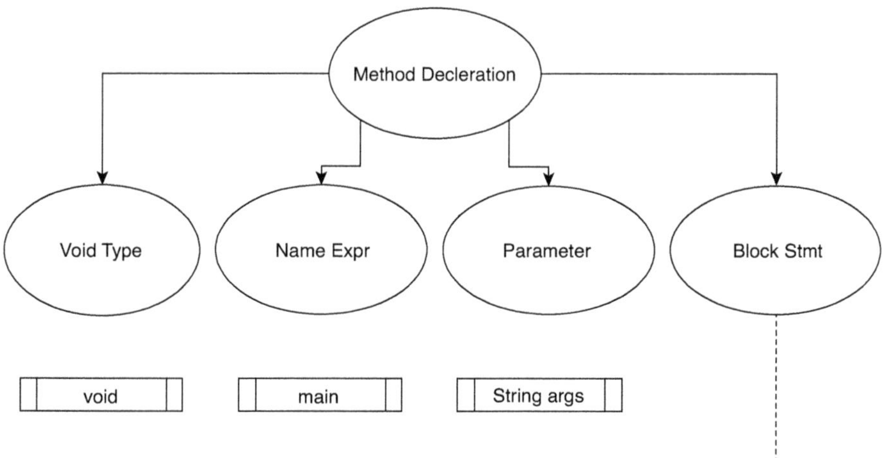

Fig. 8 Method declaration of class declaration

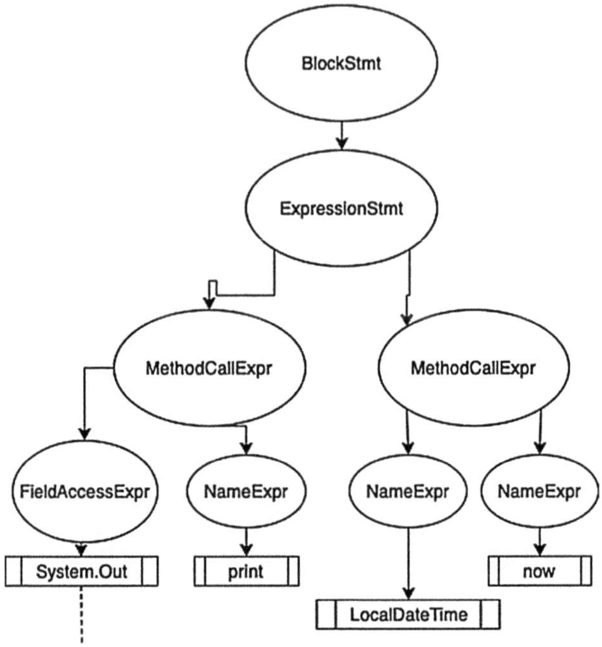

Fig. 9 Block statement of method declaration

3 Literature Review and Motivation

Observing from first principles of computer science, contracts can be viewed as programs and Law as an operating system. The organizational theorist Arthur Stinchcombe once wrote that contracts are merely organizations in miniature, and by extension all organizations are just complexes of contracts. Therefore, contract drafting and execution is a foundation of society.

The first digital morphing of contracts is named as smart legal contracts, which looked to digitize contracts into repositories and allow methods for document assembly. The new phase is termed as legal smart contracts. Legal smart contracts focus on the operational aspect of legal contracts, involving software agents on a shared ledger fulfilling obligations between parties.

Szabo [32] presented the idea of smart contracts in 1997, described as computerized transactions that execute the terms of a contract. Szabo [32] identified a vending machine as a form of smart contract, where an autonomous transfer of confectionery and funds occurs. The word "smart" is a relation to the automation of the process, and the term "contract" refers to its "tamper-proof" execution of prescribed terms.

Smart contracts are computer programs that verify and execute terms of an agreement upon occurrence of an event. Within, distributed ledgers, these events are the proposed transactions that aim to update state of the ledger, the smart contract thus validates the transaction in relation to a predefined agreement. Smart contracts are able to leverage the trust-less immutable nature of distributed ledgers to enable peer-to-peer, disinter-mediated agreements enforced automatically by code.

Distributed ledgers provide a simple framework for these contracts to operate efficiently and transfer digital property with the advantage of transparency with a synchronized ledger.

Hence, smart contracts present organizational benefits with increased efficiency, certainty, and reduction in costs, although their optimal success will come from their adaptability to existing legal frameworks regulating contracts across jurisdictions.

Currently, Blockchain is incapable of capturing nuance of contract language, and we need to provide specification of contract code, admissible to court for dispute resolutions.

Savelyev [27], Giancaspro [12], look toward exploring the legality of smart contracts, the concerns associated, and their future landscape. Giancaspro [12] concludes that when courts resolve disputes they need reference to terms of the contract, this will not be possible due to the unintelligibility of code to judges and lawyers. Reference to terms in legible linguistic form is necessary. Lawyers over the years may need to gain basic proficiency in coding to allow checking of clauses, and contractual mechanisms have been appropriately translated to relevant programming language and guide creation of smart contract code.

Hofman [19] proposes integrating a semantic legal layer, enhanced with jurisdiction specific legal ontologies, which could add the sufficient precision, flexibility and enforceability to allow smart contracts to act as full legal contracts.

The paper continues to define conditions that render a contract legally binding:

1. Arise as a result of offer and acceptance.
2. Include legally sufficient consideration.
3. Be between parties with intent to contract.
4. Be between parties with capacity to contract.
5. Comply with formal legal requirements (e.g., State of Frauds).
6. Be legal (e.g., cannot sell organs).
7. Not be void (e.g., due to non-disclosure of material facts by one party).

These are conditions that are applied at the beginning of the Smart Contract Lifecycle Management, and, once founded, the generated code can enforce itself on the Blockchain, legally.

Ugarte H. [33] proposes a vision of a Semantic Blockchain being the foundation of Web3.0 in efforts of fulfilling T. Berner Lee's vision of Web Linked Data. Ugarte, in one of his definitions, describes Semantic Blockchain as "*... a distributed database that maintains a continuously growing list of standardized data records, using Resource Description Framework (RDF), hardened against tampering and revision.*" Hofman [19] draws similarity to this definition to current paper registries that hold legal contracts and records. They propose using linked data in preserving the archival bond of the smart contracts executed on the distributed ledger. Recently, [13] described a system using blockchain architecture in combination with a peer-to-peer solution to manage and store Semantic Web graphs.

If smart contracts are written entirely in code, courts will face challenges applying contract law due to the absence of natural language specification. However, supporting smart contracts with a semantic legal layer, where parties define assets and terms as choice of laws, and conditions supported by robust linked data representing contract formation, would enable legal smart contracts.

To enable to semantics of the blockchain platform, they propose using Ontologies. The Ontology aids in establishing context, functional and procedural of the smart contract. Ontologies can also be layered to enrich the expression of context. A further benefit in using Ontologies is they can be extended and swapped per use case. This framework would establish archival bonds, which links the entry of data on the Distributed Ledger to the procedural action to establish records identity.

This is similar to how the Corda platform stores transactions. Each recorded transaction holds hash of the smart contract code is verified by. Merging this current framework with the papers proposal would require the simple addition of the semantic specification of the legal contract to link to the smart contract code, which finally links to the recorded entry. Upon dispute of a record on the ledger, the relevant smart contract and legal specification can be brought up for resolution.

Barclays presented their work on Smart Contract Templates in collaboration with UCL [7] and deployed their first prototype on the Corda platform in 2016 for a derivatives agreement [26]. Clack et al. [8] propose smart contract templates that support legally enforceable smart contracts with operational parameters that connect the code to the agreement. The code is manually developed with a specific legal documentation in mind and presented as standardized templates with configurable parameters. They focus on derivative agreement that is a financial contract between banks. They further explore the design landscape to increase use of common standardized code and long-term research.

Smart Contract Templates are legal document templates created by standards bodies and used by counter parties negotiating such contracts. These templates are defined with the following criteria:

1. Automatable—executable by machine, some parts may require human input
2. Enforceable—legal enforcement of rights and obligations
3. Structure—Ricardian contract [15] (legal prose, parameters, code)
4. Future Evolution—increasing sophistication of parameters (higher-order parameters, domain-specific languages)

Smart Contract Templates is inspired by the Ricardian Contract [15], which is a triple of a prose, parameters, and code. The parameters provide the final operational details of the code. The use of parameters would increase the standardization of code. In future, there may be an increasing use formal language, with legal prose replaced with arithmetic and logical expressions. This would lead to reduced ambiguity in legal prose and code.

Another interesting article by IBM [6], from which we derived great inspiration for this chapter, presents a novel framework for auto-generating smart contract code by translating semantic rules as constraints to code. They demonstrate the functionality successfully by applying it to the domain of clinical trials and car rentals. Their paper presents the power of using semantic web technologies to directly generate code that executes on the blockchain. However, they used predefined code templates for each domain example, anticipating functions such as createPatient, and already containing constraints that were set to empty strings. Therefore, the templates were not fully auto-generated. When a JSON array of the constraints was created from the Semantic Rules, it would manipulate the Abstract Syntax Tree (AST) of the code to fill the empty strings. Furthermore, they designed Domain-Specific Ontologies for each example and Constraints using Semantic Web Rule Language SWRL that require high expertise.

In line with the above work on exploring legal smart contracts, it is clear that specification of legal terms and assets is necessary to provide smart contracts with legal authority, since code is obfuscated with ambiguity in meaning known only to the coder. Providing the specification and code in conjunction consequently increases the adoption of smart contracts and their standardization across multiple domains. Within Corda, we can add legal prose as a specification to the deployed smart contract. However, we need to establish a direct mapping between the legal prose and the smart contract code; otherwise, there is a duplication of efforts and the presence of ambiguity. For this reason, we introduce a semantic medium that is used to represent concepts of distributed applications (smart contracts) in Corda, referred to as CorDapps. This medium acts as the source language that can represent the operational aspects of the smart contract code and the non-operational aspects of legal prose, similar to the source language described in [8]. The Ontology also acts as the formal language to describe the assets, actions, and constraints to model the agreement.

By providing this semantic medium for the legal prose and smart contract code, we enable use of Semantic Web Technologies, which can perform validation and formal verification of code and obligations defined prior to deployment of agreement to further automate the contract lifecycle.

To create this semantic medium, we model a novel Ontology, CordaO, which represents key concepts of a Corda Distributed Application. This Ontology can then be used by lawyers and engineers to jointly model legal smart contracts in a human and machine-readable syntax. The knowledge graph is defined, using the Ontology, to model an agreement as a collection of triple statements.

We also present an auto-generation tool, CordaOntoG, which extracts parameters and constraints from the triple statements to generate the CordDapp Code that executes the agreement between parties on the Corda Blockchain.

Rather than writing boilerplate code, disjoint from the legal prose, the operational actions are defined in the semantic medium, providing a modular representation of the agreement. This representation of the agreement can be evolved as per use case, standardized and extended with further Ontologies such as LKIF-Core [18], to integrate further legal semantics.

4 Methodology

To revisit, we aim to build a novel Ontology for defining Corda Distributed Applications to:

- Share a common understanding of the structure of information among people and software agents (auto-generation tool, reasoning agents).
- Enable reuse of domain knowledge (medium for lawyers and engineers to build smart contracts with legal specification).
- Made domain assumptions explicit (remove ambiguity in standalone coding of smart contracts).
- Analyze domain knowledge (enable reasoning over specification for validation).

Prior to designing the Ontology, we will go over the Key Concepts in CorDapps to be then able to extract the relevant concepts and their relationships into an Ontology. Then, to guide our design, we follow the methodology presented by Noy et al. [24]. Once our Ontology is designed, we define the auto-generation tools' pipeline. Throughout the process, to aid in the conception of the end goal, we follow the most common agreement example, issuing invoices.

4.1 Corda Key Concepts

Corda is a network made up of known peers that communicate point to point to reach agreements on updates of their ledgers.

A node on the network represents the peer. Each node maintains its own vault of states and CorDapps, same as copy of ledgers in Bitcoin and Ethereum, all peers of a shared fact share identical copies. The node abstracts away the complexities of messaging, storage, concurrency, disaster recovery, and so on from the peer client. So the client simply interacts with node through RPC (Remote Procedural Call) to execute relevant business flows.

Thus, the CorDapps form the main drivers of network transactions. CorDapps have 3 main components:

1. States—facts over which agreement is reached.
2. Contracts—guide evolution of state and define the business rules.
3. Flows—routine the node runs to propose transactions updating the ledger.

4.1.1 State

States are immutable objects representing shared facts on ledger known by one of more nodes. These facts can contain arbitrary data. Life cycle of a shared fact over time is represented by a state sequence. It evolves by creating a new version of the

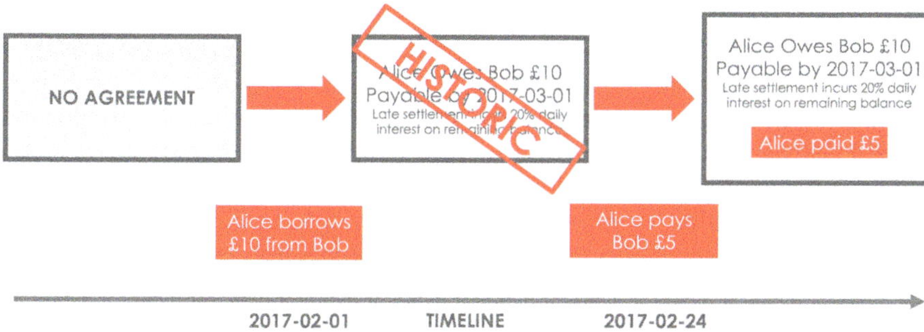

Fig. 10 States for an IOU state evolution on ledger

Fig. 11 Instance of a sample IOU state

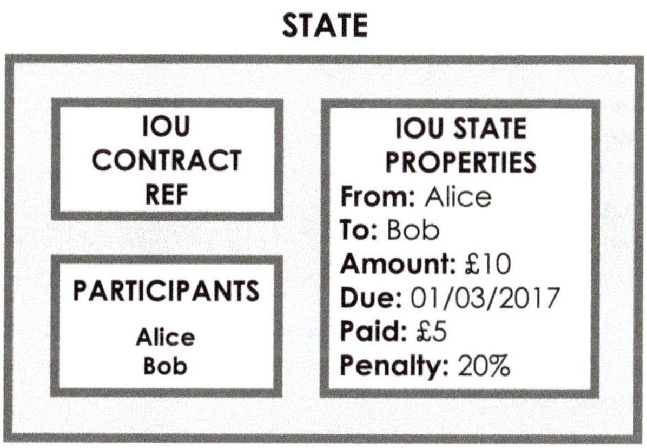

state and marking existing state as historic (consumed), and the unconsumed states present the current state of the ledger, Fig. 10.

The main characteristics of a state are:

- State properties
- Participants
- Reference to contract that governs its evolution

Figure 11 is the example of a simple invoice, loan, credit letter example. Sometimes termed as an I-Owe-You (IOU). Within this example, we see state instantiated with properties, Amount, Due, Paid, and Penalty, with participants Alice and Bob and a reference to the IOU Contract.

The behavior of states is further customized by sub-interfaces. The most common being LinearState and OwnableState.

LinearState models shared facts for which there is only one current version at any point in time. LinearState states evolve in a straight line by superseding themselves. However, OwnableState represents assets that can be freely split and merged over time. The Cash state is an example of this and is integrated within the Corda finance library.

We model all our state classes as Linearstate states since we aim to represent and manage single versions of “Paper Contracts” or “Information,” such as a Patient in a Clinical Trial, an Invoice, or a Car Rental.

As with most blockchain frameworks, they have an innate cryptocurrency that is used to drive transactions and reward miners. Although Corda does not follow proof-of-work to reward miners for computing, they do have an embedded token that can be used as digital currency, and this is the Cash State. The Corda network has an embedded Cash CorDapp within every node that handles transfer and issue flows.

We represent the integrated Cash State in Corda as an Ontology class to enable representation of monetary transactional exchange in agreements.

Transactions Transactions are built by Flows and are proposed to update the ledgers’ state to the desired state. The transaction is then verified by the relevant contract classes of the states involved in the transactions to decide validity of transaction. Notaries witness such transactions and ensure the states are never consumed more than once (no double spending).

The evolution of the states is overseen by the UTXO Model (Unspent Transaction Output Model), similar to Bitcoins model, which maintains immutability of ledger. Each transaction consumes zero or no input states on the ledger and outputs zero or more new output states and presents a single link in the states’ evolution.

Once built, the transaction is signed by parties who share the fact (Alice and Bob in the example above). The required signers only sign the transaction if following conditions hold:

- Transaction Validation—Transaction is contractually valid and signed by all required parties.
- Transaction Uniqueness—No other committed transaction that has consumed any of the inputs to our proposed transaction.

We could impose different rules of a valid transaction depending on the intent of transaction when changing a state. *Commands* allow us to indicate the transactions intent, which affects how we validate the transaction. Figure 12 shows the IOU being settled with a state change of IOU and of the Cash state. The main components of a transaction are:

- Commands
- Input states
- Output states

Additional components such as notary, attachments, and time windows can be added. But we default them for our use case to model agreement rather than infrastructure-level configuration.

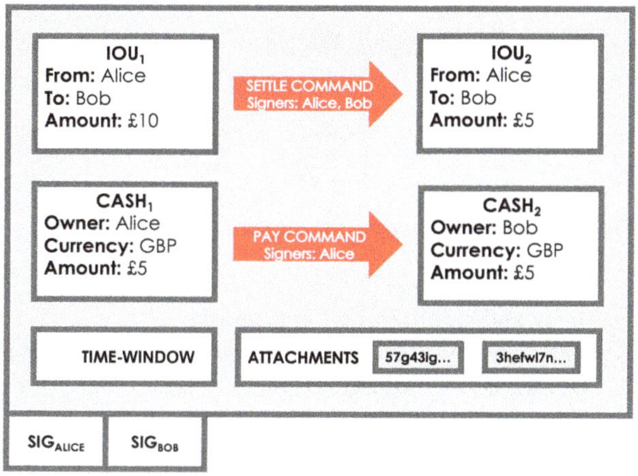

Fig. 12 Transaction proposal of settlement flow of IOU

4.1.2 Contract

Transaction validity is verified by the Contract of the CorDapp, to impose rules on the evolution of states over time. The contract takes the transaction as an input and states whether the transaction is considered valid based on the contract's rules. The combination of State and Contract forms the "Smart Contract" in Corda.

Looking at an IOU Issuance, we can pose simple constraints such as:

- Value must be non-negative.
- Lender and Borrower cannot be the same identity.
- Value must not exceed $150 (Borrow Limit).

If the transaction satisfies constraints, the Contract returns, otherwise throws an exception and rejects with the failed constraint.

4.1.3 Flow

Flows execute business logic by modelling the business processes that propose transactions and achieve specific ledger update. By installing new flows, we allow the node to handle to new business processes.

Flows consist of two classes, the Initiator and Responder, to manage transfer information between counter parties and achieve consensus. The transactions are contractually verified, signed off by nodes, and then sent to notary to check for uniqueness. Therefore, for recurrent agreements, flows automate common tasks by ensuring transactions of the agreement are contractually valid, signed off, and unique.

Taking the IOU Example, to orchestrate the issuance of an IOU, we:

- Build transaction proposal for a new IOU.

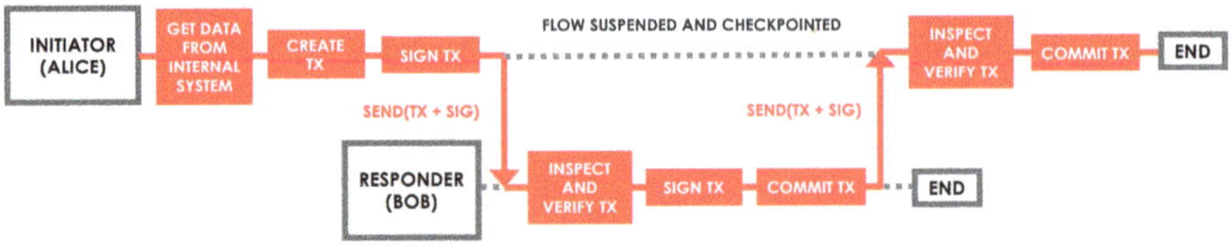

Fig. 13 Flow sequence between two parties

- Verify and sign transaction proposal.
- Send transaction to collect borrowers signature (responder flow).
- Once signed by required signers, send to notary for uniqueness.
- Once checked by notary, parties commit transaction to their vaults.

The flow can be visualized in Fig. 13. The main components are:

- Flow properties—parameters required to fulfill business logic
- Other party—party with whom we share the State (fact) with to sign and verify transaction
- Transaction—proposed transaction to achieve desired ledger state

4.2 *Ontology Development*

As mentioned, we develop the CordDapp Ontology following the methodology laid out by Noy et al. [24]. It is to be noted that there is no correct way to model a domain, and there are always viable alternatives. The best solution depends on the application we have in mind and the extensions that we anticipate. Ontology development is an iterative process and can be extended and configured as per use case. The concepts we model should be close to the logical objects and relationships in our domain of interest. The steps can be defined as follows:

1. Determine scope.
2. Consider reuse.
3. Enumerate terms.
4. Define classes.
5. Define properties.
6. Create instances.

4.2.1 Determine Scope

For this step, we answer the following questions:

Which domain should the Ontology cover? Our domain of interest is Distributed Applications that run on Corda as recurrent business processes.

What should the Ontology be used for? We aim to represent the hierarchy and relationships of an agreement as states, contracts, and flows. This representation then provides the necessary hierarchy of parameters to generate code that can execute on a Corda network. We should also be able to formally reason through and validate instances of the agreement graph to highlight inconsistencies and privacy issues.

What types of questions should be answered by the knowledge represented in the Ontology? (These are known as competence questions, and we summarize a few as follows.)

Which participants manage the evolution of the given state on the ledger?
Which properties of the state should be considered for modelling?
What contract does the state belong to?
What are the variable names of the state and their respective datatypes?
What commands are executable by the contract?
What are the constraints checked by the command?
What constraint does the condition correspond to?
What are the types and variable names in the expressed condition?
Is the type of the condition a binary or a temporal condition?
Are the variables in the condition input or output parameters?

Who will use and maintain this Ontology? Rather than writing standalone for generic recurrent business logic across an industry, standardized knowledge bases of the agreements can be designed and used by industries. These knowledge bases would be created by lawyers or engineers using the proposed Ontology and be exposed in a public regulated repository for use.

4.2.2 Reuse of Ontologies

Typically, we should reuse existing Ontologies to save cost and apply existing tools associated with them. We use existing vocabularies such as RDF and RDFs, but there were no existing smart contract Ontologies from our research. However, our Ontology shall be merged in future with existing legal Ontologies [18].

4.2.3 Enumerate Terms

In this step, we describe the main concepts and their properties, along with defining them:

- State—state name, state properties, property names and datatypes, reference to contract
- Contract—contract name, commands, constraints, constraints descriptions, condition variables, transaction inputs and outputs, size of inputs and outputs

- Flow—flow name/properties, flow property datatypes, transaction
- Transaction—input state, output state, new state, retrieved state, command, other party

4.2.4 Define Classes

The benefit of modelling a distributed application into an Ontology is the benefit of similarities in Object-Oriented Design. In a CorDapp, the main classes are State, Contract, and Flow. These classes define the collections of objects with similar properties. **Classes**[3] are:

- CorDapp—Corda Distributed Application consisting of state contract and flows.
- State—shared fact that exists on the ledger.
- Retrieved state—existing state retrieved from ledger to be used as input for new transaction.
- New state—create a new state to be used as a transaction output.
- State property—define state property name and datatype.
- Command—defines intent of transaction being proposed.
- Constraint—a business rule to validate transaction being proposed.

4.2.5 Define Properties

Define how the classes are related to each other:[4]

- amount—amount of cash we wish to transfer
- belongsTo—contract to which state belongs to
- commandName—name of command
- contractName—name of contract class
- datatype—datatype of property variable
- hasCommand—command of contract
- hasConstraint—constraint of command in contract class
- hasInputState—input state of transaction in flow
- hasTransaction—transaction proposed by flow
- otherParty—other party whose signature we require for transaction
- payee—the other party to whom we wish to transfer money to
- properties—list of properties of state
- propertyName—name of state property
- stateClass—state of new or retrieved state
- equals—binary operator—ensures x equals y. P(x,y)
- lessThan—binary operator—ensures x less than y. P(x,y)

[3] The class list is not exhaustive here.

[4] The property list is not exhaustive here.

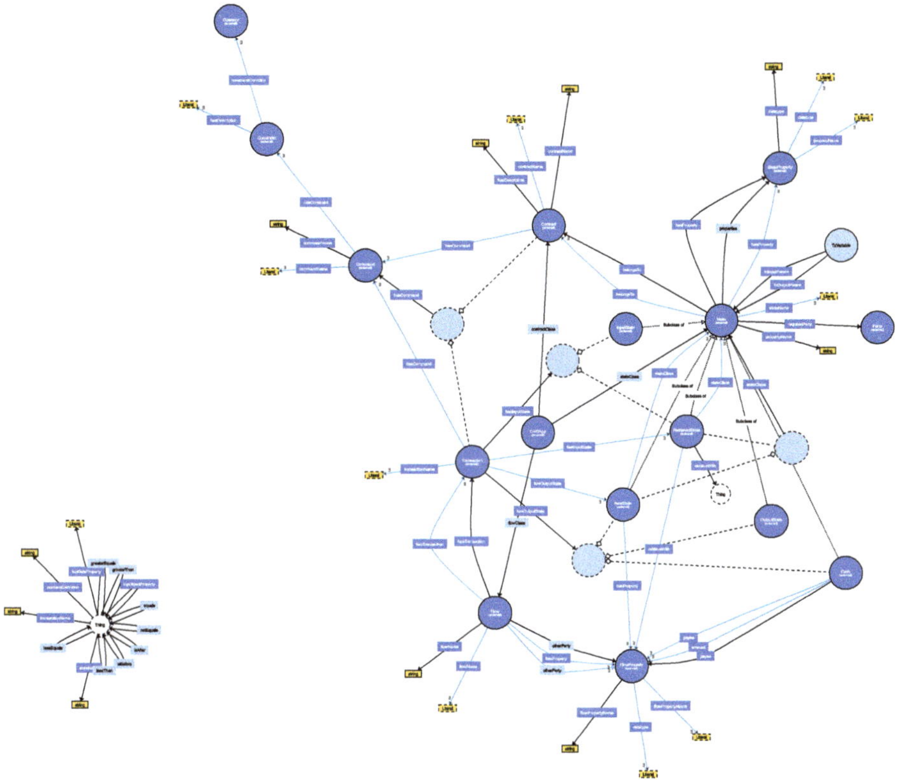

Fig. 14 Visualization of CordaO Ontology in WebOwl

4.2.6 Create the Instances

To create the instances, we used Web Protege 5.0 and visualized using WebOWL. The Ontology is displayed in Fig. 14.

4.2.7 RDF* to Annotate Constraints

Choudhury et al. [6] has implemented constraints in SWRL, a proposal presented in 2004 and still not part of the W3C standard. Although powerful, it requires high-level expertise to define conditions using SWRL[5] and more projects looked to model rules using SPIN[6] and even SP-ACT [21]. However, we wish to model simple static constraints based on the parameters of the incoming transactions' input and output

---

[5] https://www.w3.org/Submission/SWRL/.

[6] https://spinrdf.org/.

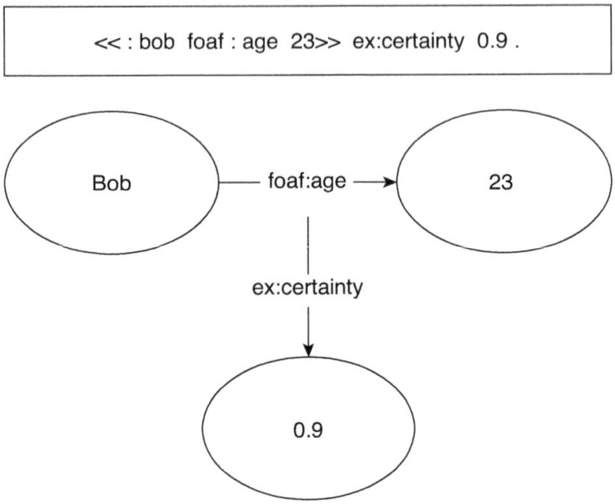

Fig. 15 Edge representation with RDF*

states. Therefore, we take advantage of a recent proposal to enable statement-level metadata in RDF* and SPARQL*,[7] and the proposal allows for a convenient way to annotate RDF triples and query such annotations in RDF. Such annotations are popular in other contemporary graph data models (BlazeGraph and AnzoGraph). With RDF*, any triple that represents metadata about another triple can contain this other triple as its subject or object. This can be visualized in Fig. 15: We want to represent certainty of the statement "Bobs age is 23" as 90%.

We represent constraints as simple triple statements and relate to them by edge properties (see Fig. 16). Therefore, rather than creating a knowledge base and separate SWRL rules, they can be defined in the knowledge base itself and queried to retrieve the variables in question and the condition operator. For example:

```
:constraint1 a :Constraint ;
    :hasDescription "Gender Must be Female."^^xsd:string .
<< :txoutsp1 :equals "Female" >> :belongsTo :constraint1 .
```

Listing 1 Constraints defined using edge triple statements and edge properties

4.3 *Auto-generation Tool*

Now that we have an Ontology that can model agreements as a collection of triple statements, to auto-generate the respective code we need to query the knowledge to get answers to our competency questions 4.2.1.

[7] https://www.w3.org/Data/events/data-ws-2019/index.html.

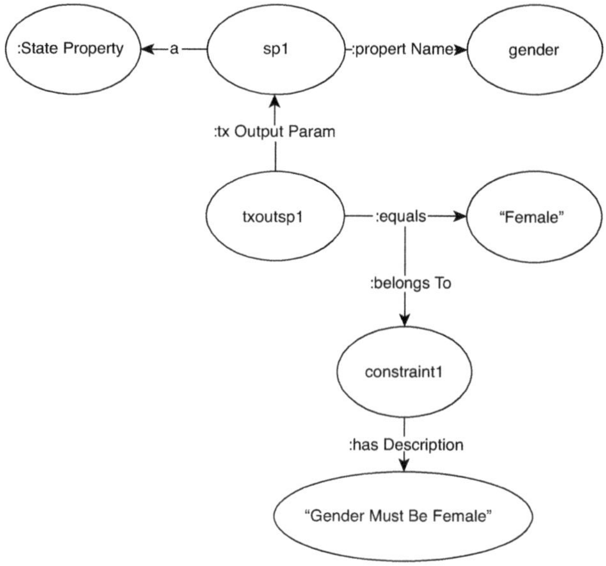

Fig. 16 Graph representation of contract constraint with RDF*

4.3.1 Architecture

The tool is sectioned into the following packages Fig. 17:

- Compilers—The compilers Package contains the Classes that query the knowledge base for parameters relating to given class being generated.
- QueryDB—Contains all the SPARQL* queries and connects to the Knowledge Base Database.
- Models—Models help in aligning the parameters with a programming data structure. For example, fields are stored in LinkedHashMaps of key(variable) and value(datatype) pairs, allowing for iteration to create their getters, setters, keys, hashCode and provide search in constant time O(1).

State Compilation See Fig. 18.

The state class is the foundation of the CorDapp. The initial step of all compilation classes is to fetch the relevant parameters for the Class and then pass them to the individual classes and methods needed to generate the final class, as shown in Fig. 19.

We define the abstract data model that would model the fact being shared between parties in the agreement in triple statements as follows:

```
# Defining State
:sp1 a :StateProperty ;
        :propertyName "value"^^xsd:string ;
        :datatype "Amount<Currency>"^^xsd:string .

:sp2 a :StateProperty ;
        :propertyName "borrower"^^xsd:string ;
        :datatype "Party"^^xsd:string ;
```

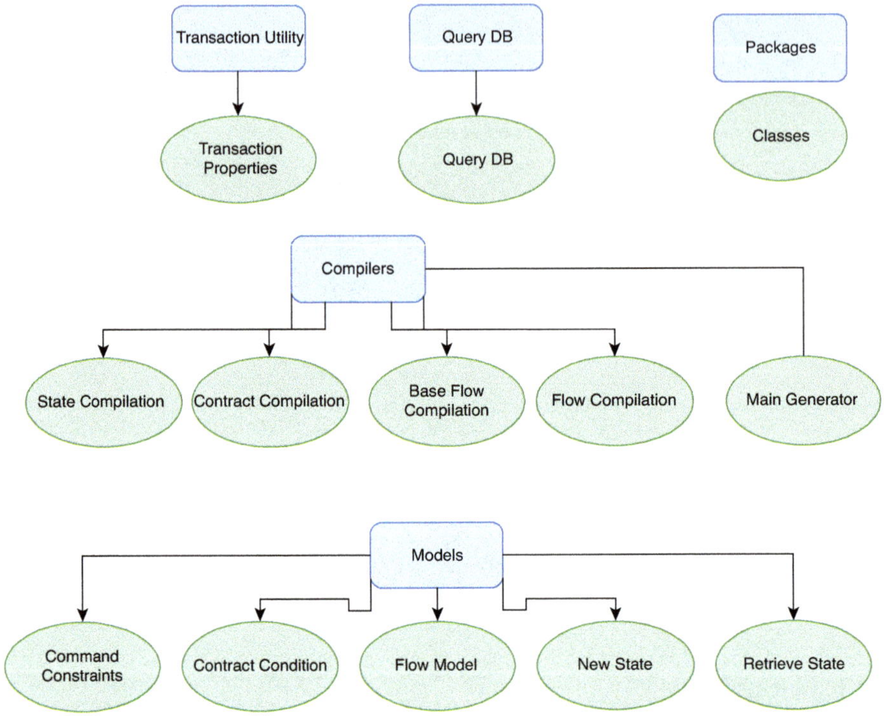

Fig. 17 CordaOntoG auto-generation tool architecture

Fig. 18 State compilation flow

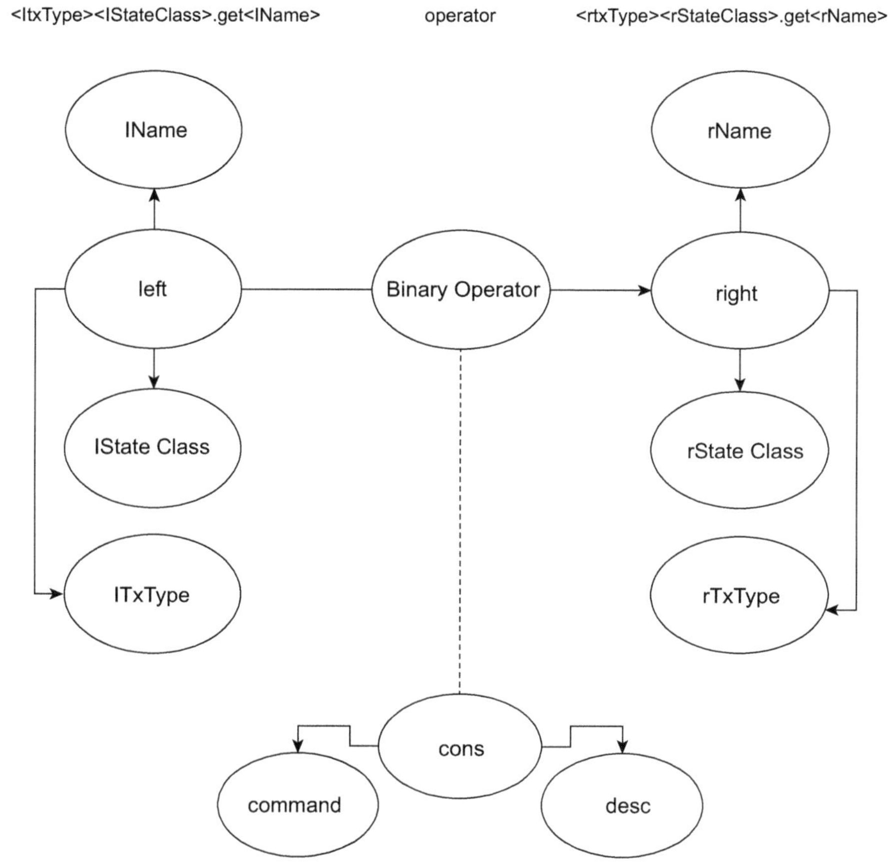

Fig. 19 SPARQL Query pattern for Listing 4

```
                :txIdentity "payer"^^xsd:string .

:sp3 a :StateProperty ;
                :propertyName "lender"^^xsd:string ;
                :datatype "Party"^^xsd:string ;
                :txIdentity "payee"^^xsd:string .

:sp4 a :StateProperty ;
                :propertyName "dateOfIssue"^^xsd:string ;
                :datatype "LocalDate"^^xsd:string .

:s1  a :State ;
         :stateName "IOUState"^^xsd:string ;
         :properties ( :sp1  :sp2  :sp3 :sp4) ;
         :hasProperty :sp1 ;
         :hasProperty :sp2 ;
         :hasProperty :sp3 ;
         :hasProperty :sp4 ;
         :belongsTo :contract1 .
```

Listing 2 IOU state defined in knowledge base

Contract Compilation As mentioned the Contract class in a CorDapp defines the business rules to validate transaction proposing. We use edge properties to relate constraints statements to a command. The following listing shows the list of constraints for the Issuance of an Invoice (IOU).

```
## Define Issue Contract Constraints
:constraint1 a :Constraint ;
    :hasDescription "The lender and borrower cannot be the same identity."^^xsd:string .

:constraint2 a :Constraint ;
    :hasDescription "A newly issued obligation must have a positive amount."^^xsd:string .

:constraint3 a :Constraint ;
    :hasDescription "A newly issued obligation must be less than \$150."^^xsd:string .

:constraint4 a :Constraint ;
    :hasDescription "No inputs should be consumed when issuing an obligation."^^xsd:string .

:constraint5 a :Constraint ;
    :hasDescription "Only one obligation state should be created when issuing an obligation."^^xsd:string .

# Conditional Constraints
<< :txoutsp3 :notEquals :txoutsp2 >> :belongsTo :constraint1 .
<< :txoutsp1 :greaterThan 0 >> :belongsTo :constraint2 .
<< :txoutsp1 :lessThan 15000 >> :belongsTo :constraint3 .
<< :txInpSize :equals 0 >> :belongsTo :constraint4 .
<< :txOutSize :equals 1 >> :belongsTo :constraint5 .
```

Listing 3 IOU contract constraints defined in knowledge base

To retrieve the variables, the following query is used.

```
SELECT ?cons ?left ?bin ?right ?desc ?leftType ?rightType ?lName ?ldatatype ?lstateClass ?commandName{
    ?c a :Command ;
            :commandName ?commandName ;
            :hasConstraint ?cons .
  ?cons a :Constraint .
  << ?left ?bin ?right >> :belongsTo ?cons .
    ?cons :hasDescription ?desc .
    ?leftType a :txPropertyType .
    ?left ?leftType ?leftProp .
    ?leftProp :propertyName ?lName .
    ?leftProp :datatype ?ldatatype .
    OPTIONAL {
    ?leftProp ^:hasProperty ?lstate .
    ?lstate :stateName ?lstateClass .}
    BIND (datatype(?right) AS ?rightType)
}
```

Listing 4 IOU contract constraints defined in knowledge base

This SPARQL Query is akin to matching the following graph pattern to the knowledge base. Then we get a list of constraints for a contract class that belong to a command class.
The following constraints are then generated in the Contracts' Issue command verify function 4.3.1, if any of these conditions fail, and IllegalArgumentException is thrown and the transaction is rejected.

```
    private void verifyIssue(LedgerTransaction tx, Set<PublicKey> signers) {
        requireThat(req -> {
            IOUState ioustateOutput = (IOUState) tx.getOutputStates().get(0);
            req.using("No inputs should be consumed when issuing an obligation.", tx.getInputStates().size()
    == 0);
            req.using("Only one obligation state should be created when issuing an obligation.", tx.
    getOutputStates().size() == 1);
            req.using("A newly issued obligation must have a positive amount.", ioustateOutput.getValue().
    getQuantity() > 0);
            req.using("A newly issued obligation must be less than \$150.", ioustateOutput.getValue().
    getQuantity() < 15000);
            req.using("The lender and borrower cannot be the same identity.", !ioustateOutput.getBorrower().
    equals(ioustateOutput.getLender()));
            return null;
        });
    }
```

Flow Compilation

Now that we have generated the smart contract, we can define the business logic that determines how these transaction proposals come about. Contracts are

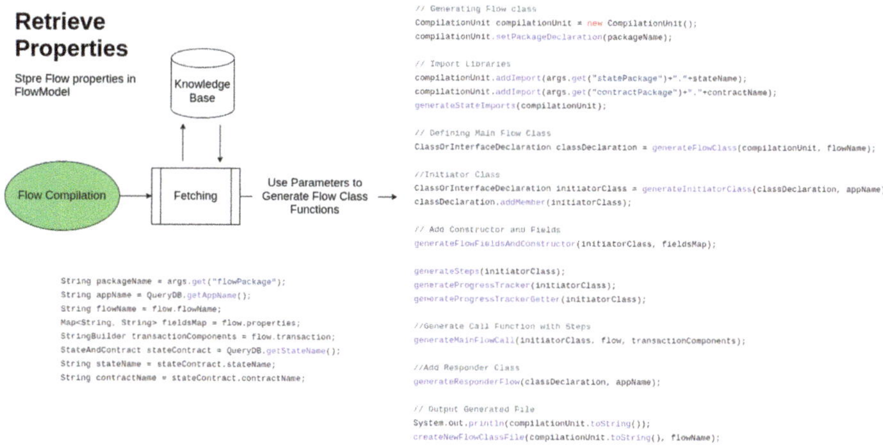

Fig. 20 Flow compilation flow

deterministic, meaning they have no access to the outside world. But, flow has access to Oracles and Services to retrieve information such as Forex Exchange Rates or Bitcoin Stock Price and so on. We stick to a simple business logic that involves two parties either issuing a new state onto the ledger, updating it or marking it historic (settlement of invoice or discharge of patient from trial). The steps for a flow involve:

- Initializing—retrieve input states and/or create new state using flow parameters.
- Building—build transaction with input/output states and command.
- Signing—verifying transaction with contract and signing.
- Collecting—transaction sent to <otherParty> to verify and sign.
- Finalizing—notary marks state uniqueness and parties commit transaction.

We generate the flow class alongside a BaseFlow Class that contains methods to get first Notary (used for transaction uniqueness) and methods to retrieve an input state with a linearId. The subsequent flow classes then extend this BaseFlow class, and the flow of the generation is shown in Fig. 20.

5 Evaluation

The Auto-generation Tool, the sample knowledge bases of the evaluation CorDapps, SPARQL Queries used, and instructions to run tool and set up Stardog Graph database are hosted at: https://github.com/QUzair/cordaOntoG. We now model three different agreements to demonstrate success generation of smart contract code from knowledge base.

5.1 *Invoice (IOU)*

First, we model a common transactional agreement between parties. An Invoice is a commercial document issued by a seller (lender) to a buyer (borrower), relating to a sale transaction of goods/services as a notice for payment. The Invoice is also termed as an IOU (I Owe You), used henceforth. Businesses do not pay for goods and services from other businesses immediately. Rather, they give a promise (agreement) for future payment once they are profitable from the goods/services they received. It helps businesses improve cash flow by having financial freedom for other businesses opportunities than if they had to wait until their customers paid their balances in full to repay amount due.

Defining the State Our state will be agreement that contains:

- Parties: lender and borrower.
- Amount: Value of the IOU due to the borrower. This can be represented as an amount of any currency.

Defining the Contract For this agreement, we can imply two commands: issue and settle. The following Fig. 21 is a state machine view that helps in visualizing how

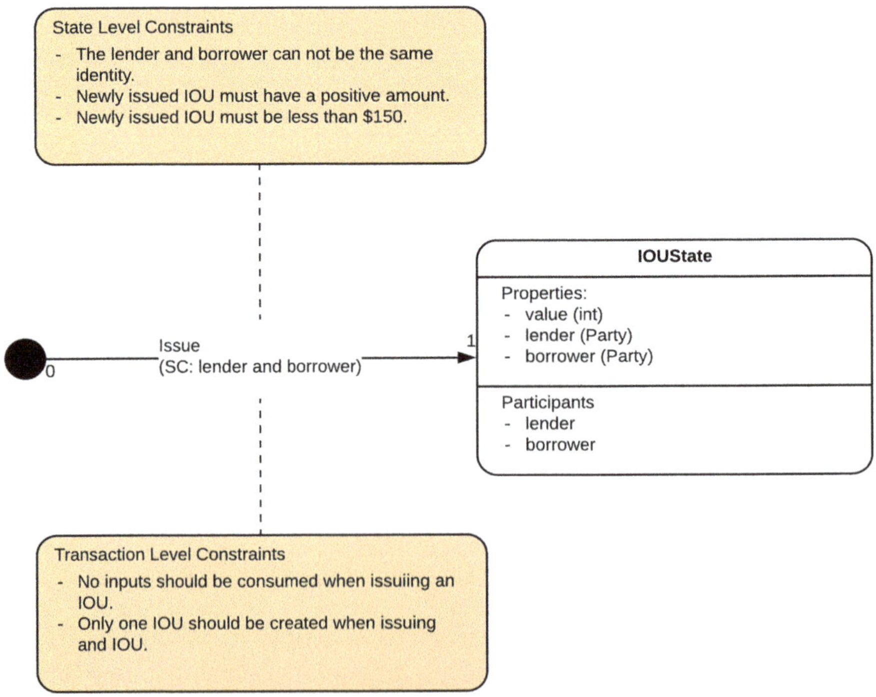

Fig. 21 IOU issue state machine view

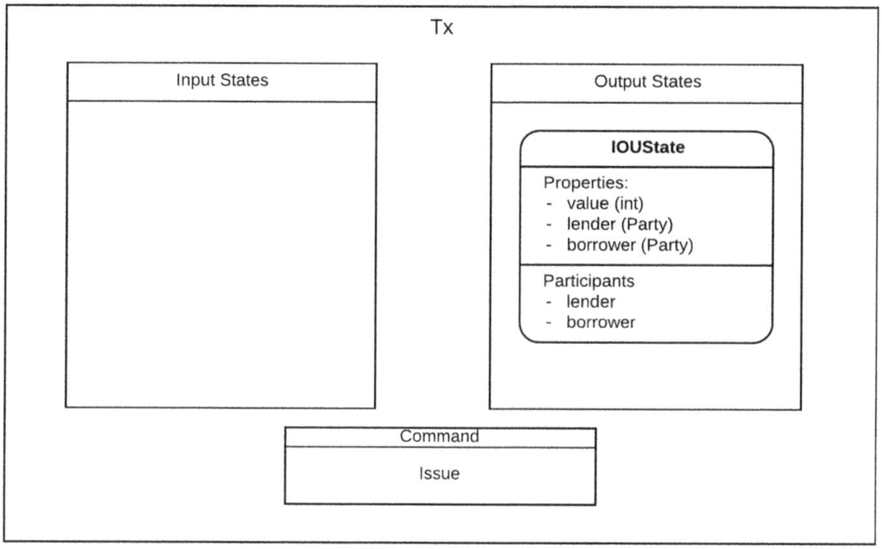

Fig. 22 Transaction proposal by IssueFlow for IOU

the state evolves within a specific flow. When issuing an IOU, we require no states to be consumed and require only one IOU to be outputted. We also require the amount to not be greater $150 (currency is multiplied by 100 to represent a token, therefore, 1500). Finally, we require that the lender and the borrower be different parties.

When settling an IOU, we wish there to be one input IOU state and no output IOU state. We also wish to check that there is an output cash state and that the amount settled to lender equals the amount in the input (current) IOU state.

Defining the Flows For our Issue business logic, we want to propose a transaction that yields in a new IOU state being created between two parties Fig. 22 and committed to the ledger for future settlement. For our Settle business logic, we want to propose a transaction that yields in the current IOU being consumed and with the only output being a Cash state transferred from the borrower to the lender Fig. 23. To generate the classes, we place the database URL in *src/main/java/query-DB/QueryDB.java* → *dbUrl* and run *src/main/java/compilers/MainGenerator.java* main class. The corresponding IOU classes are generated: State Class, Contract Class, Base Flow Class, Issue Flow Class, and Settle Flow Class. We can then deploy these contracts to an existing networks node. For testing purposes, we pass them to simulated network. The following project should be cloned, CordaMockNetwork, to place the generated state contract and flow classes. When we run the simulated network, we deploy a Corda network with one notary service and two parties. Both parties have the CorDapps deployed in their nodes. We can then connect to the nodes

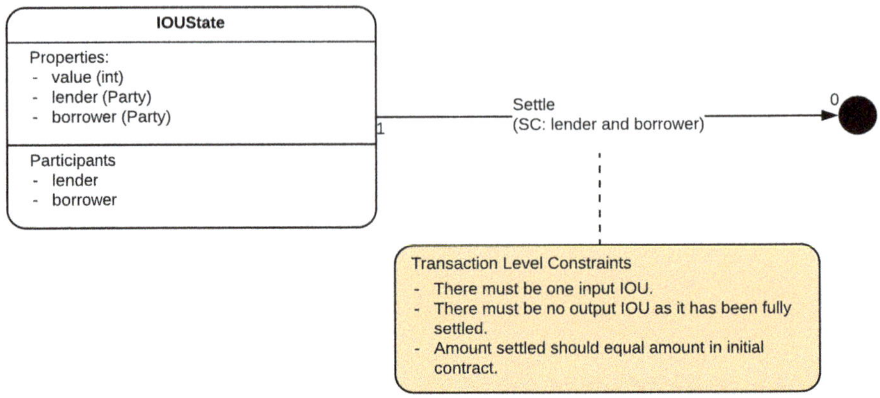

Fig. 23 IOU settle state machine view

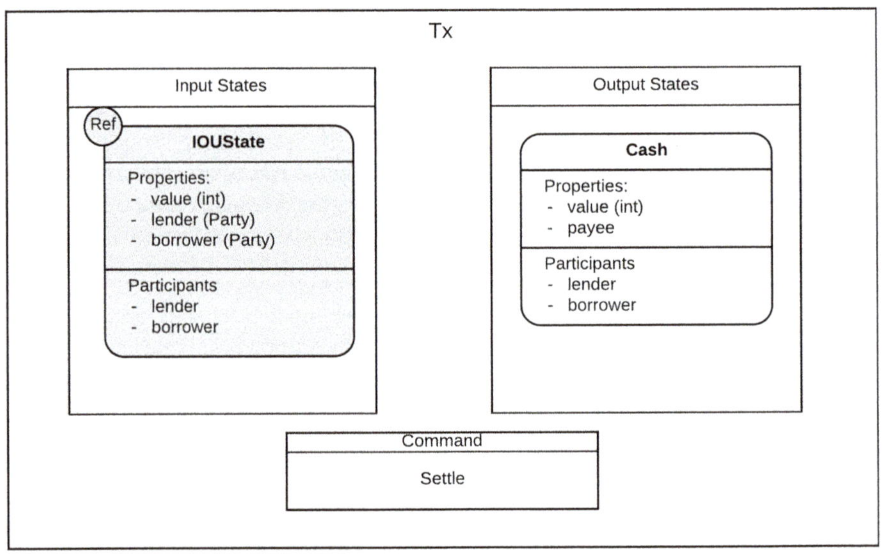

Fig. 24 Transaction proposal by SettleFlow for IOU

and execute the flows. Below are some details together with examples for which the infrastructure returns errors when arguments are infringing the constraints (Fig. 24).

Successful Issue of IOU The following flow is executed by PartyBs node. We successfully issue an IOU with PartyB as the Borrower and PartyA as the lender.

```
Sun Apr 26 21:32:22 IST 2020>>> start IssueFlow$Initiator amount: $10, lender: "O=PartyB,L=New York,C=US",
borrower: "O=PartyA,L=London,C=GB" , externalId: "1234"

✓   Starting

✓   Starting
✓   Performing Initial Steps.
✓   Building Transaction.
✓   Signing transaction.
✓   Collecting counterparty signature.
    ✓   Collecting signatures from counterparties.
    ✓   Verifying collected signatures.
✓   Finalising transaction.
    ✓   Requesting signature by notary service
        ✓   Requesting signature by Notary service
        ✓   Validating response from Notary service
    ✓   Broadcasting transaction to participants
-   Done
Flow completed with result: SignedTransaction(id=
02F5F5F0CBA8EB231B2A7D72F694570A8AFEE394DD60ABB175BF5D85FE50BE95)
```

Once issued we can view the fact in both Party A and B's vault.

```
Sun Apr 26 21:35:45 IST 2020>>>  run vaultQuery contractStateType: com.template.states.IOUState
states:
- state:
    data: !<com.template.states.IOUState>
      value: "10.00 USD"
      borrower: "O=PartyA, L=London, C=GB"
      lender: "O=PartyB, L=New York, C=US"
      linearId:
        externalId: "1234"
        id: "c37c72a1-c536-4f4e-b454-3dc728b006aa"
    contract: "com.template.contracts.IOUContract"
    notary: "O=Notary, L=London, C=GB"
    encumbrance: null
    constraint: !<net.corda.core.contracts.SignatureAttachmentConstraint>
      key:
"aSq9DsNNvGhYxYyqA9wd2eduEAZ5AXWgJTbTEw3G5d2maAq8vtLE4kZHgCs5jcB1N31cx1hpsLeqG2ngSysVHqcXhbNts6SkRWDaV7xNcr6
MtcbufGUchxredBb6"
  ref:
    txhash: "4C0C693EB425158A9487D01E15E7F12C49388E16B50BC88089D7C48DAE453ABA"
    index: 0
- state:
    data: !<com.template.states.IOUState>
      value: "10.00 USD"
      borrower: "O=PartyA, L=London, C=GB"
      lender: "O=PartyB, L=New York, C=US"
      linearId:
        externalId: "1234"
        id: "5c4912e1-7f1b-4111-83d4-20481633cf82"
    contract: "com.template.contracts.IOUContract"
    notary: "O=Notary, L=London, C=GB"
    encumbrance: null
    constraint: !<net.corda.core.contracts.SignatureAttachmentConstraint>
      key:
"aSq9DsNNvGhYxYyqA9wd2eduEAZ5AXWgJTbTEw3G5d2maAq8vtLE4kZHgCs5jcB1N31cx1hpsLeqG2ngSysVHqcXhbNts6SkRWDaV7xNcr6
MtcbufGUchxredBb6"
  ref:
    txhash: "77B401542228126AB5AFFA7DF269B89BEAA02A0707612B8F56F3AB66F7BFC0BC"
    index: 0
```

Same Party as Borrower and Issuer If we try to execute flow from PartyA with PartyA as borrower, we fail because you cannot send the flow to oneself to sign, because we have defined the other Party as being the "borrower," and therefore only the lender can Issue IOUs.

```
Sun Apr 26 21:34:15 IST 2020>>> start IssueFlow$Initiator amount: $10, lender: "O=PartyB,L=New York,C=US",
borrower: "O=PartyA,L=London,C=GB" , externalId: "1234"

✓   Starting
✓   Performing Initial Steps.
✓   Building Transaction.
✓   Signing transaction.
✓   Collecting counterparty signature.
  -   Collecting signatures from counterparties.
  ⊘   Verifying collected signatures.
⊘   Finalising transaction.
  ⊘   Requesting signature by notary service
      ⊘   Requesting signature by Notary service
      ⊘   Validating response from Notary service
  ⊘   Broadcasting transaction to participants
⊘   Done
*   java.lang.IllegalArgumentException: O=PartyB, L=New York, C=US is a required signer, but no session has
been passed in for them
```

Negative Amount Because the amount is not a simple int, when passing negative amount $-10, the node views it as an invalid argument not from the contract, but from the flow input itself. But the constraint in the contract shall be placed for full proof purposes.

```
Sun Apr 26 21:35:17 IST 2020>>> start IssueFlow$Initiator amount: -$145, lender: "O=PartyB,L=New York,C=US",
borrower: "O=PartyA,L=London,C=GB" , externalId: "1234"
No matching constructor found:
- [net.corda.core.contracts.Amount<java.util.Currency>, class net.corda.core.identity.Party, class
java.lang.String, class net.corda.core.identity.Party]: Could not parse as a command: Did not recognise the
currency in -$145 or could not parse

Sun Apr 26 21:35:19 IST 2020>>> start IssueFlow$Initiator amount: $-145, lender: "O=PartyB,L=New York,C=US",
borrower: "O=PartyA,L=London,C=GB" , externalId: "1234"
No matching constructor found:
- [net.corda.core.contracts.Amount<java.util.Currency>, class net.corda.core.identity.Party, class
java.lang.String, class net.corda.core.identity.Party]: Could not parse as a command: Could not parse $-145
as a currency
```

Successfully Settle Once the IOU is in our ledger as unconsumed, we are meant to pay for it and settle the agreement. This is done through the settle command by passing the id of the IOU and the amount to pay.

```
Sun Apr 26 21:36:20 IST 2020>>> flow start SettleFlow$Initiator linearId: "c37c72a1-c536-4f4e-
b454-3dc728b006aa", amount: "$10", lender: "O=PartyB,L=New York,C=US"

✅  Starting
✅  Performing Initial Steps.
✅  Building Transaction.
✅  Signing transaction.
✅  Collecting counterparty signature.
   ✅  Collecting signatures from counterparties.
   ✅  Verifying collected signatures.
✅  Finalising transaction.
   ✅  Requesting signature by notary service
      ✅  Requesting signature by Notary service
      ✅  Validating response from Notary service
   ✅  Broadcasting transaction to participants
-  Done
Flow completed with result: SignedTransaction(id=
248247EA50E4B0B4695742A83844E840780380B6BFB328DF5BD2522EB5AEAD6F)
```

Non-existent IOU If we pass an id that does not exist in the ledger or is consumed already, the transaction fails.

```
Sun Apr 26 21:36:24 IST 2020>>> flow start SettleFlow$Initiator linearId: "c37c72a1-c536-4f4e-
b454-3dc728b006aa", amount: "$10", lender: "O=PartyB,L=New York,C=US"

✅  Starting
-  Performing Initial Steps.
🚫  Building Transaction.
🚫  Signing transaction.
🚫  Collecting counterparty signature.
   🚫  Collecting signatures from counterparties.
   🚫  Verifying collected signatures.
🚫  Finalising transaction.
   🚫  Requesting signature by notary service
      🚫  Requesting signature by Notary service
      🚫  Validating response from Notary service
   🚫  Broadcasting transaction to participants
🚫  Done
✘  IOUState with id c37c72a1-c536-4f4e-b454-3dc728b006aa not found.
```

With the following results, we conclude our generated CorDapp successfully models the business logic (Flow) and the business rules (Contract) of an Invoice Agreement.

5.2 Car Rental

The Car Rental Application is related to one of the domains evaluated by Choudhury et al. [6]. In this scenario, we model an agreement between drivers and a car rental company. Whenever a driver wishes to rent a car, they must satisfy certain conditions related to their personal information. As constraints, all drivers at the time of rental must:

- Meet the renting location's minimum age requirement (25).

- Have a valid driver's license.
- A valid driving record.
- Credit card in their name.

The corresponding[8] Car Rental Classes are generated: State Class, Contract Class, Base Flow Class, and Register Rental Flow Class.
Driver Age Restriction:

```
Sun Apr 26 21:49:27 IST 2020>>>start com.template.flows.RegisterRentalFlow$Initiator age: 25, recordStatus:
"clean", rentee: "O=PartyA,L=London,C=GB", rentalCompany: "O=PartyB,L=New York,C=US", name: "Uzair",
cardCompany: "Visa", cardNum: "123456678", licenseStatus: "valid", externalId: "1568"

     Starting
     Performing Initial Steps.
     Building Transaction.
 -  Signing transaction.
    Collecting counterparty signature.
        Collecting signatures from counterparties.
        Verifying collected signatures.
    Finalising transaction.
        Requesting signature by notary service
            Requesting signature by Notary service
            Validating response from Notary service
        Broadcasting transaction to participants
    Done
    Contract verification failed: Failed requirement: Age of rentee must be greater than 25., contract:
com.template.contracts.RentalContract, transaction:
38016CB523F0506EC05B4752B8F8E7179253F177CD6C91F8A983F99F9E046094
    java.lang.IllegalArgumentException: Failed requirement: Age of rentee must be greater than 25.
```

Invalid License Driver:

```
Sun Apr 26 21:50:45 IST 2020>>> start com.template.flows.RegisterRentalFlow$Initiator age: 25, recordStatus:
"clean", rentee: "O=PartyA,L=London,C=GB", rentalCompany: "O=PartyB,L=New York,C=US", name: "Uzair",
cardCompany: "Visa", cardNum: "123456678", licenseStatus: "invalid", externalId: "1568"

     Starting
     Performing Initial Steps.
     Building Transaction.
 -  Signing transaction.
    Collecting counterparty signature.
        Collecting signatures from counterparties.
        Verifying collected signatures.
    Finalising transaction.
        Requesting signature by notary service
            Requesting signature by Notary service
            Validating response from Notary service
        Broadcasting transaction to participants
    Done
    Contract verification failed: Failed requirement: License Status should be valid., contract:
com.template.contracts.RentalContract, transaction:
CD73A36ED245511CC628F6553C37BC8600C4C5BEB3F322C389D17BEE155AE7EC
    java.lang.IllegalArgumentException: Failed requirement: License Status should be valid.

start com.template.flows.RegisterRentalFlow$Initiator age: 25, recordStatus: "probation", rentee:
"O=PartyA,L=London,C=GB", rentalCompany: "O=PartyB,L=New York,C=US", name: "Uzair", cardCompany: "Visa",
cardNum: "123456578", licenseStatus: "valid", externalId: "1568"
```

Successfully Register Car Rental:

```
Sun Apr 26 21:49:03 IST 2020>>> start com.template.flows.RegisterRentalFlow$Initiator age: 26, recordStatus:
"clean", rentee: "O=PartyA,L=London,C=GB", rentalCompany: "O=PartyB,L=New York,C=US", name: "Uzair",
cardCompany: "Visa", cardNum: "123456678", licenseStatus: "valid", externalId: "1568"
     Starting
     Performing Initial Steps.
     Building Transaction.
     Signing transaction.
     Collecting counterparty signature.
         Collecting signatures from counterparties.
         Verifying collected signatures.
     Finalising transaction.
         Requesting signature by notary service
             Requesting signature by Notary service
             Validating response from Notary service
         Broadcasting transaction to participants
 -   Done
Flow completed with result: SignedTransaction(id=
32EDAABA2F1E1AA87AF946CA22F4EA9471E1F834616CBE35A7FD941D4E279283)
```

[8] The Knowledge Base is available from: Car Rental CorDapp Knowledge Base.

5.3 Clinical Trial Patient

The other domain modeled by Choudhury et al. [6] is a Clinical Patient Trial agreement. This agreement is definitely something that requires regulatory backing, since it concerns public health and is high risk. For this domain,[9] we model the two parties as the Investigator (one carrying out the trial) and the Regulator (one overseeing the trial). Having the Clinical Trial patients on a distributed ledger benefits the Regulator greatly as they can audit the patients and have a synchronized copy of all new patients. The following restrictions were placed for registering a new Clinical Trial Patient:

- Female patients of age 6 and older are eligible.
- Patients with a score ≥ 1 on the Sino-Nasal questionnaire (SNQ).
- Conducted on visit 1 (V1), between Feb. 1st and Feb. 15th, 2010.

All the mentioned requirements must be fulfilled, if even one requirement is not met we reject taking on the patient (ledger not updated). The corresponding Classes are generated: Patient State Class, Patient Contract Class, Patient Base Flow Class, and Patient Register Flow Class.
Patient Age:

```
Failed Age

Sun Apr 26 21:43:59 IST 2020>>> start Register$Initiator age: 5, regulator: "O=PartyB,L=New York,C=US",
externalId: "1234", gender: "Female", snq: "1", visit: "1", investigator: "O=PartyA,L=London,C=GB",
visitDate: "2010-02-02"

 ✓   Starting
 ✓   Performing Initial Steps.

 ✓   Starting
 ✓   Performing Initial Steps.
 ✓   Building Transaction.
-   Signing transaction.
⊘    Collecting counterparty signature.
   ⊘    Collecting signatures from counterparties.
   ⊘    Verifying collected signatures.
⊘    Finalising transaction.
   ⊘    Requesting signature by notary service
      ⊘    Requesting signature by Notary service
      ⊘    Validating response from Notary service
   ⊘    Broadcasting transaction to participants
⊘    Done
*   Contract verification failed: Failed requirement: Patient Age should be greater than or equal to 6.,
contract: com.template.contracts.PatientContract, transaction:
ECC9A15F106ECE8CFF8039AD1787A493EDADA7A274BD6085930E08A0E40B8750
*   java.lang.IllegalArgumentException: Failed requirement: Patient Age should be greater than or equal to
6.
```

Invalid Date:

```
Sun Apr 26 21:40:33 IST 2020>>> start Register$Initiator age: 15, regulator: "O=PartyB,L=New York,C=US",
externalId: "1234", gender: "Female", snq: "1", visit: "1", investigator: "O=PartyA,L=London,C=GB",
visitDate: "2015-02-02"

-   Starting
    Performing Initial Steps.

 ✓   Starting
 ✓   Performing Initial Steps.

 ✓   Starting
 ✓   Performing Initial Steps.
 ✓   Building Transaction.
-   Signing transaction.
⊘    Collecting counterparty signature.
   ⊘    Collecting signatures from counterparties.
   ⊘    Verifying collected signatures.
⊘    Finalising transaction.
   ⊘    Requesting signature by notary service
      ⊘    Requesting signature by Notary service
      ⊘    Validating response from Notary service
   ⊘    Broadcasting transaction to participants
⊘    Done
*   Contract verification failed: Failed requirement: SNQ must have been conducted before February 15th
2010., contract: com.template.contracts.PatientContract, transaction:
AD7D21D54F43C8A8B280CDFE153A4649F3E78D2D27AD54FE286876C6909CCD80
*   java.lang.IllegalArgumentException: Failed requirement: SNQ must have been conducted before February
15th 2010.
```

[9] The associated knowledge base is available at: Patient CorDapp Knowledge Base.

Patients Visit Occurrence:

```
Failed Visit
Sun Apr 26 21:42:31 IST 2020>>> start Register$Initiator age: 15, regulator: "O=PartyB,L=New York,C=US",
externalId: "1234", gender: "Female", snq: "1", visit: "2", investigator: "O=PartyA,L=London,C=GB",
visitDate: "2015-02-02"

-   Starting
    Performing Initial Steps.

 ✓    Starting
 ✓    Performing Initial Steps.
 ✓    Building Transaction.
-   Signing transaction.
⊘    Collecting counterparty signature.
   ⊘    Collecting signatures from counterparties.
   ⊘    Verifying collected signatures.
⊘    Finalising transaction.
   ⊘    Requesting signature by notary service
      ⊘    Requesting signature by Notary service
      ⊘    Validating response from Notary service
   ⊘    Broadcasting transaction to participants
⊘    Done
☠   Contract verification failed: Failed requirement: SNQ must have been conducted before February 15th
2010., contract: com.template.contracts.PatientContract, transaction:
25284C8A477FF75E29214D146D107BE5BB37FD2256D336A59F89BD15A49A3603
☠   java.lang.IllegalArgumentException: Failed requirement: SNQ must have been conducted before February
15th 2010.
```

Successfully Registering Patient for Clinical Trial:

```
Sun Apr 26 21:35:49 IST 2020>>> start Register$Initiator age: 15, regulator: "O=PartyB,L=New York,C=US",
externalId: "1234", gender: "Female", snq: "1", visit: "1", investigator: "O=PartyA,L=London,C=GB",
visitDate: "2010-02-02"

 ✓    Starting
 ✓    Performing Initial Steps.
 ✓    Building Transaction.
 ✓    Signing transaction.
 ✓    Collecting counterparty signature.
    ✓    Collecting signatures from counterparties.
    ✓    Verifying collected signatures.
 ✓    Finalising transaction.
    ✓    Requesting signature by notary service
       ✓    Requesting signature by Notary service
       ✓    Validating response from Notary service
    ✓    Broadcasting transaction to participants
-   Done
Flow completed with result: SignedTransaction(id=
47033DC0E6ABC0CAEDA9F19EAF61F619FE2B2A19440A8CE9C52E66DEB4900202)
```

6 Conclusion and Future Work

Within this section, we provide in general an analysis of the framework carried out in this effort to enable a semantic medium between operational code on the Corda Blockchain and Legal Agreements.

6.1 Conclusion

The chapter started with proposals backed from legal experts to assess how future business agreements and processes may exist on the Blockchain. Most proposals from these papers and recent work focused on abstracting code to provide specification of the code. To be able to abstract the operational aspects of distributed applications on Blockchain, we turned to the Semantic Web. Semantic Web Technologies, with decades of expertise behind it, provided a robust framework for representing, standardizing, and reasoning knowledge on the Web. We further assessed different Blockchain platforms, noticing the vision of Public Blockchains is severely hindered by anonymity. Corda seemed to provide the best integration for the vision of legal smart contracts. It is clear that domain-specific knowledge of the Blockchain platform is necessary to build these ontologies. As of yet, the smart contract implementation within the different platforms does vary significantly, so a single hybrid Ontology for all platforms is not yet possible. To progress the field of legal smart contracts, we require expertise from all domains of Law, Blockchain, and the Semantic Web. Therefore, within this chapter, we hoped to present an approach for modelling lightweight semantic agreements that can directly map to smart contract code and execute on a distribute ledger. To move toward autonomous industries, the Web and Blockchain will play a huge role, as it should provide a meaningful context for software agents, such as IOT devices in a Supply Chain, to make decisions and transact in a secure and distributed manner.

Although most lawyers are not experts in building knowledge bases, with the coming paradigm of legal smart contracts, lawyers will soon need to gain proficiency in programming languages to ensure obligations are modelled correctly in legally binding automatable code.

6.1.1 Future Work

It must be noted that we were not able to fully integrate a legal Ontology or an appropriate reasoning capability that could validate the statements prior to deployment to the distributed ledger.

Heavyweight Ontology As the current Ontology is relatively lightweight, we hope to add further constraints and restrictions to be able to formally verify the constraints and relationships between classes such as highlighting contradicting or redundant constraints in a command. We should also be able to add non-operational aspects such as articles of Law to constraints and regulatory bodies concerned with the dispute of a CorDapp.

Supply Chain Pricing Contracts One of the main purposes for this effort was to look toward enabling the transition of manual agreements to smart contracts to enable autonomous industries. Supply chains currently are notoriously inefficient because they are manually managed and contain so many interconnected parties in the chain, aggregating inefficiencies in multitude. A prominent research field in Economics is Supply Chain Coordination, which focuses on building Game Theory Models for Supply Chain efficiency. These models represent the individual parties as mutually distrusting and build a mathematical model to maximize profits of the system and react to changes in distribution of supply and demand. The most famous of these types of contracts are the Buy-Back contract and the Revenue-Sharing contract. However, these contracts are not practical in real life since human decisions render them inefficient, and it becomes very hard to ensure that the other party is also acting in your best interest and following the protocol. Therefore, considering that supply chain actors agree to follow the same protocol (enforced by the smart contract), the whole system gains greater profit and surges or declines in Supply Chain can be dealt with quicker. Therefore, an interesting research path would be to build the Buy-Back, Revenue Sharing, and WholeSale Pricing agreements, using the Ontology, and simulate a network of nodes for different Supply Chain networks and evaluate the system and individual profit, comparing to the math simulations.

Agreement as a Template of Blocks To abstract the building of the agreement using the Ontology further, it could be beneficial to have a template of blocks that represent the classes with parameters of the properties and commands that can be filled and interconnected. This would function similar to a lot of existing document assembly tools that generate legal prose, but now it would also generate the business rules and logic as code executable on a Corda Blockchain.

References

1. E. Androulaki, A. Barger, V. Bortnikov, C. Cachin, K. Christidis, A. De Caro, D. Enyeart, C. Ferris, G. Laventman, Y. Manevich, et al. Hyperledger fabric: a distributed operating system for permissioned blockchains, in *EuroSys* (2018), pp. 1–15
2. N. Atzei, M. Bartoletti, T. Cimoli, A survey of attacks on Ethereum smart contracts (SoK), in *International Conference on Principles of Security and Trust* (Springer, Berlin, 2017), pp. 164–186
3. T. Berners-Lee, M. Fischetti, *Weaving the Web: The Original Design and Ultimate Destiny of the World Wide Web by Its Inventor* (DIANE Publishing Company, Collingdale, 2001)
4. T. Berners-Lee, J. Hendler, O. Lassila, The semantic web. Sci. Am. **284**(5), 34–43 (2001)
5. V. Buterin et al., A next-generation smart contract and decentralized application platform. White Paper **3**(37), 1–2 (2014)
6. O. Choudhury, N. Rudolph, I. Sylla, N. Fairoza, A. Das, Auto-generation of smart contracts from domain-specific ontologies and semantic rules, in *2018 IEEE International Conference on Internet of Things (iThings) and IEEE Green Computing and Communications (GreenCom) and IEEE Cyber, Physical and Social Computing (CPSCom) and IEEE Smart Data (SmartData)* (IEEE, Piscataway, 2018), pp. 963–970
7. C.D. Clack, Smart contract templates: legal semantics and code validation. J. Digital Banking **2**(4), 338–352 (2018)
8. C.D. Clack, V.A. Bakshi, L. Braine, Smart contract templates: foundations, design landscape and research directions (2016). Preprint. arXiv:1608.00771
9. S.A. El-Seoud, H.F. El-Sofany, O.H. Karam, The semantic web architecture and its impact on e-learning systems development. Int. J. Emerg. Technol. Learn. **10**(5), 29–34 (2015)
10. I. Eyal, E.G. Sirer, Majority is not enough: Bitcoin mining is vulnerable, in *International Conference on Financial Cryptography and Data Security* (Springer, Berlin, 2014), pp. 436–454
11. J. Frankenfield, Double-spending (2020). https://www.investopedia.com/terms/d/doublespending.asp (last accessed December 2023)
12. M. Giancaspro, Is a 'smart contract' really a smart idea? insights from a legal perspective. Comput. Law Secur. Rev. **33**(6), 825–835 (2017)
13. D. Graux, S. Mahmoodi, A fully decentralized triplestore managed via the Ethereum blockchain, in *Further with Knowledge Graphs* (IOS Press, Amsterdam, 2021), pp. 183–197
14. D. Graux, G. Sejdiu, H. Jabeen, J. Lehmann, D. Sui, D. Muhs, J. Pfeffer, Profiting from Kitties on Ethereum: Leveraging blockchain RDF data with SANSA, in *SEMANTiCS Conference* (2018)
15. I. Grigg, The Ricardian contract, in *Proceedings of the First IEEE International Workshop on Electronic Contracting, 2004* (IEEE, Piscataway, 2004), pp. 25–31

16. T.R. Gruber et al., A translation approach to portable ontology specifications. Knowl. Acquis. **5**(2), 199–221 (1993)
17. M. Hearn, Corda: a distributed ledger. *Corda Technical White Paper, 2019* (2019)
18. R. Hoekstra, J. Breuker, M. Di Bello, A. Boer, et al., The LKIF core ontology of basic legal concepts. LOAIT **321**, 43–63 (2007)
19. D.L. Hofman, Legally speaking: smart contracts, archival bonds, and linked data in the blockchain, in *ICCCN* (IEEE, Piscataway, 2017), pp. 1–4
20. B. McBride, Jena: a semantic web toolkit. IEEE Int. Comput. **6**(6), 55–59 (2002)
21. G. Meditskos, S. Dasiopoulou, V. Efstathiou, I. Kompatsiaris, SP-ACT: a hybrid framework for complex activity recognition combining OWL and SPARQL rules, in *PERCOM Workshops* (IEEE, Piscataway, 2013), pp. 25–30
22. J. Mogan, Quorum. Advancing blockchain technology [En línia] (2018). Available: https://www.jpmorgan.com/country/US/EN/Quorum
23. S. Nakamoto, Bitcoin: a peer-to-peer electronic cash system. Technical report, Manubot (2019)
24. N.F. Noy, D.L. McGuinness, et al., Ontology development 101: a guide to creating your first ontology, as a Stanford knowledge systems laboratory technical report (2001)
25. M.-J. Park, J. Lee, C.-H. Lee, J. Lin, O. Serres, C.-W. Chung, An efficient and scalable management of ontology, in *International Conference on Database Systems for Advanced Applications* (Springer, Berlin, 2007), pp. 975–980
26. R3. Barclays' smart contract templates – presented by Dr Lee Braine, introduced by Brad Novak (2016)
27. A. Savelyev, Contract law 2.0: 'smart' contracts as the beginning of the end of classic contract law. Inform. Commun. Technol. Law **26**(2), 116–134 (2017)
28. A. Schätzle, M. Przyjaciel-Zablocki, S. Skilevic, G. Lausen, S2RDF: RDF querying with SPARQL on Spark (2015). Preprint. arXiv:1512.07021
29. N. Smith, D. van Bruggen, F. Tomassetti, JavaParser: visited. *Leanpub, Oct. de* (2017)
30. R. Studer, V.R. Benjamins, D. Fensel, Knowledge engineering: principles and methods. Data Knowl. Eng. **25**(1–2), 161–197 (1998)
31. D. Sui, G. Sejdiu, D. Graux, J. Lehmann, The hubs and authorities transaction network analysis using the SANSA framework, in *SEMANTiCS Posters & Demos* (2019)
32. N. Szabo, Smart contracts: building blocks for digital markets. EXTROPY J. Transhumanist Thought, (16) **18**, 2 (1996)
33. H. Ugarte, A more pragmatic Web 3.0: Linked blockchain data (2017). https://doi.org/10.13140/RG.2.2.10304.12807/1
34. M. Valenta, P. Sandner, *Comparison of Ethereum, Hyperledger Fabric and Corda* (Frankfurt School Blockchain Center, Hessen, 2017), pp. 1–8

Blockchain and Smart Contracts for Telecommunications: The Whys and Wherefores

Meroua Moussaoui, Emmanuel Bertin, and Noel Crespi

Abstract Telecommunications represents a fertile field for blockchain and smart contracts to harvest innovative use cases. The development of 5G and the impending 6G have prompted the research and industry sectors to look into blockchain's key benefits of decentralization, distribution, trust, immutability, security, and traceability to address some identified challenges or to develop blockchain-based systems that natively inherit from these benefits. In this chapter, we explore the potential paths and research directions where blockchain and smart contracts can be leveraged to serve telecommunications. We start the chapter off by defining telecommunications and the major improvements brought to the sector through the different generations. We then present several areas where blockchain and smart contracts can bring an added value. These areas are: spectrum management, network resource management, business, billing and service level agreements (SLAs), as well as mobility and signaling. We later present the raising movement of blockchain-based wireless networks known as "Decentralized Wireless (DeWi)." We also briefly present the main business and regulatory transformations that should follow the migration towards blockchain-based systems in telecommunications. We wrap the chapter with a summary of the main challenges and research directions ahead of the utilization of blockchain and smart contracts to support telecommunications applications.

Keywords Telecommunications · 5G · 6G · Blockchain · Smart contracts · Spectrum management · Network resources · Service level agreements · Mobility · Signaling

---

M. Moussaoui (✉) · E. Bertin
Orange Innovation, Caen, France

SAMOVAR, Telecom SudParis, Institut Polytechnique de Paris, Palaiseau, France
e-mail: meroua.moussaoui@orange.com; emmanuel.bertin@orange.com

N. Crespi
SAMOVAR, Telecom SudParis, Institut Polytechnique de Paris, Palaiseau, France
e-mail: noel.crespi@it-sudparis.eu

N. El Madhoun et al. (eds.), *Blockchain and Smart-Contract Technologies for Innovative Applications*, https://doi.org/10.1007/978-3-031-50028-2_9

1 Introduction

1.1 The Evolution of Telecommunications

The word "telecommunication" is made up of the two parts "tele" and "communication." "Tele" stands for "at distance" and "communication" refers to the transmission of information between a transmitter and a receiver across a medium. Telecommunication, then, is any form of exchanging information over a specific distance while utilizing a particular medium. This broad definition encompasses a variety of at-distance-communication forms ranging from pigeon posts to instant chat messages. However, as you might have suspected it, we are not going to discuss about the many carrier pigeon training techniques (especially that blockchain did not exist in that era) but rather the digital forms of telecommunication. More specifically, we focus on "mobile telecommunications," leveraging cellular networks; nevertheless, we will occasionally highlight how blockchain can be used in conjunction with other types of communications networks.

One could never have predicted how Information and Communication Technologies (ICT) would change the world around us. The first generation of mobile telecommunications (1G) was an analogue voice-only technology with no data communication. It represents a significant technological milestone because it marks the beginning of the 10-year periodic evolution of mobile telecommunications. With each increment, a new generation is introduced, with the aim of responding to emerging use cases with significant technological advancements in data-carrying capacity and latency reduction. In 1991, the second generation (2G) was launched as a digital voice and data transmission technology supporting low data rates. Later, the third generation of mobile telecommunications (3G) was introduced in 2000 as a set of services combining high-speed mobile access with Internet Protocol (IP)-based services, following the International Mobile telecommunications-2000 (IMT-2000) standards. The fourth generation of mobile telecommunications (4G) came to light in 2009, and was centered on a collection of technologies and protocols that were expected to provide a complete and secure all IP-based packet-switched networks, optimized for data, as defined by the IMT-Advanced standard. The fifth generation of mobile telecommunications (5G) was released in 2017, and it meets the IMT-2020 requirements. It is an evolution of its predecessor (4G), and it supports three types of services: enhanced Mobile BroadBand (eMBB), ultra-Reliable Low Latency Communications (uRLLC), and massive Machine Type Communications (mMTC). While the world is still witnessing the rollout of 5G technologies, early stages of research about 6G have already started. Many working groups have been created, gathering academia and industry, with the purpose of tracing the first lines of this future era. 6G is expected to leverage recent advancements in machine learning (ML), artificial intelligence (AI), and virtual reality to efficiently integrate these technologies into verticals (industry automation, healthcare, transportation, etc.). Many supporting technologies, including a greater reliance on satellite communications, the use of higher frequency bands (Terahertz bands) and Intelligent Reflective Surfaces (RIS) will come into play. Blockchain is also expected to be a major 6G enabler that would be leveraged to overcome the 5G limitations and shortcomings to shape 6G networks [1].

1.2 The Structure of the Chapter

The hype surrounding bitcoin and other cryptocurrencies led to the first applications of blockchain technology in the financial sector. However, the advent of smart contracts in 2010 revealed novel uses for this technology that later led to the development of distributed applications (DApps), transforming blockchain into a fully automated self-managing system. This has made blockchain a possible contender for many Real-Time Business use cases [2]. This chapter will present the major contribution opportunities for blockchain and smart contracts in four main areas of telecommunications:

- Spectrum management
- Network resource management
- Service Level Agreements (SLAs), billing and business-related agreements
- Mobility and signaling

After this presentation, we briefly introduce the novel generation of blockchain-based wireless networks known as "Decentralized Wireless (DeWi)." The chapter also discusses some business and regulatory changes that should go with the introduction of blockchain and smart contract in the telecommunications sector, followed by a summary of the most prominent challenges and future directions. The chapter is wrapped by a conclusion that recalls the main contributions of blockchain and smart contracts in telecommunications.

Figure 1 illustrates the structure of this chapter.

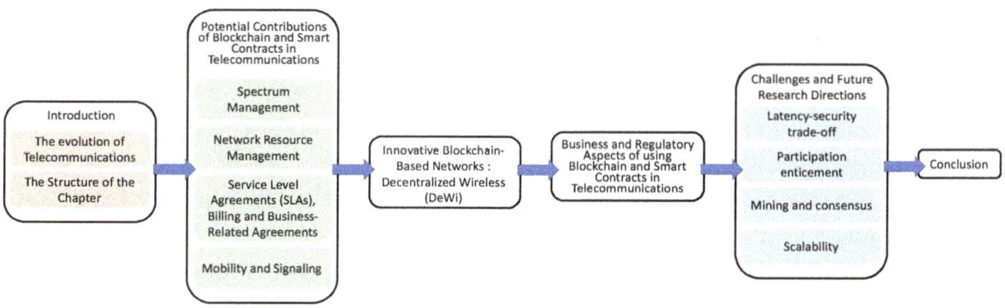

Fig. 1 The structure of the chapter

2 Potential Contributions of Blockchain and Smart Contracts in Telecommunications

2.1 Spectrum Management

Spectrum is regarded as a national asset that is subject to sovereign laws in the same way that other assets (such as water, land, etc.) are. Spectrum management is subject to a three-tier framework that administers this asset on: national, regional, and global (via the International Telecommunication Union-Radiocommunication Sector (ITU-R)) levels [3]. Radio spectrum bands in telecommunications are divided into two big categories: unlicensed spectrum bands, like Bluetooth and Wireless Fidelity (Wi-Fi). These bands are open to the large public without the requirement for any exclusive rights or payment. The second category is licensed spectrum bands, which are attributed to demanders, via multiple approaches, for exclusive usage rights in exchange of payment. The most known approach for spectrum licensing is "auction" where the spectrum is awarded to the bidder proposing the highest price. Another approach is "administrative procedures," which are primarily employed for the licensing of governmental entities and crucial first responder services. A third approach is a hybrid of these two first approaches, where the regulator holds an auction among a few candidates previously selected via administrative procedures [3].

A common approach to spectrum management is centralized spectrum management, in which an authoritative institution, generally the government of a country, bears responsibility for the whole spectrum management system. This includes supervising essential spectrum management functions such as spectrum licensing, strategic spectrum allocation, spectrum policy and regulation formulation, enforcing compliance with national, regional, and international spectrum regulatory frameworks, and vigilant monitoring of spectrum sharing activities. So far, this centralized method has yielded satisfactory spectrum management in terms of prudent resource allocation and interference limitation between parties. Nonetheless, with the widespread use of the Internet and the substantial use of digital services, certain limitations have become apparent with this approach. These include the time-consuming and costly centralized licensing procedures, which are unsuitable for the rapidly changing spectrum utilization needs. Furthermore, centralized spectrum management raises the chance of "Spectrum market monopoly" by a few firms, increasing the risk of spectrum market manipulation and unfair pricing of spectrum resources. Decentralization of the spectrum management system offers a promising possibility to solve these limits while also aligning with the evolving dynamic nature of spectrum demand.

The centralized spectrum management system relies on a centralized database that can be decentralized by coupling blockchain and a distributed database system (such as Inter Planetary File System (IPFS) – distributed offline storage). In this approach, blockchain is coupled with an offline storage to ensure the scalability and security of the storage system, as one of the major constraints of blockchain technology is its storage capacity. Each node in the blockchain network must keep a local copy of the distributed ledger. Scalability becomes an issue when the ledger's stored information grow in size, as nodes would require more storage space. This can be especially problematic for specific devices, such as mobile devices often used by telecommunications users, which may lack the ability to meet these increasing storage demands. Furthermore, in the case of public blockchains, the information held within the blockchain blocks is accessible, albeit encrypted, to all participants. This raises serious privacy and security concerns, as the encrypted sensitive data may be available to anybody with blockchain access. While the immutability of blockchain storage is frequently regarded as a significant advantage since it allows verification of system users' honesty and ensures transaction traceability and history, it may also be viewed as a constraint. The inability to edit stored data limits the functionality of blockchain storage, rendering it unsuitable for situations requiring data updates or corrections. In such instances, an additional storage solution that allows for editable data would be required. This emphasizes the importance of integrating blockchain with an offline distributed storage system to provide data distribution, security, and confidentiality. Furthermore, it would improve the overall scalability of the storage infrastructure and enable a variety of functionalities and operations on the stored data, with a special emphasis on data modification capabilities [4]. Confidential and sensitive information can be stored in the distributed database, and its corresponding hash can be stored in the blockchain, as well as other data that require peer verification and immutability. Blockchain technology is used in this situation primarily for its capacity to function effectively in an untrustworthy environment involving various actors, such as spectrum regulators and spectrum demanders. Its primary strengths are identity management and access control properties, which allow for secure and authenticated interactions between participants. Furthermore, blockchain's ability to facilitate contract administration and enforce spectrum management policies via smart contracts, reinforces its importance in this context.

Centralized spectrum management can result in multiple performance and security problems, such as: Single-Point-Of-Failure (SPOF), privacy leakage, identity spoofing, and processing delays [5]. Other issues can also result from the process of buying licenses such as: long licensing delays, unfair and/or biased license attribution, corrupted auctioneers manipulating the bid for their advantage, identity spoofing, among others. To overcome these issues, one solution is to use a permissioned blockchain in conjunction with smart contracts to accomplish decentralized spectrum management. In spectrum management scenarios, a permissioned blockchain is preferable to a public blockchain, owing to the latter's longer block verification delays and lower throughput, which unavoidably hamper the spectrum management system's efficiency. The permissioned blockchain's applicability originates from the lower number of participants as compared to a public blockchain, as well as its optimized consensus algorithms, which adds to faster block verification times. Furthermore, the identities of participating nodes and validators are thoroughly verified before getting access to the blockchain, lowering block validation time even further [5].

The use of a permissioned blockchain eliminates the need for a central administration and coordination body, reducing the risk of Single Point of Failure (SPOF). Furthermore, the interaction with a central authority, which might result in privacy leakages is avoided because all transactions are signed with a private/public key pair and then stored in an immutable ledger. This ensures the integrity, security, and privacy of information. In the context of spectrum management, the use of a permissioned blockchain eliminates the threat of identity spoofing by requiring certificate authentication (CA) of all participating nodes before providing them access to the system [5]. As for the licensing procedure, the implementation of smart contracts automates the licensing processes, allowing for the elimination of costly and time-consuming processing delays and bureaucratic constraints. The possibility for corruption that could result in unfair or biased license allocation is prevented by replacing the central licensing body with a decentralized tamper-proof governance structure. Furthermore, because the fundamental design of blockchain prohibits a majority of participating nodes from being corrupted, the use of blockchain-based bids for spectrum auctioning reduces the possibility of bid manipulation.

Spectrum licenses usually last for long periods; however, the market demand for this scarce asset is growing dynamically. Trading spectrum assets via "Secondary Spectrum Markets" allows the spectrum licensees to value this asset by fully or partially leasing it to other actors. This results in more flexibility and efficiency in spectrum usage. Another approach that allows efficient and optimal spectrum usage is the establishment of "spectrum-sharing cognitive radios." This generation of radios allows dynamic and opportunistic sharing of spectrum bands between licensees (called *primary users (PUs)*, who hold spectrum licenses for the corresponding band) and other users who do not hold the shared band license (called *secondary users (SUs)*). The SUs use intelligent and dynamic spectrum sensing algorithms to discover vacant spectrum channels, known as spectrum holes. These spectrum holes are the portions of the spectrum band where the SUs are permitted to send packets, without interfering with neither the PUs nor the other SUs.

To achieve secure, automated, and reliable spectrum sharing processes among various entities, the auction method used by authorities for attributing spectrum licenses can also be mimicked by PUs to share their spectrum with SUs. In this method, PUs could make use of blockchain and smart contract to launch an auction between the various SUs and share portions of their idle spectrum with the highest bidder or with the auctioneer that maximizes a specific function such as the system's utility [6]. Digital tokens could be used to pay for spectrum access as a currency, but also as an indicator of some specific conditions, for example to indicate access priority or encourage miners. Additionally, blockchain also allows the elimination of intermediaries and third parties in spectrum licensing and trading procedures, by implementing the corresponding constraints, regulations, and contracts as smart contracts. This allows the automation of these procedures and thus an immense gain in time, cost, and efforts.

Spectrum sharing in "cognitive radios" allows opportunistic spectrum access. For this end, a phase of "spectrum sensing" should precede spectrum attribution. In this phase, SUs collaborate to periodically check the band availability, and store the sensing information in a centralized fusion center; this information is then analyzed through fusion methods to determine spectrum availability and thus attribute it to the demanders. The centralized nature of this system's architecture and its reliance on a centralized database that keeps record of the sensing data make the system vulnerable to security problems such as SPOF and corrupted sensing information reporting among others. Fraudulent SUs may purposefully inject false sensing data in order to corrupt the system and alter it to their own advantage so that they can benefit of a maximum of white spaces. To counter this type of attacks, blockchain and smart contracts can be used to implement a reputation system, where the history of SUs reported data is kept in the distributed ledger, allowing the transparency and auditability of the SUs history. This data can be later analyzed to detect any abnormal or suspicious behavior. In the current centralized system, SUs perform sensing tasks without any incentive, however, using blockchain with a suitable consensus algorithm, SUs, which represent the system's miner, would be incentivized for achieving truthful sensing tasks; this will encourage them to work with optimal performance, and benefit from a digital token wallet, which can be used to buy spectrum white spaces. This would also achieve better fairness in spectrum attribution, where SUs benefit from spectrum resources proportional to their contribution in the sensing task. However, this reputation system can also be vulnerable to "collusion attacks," where malicious SUs promote each other's reputations, in order to increase their chances in benefiting from spectrum resources. To resolve this problem, two types of reputation can be used: "active reputation," which represents the number of tokens gained when uploading sensing information in a given time, and "passive reputation," which represents the number of tokens inside the SU's digital wallet (historical reputation) [7]. It is important to note that incorporating blockchain into the sensing phase could extend this phase's duration and slow down the access procedure.

In response to the increasing demand for spectrum, several governments have taken the initiative to open additional frequency bands for public use. For instance, the Citizens Broadband Radio Service (CBRS) band (between 3.55 and 3.7 GHz) was made available to the general public by the USA in 2017. This band was initially intended for military use. Three different access types currently share this band according to hierarchical priorities. Incumbent Access (IA) is given first priority (US navy), followed by Priority Licensed Access (PLA), which is typically granted for network operators who purchase licenses to this band and thus benefit from interference protection, and finally General Authorized Access (GAA), where users are given opportunistic access to the spectrum and are granted access to portions of the band so long as they are not used by the two first priorities. The CBRS band is controlled and managed by the Spectrum Access System (SAS). This system has a centralized architecture, where spectrum access and sharing information are stored in a central database. This makes SAS prone to SPOF, administrator corruption, and security and privacy issues. A solution for this is decentralizing the SAS using blockchain, which can be used manage spectrum access and enforce spectrum assignment to parties using smart contracts. Another solution proposes using different local blockchains that handle spectrum assignment, as a blockchain-based SAS, in specific geographical zones. These local blockchains are connected to a global blockchain that coordinates their activities, as well as ensures their conformity to rules by including the regulator and other entities that frame the activities between the different SASs, as participating nodes [8]. Blockchain and smart contracts can also be used to share spectrum between PA and GAA. This can be achieved by creating smart contracts that grant GAA access to available CBRS spectrum bands and unused PA bands by managing access demands and interference conditions. To reduce administrative and operational costs, novel consensus algorithms can also be implemented by exploring the systems' particularities, such as minimizing interference, increasing individual and global revenue, maximizing spectrum occupancy, etc. It is also worth noting that a profitable blockchain-based incentivization system can be beneficial to encourage PA users to share their available spectrum with GAA users.

While 5G is still in its early deployment phases, Dynamic Spectrum sharing (DSS) allows 4G Long-Term Evolution (LTE) and 5G New Radio (5G-NR) to coexist in the same band. Blockchain can be used in this case to manage the spectrum sharing using a reputation mechanism in order to reduce interferences and protect the system from fraudulent access. As stated before, spectrum is a scarce asset which requires responsible and meticulous usage. Moreover, licensed spectrum requires costly and time-consuming licensing procedures, which has pushed many actors, especially in the 5G context, to reconsider the usage of unlicensed spectrum for the broadcast of their services. Blockchain along with novel consensus procedures can be leveraged to allow safe and efficient sharing of unlicensed bands between the different access technologies (Wi-Fi, cellular, etc.) and limit access overhead.

2.2 *Network Resource Management*

Radio Access Network (RAN) is the portion of the mobile network that uses base stations to transmit signals linking end devices to the operator's core network and other services. Leveraging blockchain and smart contracts in RAN gave birth to the concept of "Blockchain-based RAN (BRAN)," which is a novel RAN architecture that relies on blockchain technology and smart contracts for process automation, and the creation of a collaborative RAN infrastructure by establishing trust among inherently untrusted RAN infrastructure owners and operators. This novel architecture was first introduced in [9]; it decentralizes RAN management and control, where blockchain handles the access control and authentication among the different untrusted actors, reducing centralization overhead and security threats (SPOF, DoS, etc.), facilitating cross-network roaming, and lowering the cost of RAN services.

In order to accommodate scalable and latency-sensitive RAN operations, a novel voting-based consensus algorithm, called Proof-of-Device (PoD), has been proposed by [9] to complete the BRAN architecture. It leverages the uniqueness of IT hardware identifiers (such as mobile devices' International Mobile Equipment Identity (IMEI) and base stations' Cell Global Identity (CGI)). This novel consensus is less complex compared to puzzle-solving-based Proof-of-Work (PoW) consensus, and thus requires less computation and mining expanses, hence lowering latency and energy consumption. In this algorithm, the participating devices, called miners, evaluate the following hash function:

$$\text{Hash}\,(\text{previous hash} + \text{data} + \text{device ID} + \text{timestamp}) \leq \text{Given value}$$

Where the device whose identifier (ID) satisfies this hash function is chosen as a winner. Meaning that the device who achieves a hash value less than or equal to a randomly assigned "Given value" is granted the privilege to add the next block to the blockchain. Due to the uniqueness of the device's ID, a device can only calculate the hash function once per timestamp, in contrast to PoW, where a device does numerous trials to figure out the nonce value that solves the puzzle. PoD, thus, decreases the required computation and energy for solving the puzzle, as well as enhances the system's fairness by giving the miners equal opportunities to insert a block. The variables constituting the aforementioned hash function are:

- *Hash*: the used hash function, such as MD5, SHA-1 and SHA-2.
- *Previous hash*: the hash of the previously inserted blockchain block.
- *Data*: the data contained in the block (random information, transactions, smart contracts, or any other type of data) that must be recorded and verified on the blockchain.
- *Device ID*: a unique serial number or an alphanumeric value that uniquely identifies the participating device (miner).
- *Timestamp*: the block's construction date and time.
- *Given value*: a random value chosen by the system, and which changes with every timestamp.

Blockchain has the potential to transform RANs into decentralized, scalable, and self-organized mutualized networks. It also democratizes RAN and lowers the entry-barrier to small and low-capital actors. Moreover, it maximizes the occupation of the RAN infrastructure, as RAN owners may share their idle infrastructure with other players, thus maximizing the infrastructure utility and generated revenue. Another major advantage of BRAN is that it does not requires any extra radio infrastructure, as all the management and control procedures are achieved on the software level via blockchain and smart contract. Furthermore, smart contracts can be used to automatically manage the relationships among the different actors in procedures such as RAN resource pooling, trading, access control, and authentication.

BRAN can also be seen as a platform that groups multiple actors with simple market access procedures, allowing any actor to integrate the platform and, eventually, leave it when suitable. This platform grants equitable and unbiased repartition of revenue among the participants, in addition to the fair pricing of services intended for customers. For instance, Mobile Network Operators (MNOs) can also leverage blockchain to create dedicated platforms for trading infrastructure resources and other network resources among Mobile Virtual Network Operators (MVNOs). Moreover, BRAN is open to different categories of infrastructure owners and operators (Tower Companies (TowerCos), Infrastructure Companies (InfraCos), micro-operators, Small-cell owners, etc.), and does not require the customers to have any long-term engagement with any of these RAN providers. Another approach could be to take the platformization paradigm to the next level by offering BRAN-as-a-Service (BRANaaS), eliminating the need for active involvement from both the service provider and the customer. The use of smart contracts to manage business conditions between the parties can ensure the automation, adaptability, and dynamicity of BRANaaS.

BRAN offers flexibility by efficiently enhancing RAN services within dynamic network environments, hence catering to a wide range of device categories and service requirements. Furthermore, the open nature of BRAN allows for the integration of equipment from many manufacturers, as well as various protocols and policies in both public and private settings. This method creates a transparent, distributed, and democratized environment, encouraging collaboration among different service providers and catering to various customer segments. Furthermore, BRAN showcases remarkable scalability, by seamlessly managing an increasing number of participants. As a result, the BRAN architecture operates optimally in small-scale deployments such as private networks as well as in large-scale scenarios such as national coverage networks, all while maintaining superior levels of performance and security. It is important to highlight that BRAN effectively demonstrates the concept of the "network effect." Participants in this approach initially join the network for an initial incentive, and as they become active contributors, they produce additional incentives. These newly established incentives, in turn, entice other participants to join the network, generating a positive feedback loop of increased participation and rewards [10].

Blockchain is a potential candidate to the establishment of RAN resource pooling, where RAN infrastructure owners and operators create a shared pool of their resources. These resources are then used by the participants following some specific policy (proportional to their participation, in exchange of fees, etc.). In this use case, blockchain can be used in two levels: the first blockchain would be used to manage the pooling procedure, and the second would be used to share the pool resources among the participants. The pooled resources can also be leased to new actors that have not participated in the creation of the shared pool, in exchange of revenue.

To solve the scalability issues that stem from the geographical dispersion of the participating RANs, multiple blockchains can be used. This would increase the system's scalability, flexibility and would help reduce considerable delays caused by connecting geographically dispersed nodes to a single blockchain. Instead, nodes in the same geographical area are linked to their respective local blockchain, and each local blockchain is then linked to a central main blockchain. Moreover, smart contracts have the potential to implement efficient network orchestration and management via synchronizing processes between the different local blockchains and enforcing regulations and policies [10].

Another type of networks where blockchain can be introduced is relay networks, used to improve network coverage via the usage of multiple intermediary structures that may or may not belong to the same owner or operator [11]. Blockchain can be beneficial to record the usage history of the different relays and to track the ownership of the relay services. Furthermore, it would effectively handle a large number of relay providers, regardless of whether they have mutual trust. This mechanism would also create a distributed and democratized environment, providing equitable incentives to all participants, allowing the network to expand even further. This would improve network coverage and thus achieve one of the main goals of 5G and 6G networks, which achieving ubiquitous connectivity. 5G and 6G networks are also expected to make great use of softwarization, virtualization, and cloudification technologies. Blockchain can also contribute to the improvement of virtual and cloud network resources' management and usage, and thus, laying the ground for a multi-operator multi-vendor ecosystem, by leveraging the concept of openness that spans from RAN to business level. Moreover, various players can collaborate for service provision to end users by pooling their cloud and edge infrastructure as well as their other network resources.

In an end-to-end network, the use of multiple blockchains becomes practical for effectively managing various resources and services, each adapted to its individual requirements by implementing the proper consensus mechanism. Customization of blockchain technology is required to be consistent with the intended use case. For example, certain services may require low latency, whilst others emphasize scalability over latency concerns. According to this approach, blockchain-managed services can be advertised to customers as black boxes, each delineated by specific attributes (such as the chosen consensus algorithm, scalability capabilities, message latency, and so on) tailored to precisely align with the individual customer's needs and requirements. Customers might then use these services to build their own private and customized network.

It is worth noting that all solutions relying on blockchain for the purpose of resource sharing need to be suitable to users' hand-held devices, which are usually limited in terms of computation resources, storage space, and energy.

2.3 Service Level Agreements (SLAs), Billing, and Business-Related Agreements

A Service Level Agreement (SLA) is a legal agreement wherein the service provider commits to delivering its service to its clients in accordance with the agreed values for the specific service metrics (Quality of Service (QoS)/Quality of Experience (QoE) among others).

In the 6G era, customers may no longer enter into long-term agreements for connectivity provision with a particular MNO. Smart-contract-based SLAs can be used to allow end users to dynamically and seamlessly switch operators while using services. In this situation, using cryptocurrency to pay for the utilized connectivity is also a viable option. The use of blockchain for this purpose can be improved further by giving users a real-time overview of their QoS/QoE while also giving them the option to choose between several offers, or by allowing users to priorly set their necessary QoS/QoE and price conditions, leaving it to the smart contracts to choose the appropriate MNO and the specific offer to adhere to. The smart contracts used in this case can also be further polished to choose offers that realize the best QoE/QoS-price tradeoff [12].

SLA violation occurs when a service/connectivity provider fails to offer service/-connectivity that meets the agreed-upon standards for a certain amount of time. In this situation, the service/connectivity provider has to compensate the customers; however, the current compensation processes lacks transparency and disadvantages the customers since the service/connectivity provider is responsible for identifying SLA violations and initiating the compensation process without negotiating with the clients. If not, the client must provide evidence, which is frequently difficult to get because the offense has occurred in the past. Corrupted service/connectivity providers can take use of this point to violate SLA more frequently without facing legal repercussions. In order to preserve users' rights, a mechanism relying on Trusted Third Parties (TTPs) is put in place to collect SLA violation proofs and guarantee user compensation. Nevertheless, reliance on intermediaries renders the procedure more cumbersome, expensive, and time-consuming. To address these issues, blockchain can be used for its transparent and tamper-proof storage, in order to keep record of the SLA violation incidents. Smart contracts can be used to implement the SLA terms and the compensation process, allowing fair compensation of users without over- or underestimation of the compensation amount. By translating SLAs into smart contracts, it is possible to avoid drafting errors that could lead to ambiguous SLA clauses, and thus result in different (and maybe dishonest) interpretations. This increases the legal protection available to both the service provider and the client. Additionally, smart contracts enable continuous SLA monitoring, making the system more trustworthy by catching every instance of violation, ensuring that none goes unnoticed. Cryptocurrency can also be leveraged for SLA billing and SLA violation compensation of the users. The usage of blockchain and smart contracts for SLA violation automates the procedures and renders it less expensive and less time-consuming by eliminating the intervention of third parties [13]. By limiting access to information about SLA violations to participants only, it also ensures the system's secrecy.

Despite the fact that cloud environments are extremely dynamic, the SLAs between clients and service providers who offer services hosted in the cloud are statistically issued. These SLAs are inflexible and restrict fine-grained customization of the offered services, pushing the customer to agree to contracts that do not precisely meet their intended QoS requirements. Additionally, current SLA solutions for cloud services tend to be centralized, and the SLA terms do not adjust their parameters in accordance with the network status to reflect the dynamic nature of the network. Blockchain and smart contracts can be leveraged to address the aforementioned issues. A system built on two blockchains can be put in place.

A public blockchain can be used for dynamic SLA management where SLAs are implemented as smart contracts. A private blockchain can be used for cloud environment sensing, where the participant nodes, called "oracles," detect and record different events and signal them to the public blockchain. These events can be SLA violation incidents or changes in the network parameters, so that the SLAs adapt their terms accordingly [14].

Multicloud systems contain multiple systems that belong to multiple services. In their turn, the services belong to multiple providers. This renders the SLA monitoring procedure very complicated to achieve, especially if the component services are heterogeneous and tangled through multiple dependencies on different levels. A problem that stems from this hierarchical organization of a multicloud environment is that a specific service is only aware of the services that are directly related to it. However, when an SLA violation occurs in a service it may be caused by (a) that specific services, (b) services directly related to it, or (c) other services that are not directly connected to it. Add to that, SLAs of the services in the hierarchy are not openly shared with all participants due to the confidential nature of the document. This makes SLA monitoring and violation very hard and complicated to achieve as it renders the detection of the root cause a very complex process that lacks automation and optimization. The usage of a multichannel blockchain is a suitable contender to address this issue. This blockchain type allows the creation of a private channel that is visible and accessed only by a subgroup of the blockchain participants, providing scrupulous confidentiality even among the participants of the network. The participating nodes are the service providers and their clients (service providers can also be clients if they use services provided by other service providers). All parties participating in a service participate in monitoring its SLA. A smart contract is used to register the SLA monitoring data into a wide channel accessed by all participants, called "the monitoring channel." On the other hand, private channels are used to detect and record the SLA violation incidents between specific parties. To detect the root service causing the SLA violation, another wide channel is used. This channel is managed by a smart contract that receives an SLA violation detection record from the service provider, and detects correlations with other SLA violation incidents to finally come up with the root cause service [15].

In a context of extensive dependance on cloud technologies, decentralized and open service marketplaces are needed to facilitate the service adherence procedure between service providers and clients. These marketplaces can be equipment with an SLA-violation system in order to improve their trustworthiness. In such dynamic ecosystem, utilizing witness-enhanced SLA monitoring is a further option. Along with customers and service providers, "witnesses" can also be added as blockchain participants. They are in charge of continuously monitoring SLA compliance and, eventually, violations in exchange for rewards. If they report inaccurate information, they run the danger of being fined. By introducing this new role, the system's reliability and SLA compliance are enforced [16].

Network slicing SLAs are agreements made between slice providers and customers that outline the performance metrics that the slice promises to meet (QoS, latency, availability, data rate, ...), its priorities, as well as business-related requirements like service pricing and usage duration. Traditionally, slice SLA auditing relies on third parties. This introduces the possibility that these third parties can be corrupted and may manipulate the results for their own financial benefits. Additionally, blockchain can cater to this by eliminating the need for a third party and establishing trust between the different parties. However, a persistent problem in SLA auditing is "privacy leakage." To solve this problem, SLA auditing smart contracts can be introduced. These smart contracts handle the insertion of customers slice requirements into the blockchain, as well as the slice provider's slice features. These information are kept encrypted inside the blockchain which ensures their non-tampering. Once the contract is established between the two parties, both the slice provider and customer continuously register SLA monitoring information inside the blockchain. These information are encrypted using a specific smart contract that also handles their comparison with the SLA terms in order to detect violations. In case of SLA violation, the smart contract proceeds in compensating the damaged party [17]. Specific smart contracts can be used to achieve the service billing procedure, using dedicated cryptocurrency. This renders the billing procedure more transparent to the clients and thus less difficult to tamper with by malicious parties.

2.4 Mobility and Signaling

The process known as roaming enables Home Network Operator (HNO) subscribers having an International Mobile Subscriber Identity (IMSI), to continue accessing mobile service and data even after leaving their operator's service area. This is made possible through roaming agreements previously established between the HNO and the Visited Network Operator (VNO). However, this procedure still suffers from delays and extra charges for the subscribers.

In the current system, scalability represents a major issue as peer-to-peer roaming agreements can rapidly escalate in number and thus complicate its management. Smart contracts allow the automated management and monitoring of peer-to-peer roaming contracts, and thus make the system more scalable. Another solution would be to create blockchain-roaming hubs where all the participating peers agree on one single roaming contracts instead of multiple peer-to-peer contracts.

The traditional roaming registration process requires lengthy delays to register the roaming user in the VNO. To ensure that customers are properly charged for the services they receive, user authentication process is completed in order to confirm users' identities and their eligibility to utilize roaming services provided by the VNO. These delays do not go in line with the stringent latency requirement of next-generation networks. To address this issue, blockchain records can be used to share roaming data between the different operators. Smart contracts can be used to manage contracts between users and VNOs. These contracts enforce the existing policies and regulations, as well as determine permissions over roaming data to both the HNO and VNO and propose different roaming offers to the users. In this approach, the various MNOs that wish to take part in roaming operations register as participating nodes on the blockchain, and several smart contracts are issued to define peer-to-peer roaming agreements between the MNOs. The roaming conditions that the HNO and VNO agree upon are contained in these smart contracts. A user instantly has the option to roam to all other MNO networks that have entered into roaming agreements with a specific participating MNO once they become a client of that MNO. Different smart contracts are run when a user registers with their HNO, and during this time, the user chooses the roaming agreements that best meet their requirements (price, QoS, etc.). Due to the fact that the authentication and registration processes have been completed in advance and once for all, this will eliminate the authentication delay when the user begins a roaming procedure [18]. This system allows billing transparency as the user gets to select among multiple roaming options and thus choose the best QoE/price trade-off. Another approach would be to distribute the storage of user subscription data on an off-chain distributed database (e.g., IPFS) and to establish blockchain-based access control and authentication. There would be two different sorts of smart contracts. The first smart contract is in charge of enforcing access control regulations, and the second smart contract is in charge of distributing the subscribers' data by keeping a hash of it in the distributed ledger of the blockchain. When the BC comes to a consensus, the subscriber's data can then be easily exchanged between the HNO and VNO. This guarantees tamper resistance, secure access, and authentication of subscription data. This solution eliminates roaming latency and storage costs since the data is hashed just once [19].

International roaming is a centralized procedure that includes two main processes: roaming agreements and clearinghouses. Roaming agreements outline the specific terms of international roaming for both the HNO and the VNO, meanwhile, clearinghouses provide assurances of adherence to those rules. This current international roaming approach is, however, expensive and could result in performance degradation and considerable delays due to the reliance on intermediaries for clearinghouses as well as the constant regulatory changes and conflicts in roaming agreements. As a result, the operators will transfer the expenses all the way down to the end consumers. To deal with this, a consortium blockchain has the potential to handle roaming agreements and clearinghouses, with MNOs endorsing one another to register to it. Users turn on the roaming option on their handsets when they want to travel abroad, which starts a smart contract that details the user's identity, the required service QoS, and the roaming fees that the HNO and VNO had previously discussed, when endorsing each other. The user is granted access to the VNO network once the contract has been completed and the transaction has been verified by the peers' consortium. This solution reduces delays, ensures compliance to roaming agreement without the intervention of third parties, and reduces expanses [20].

An issue that operators face in roaming procedures is "roaming fraud." Due to synchronization delays, this occurs when the roaming user continues to access VNO services after the session has ended. However, the HNO is charged for the use of their subscribers' services, by the VNO, without having the ability to bill those same subscribers back. To solve this problem, data exchange delays can be reduced by leveraging smart contracts to automate the billing procedure right at the service expiration time [21].

Increasing coverage and reducing latency are two of the main objectives of 5G. To achieve these objectives, 5G leverages small-cell technology. Such networks, however, call for higher cell densification, causing frequent user mobility between the various heterogeneous cells. As a result, the overall network performance and user experience are negatively impacted by frequent and inefficient handover operations, which can conflict with 5G's requirement for lower latency. Combining Software-Defined Networking (SDN) and blockchain technology can eliminate the need for extra reauthentication steps while switching cells. A public blockchain with an upgraded Delegated Proof-of-Stake (D-PoS) is used in this solution. Several SDN controllers take part as participating nodes, where they delegate a smaller group of SDN controllers to handle the authentication and payment processes. Blockchain is used for its distributed and immutable storage as well as for its encryption capabilities in order to generate encryption keys and identifications of devices. The blockchain allows for the participation of all end devices, and they all gain access to a public/private key for cell identification. Blockchain manages the transmission of the user's necessary authentication data via encrypted packets among the multiple SDN controllers, removing the need for additional authentication steps and shortening processing times [22].

3 Innovative Blockchain-Based Networks: Decentralized Wireless (DeWi)

Decentralized Wireless (DeWi) networks are a new generation of blockchain-powered wireless mesh systems that connect numerous nodes across various geographical locations. Individuals strategically deploy and operate these nodes to enable public connectivity and wireless service access in exchange of fees. These networks deploy a distributed, decentralized, and programmable infrastructure that is collaboratively owned by individuals by using the capabilities of blockchain technology. The fundamental economic framework behind DeWi networks is based on cryptoeconomics, with crypto-incentives playing an important role in bootstrapping the network and stimulating active participation from stakeholders [23].

The DeWi movement emerged in 2013 with the development of the Helium network, which became known as the "People's network." Initially, this network was designed to empower individuals to provide Long-Range Wireless Area Network (LoRaWAN) coverage for Internet-of-Things (IoT) devices. In 2020, the DeWi Alliance was founded to serve as a supportive stakeholder for Helium networks. Its principal goal was to achieve ubiquitous and democratized wireless coverage while providing equitable incentives for all stakeholders and fair pricing of resources, guided by the motto of "connecting to the internet everything that can be connected to the internet" [24]. This alliance expanded over time into The Helium Foundation, which now supervises all Helium-based networks (IoT and Mobile). With all the hype surrounding the success that Helium witnessed, several other firms have entered the domain, replicating Helium's approach to establish their own blockchain-powered networks, with their own currencies and governance politics. These networks include a wide range of wireless technologies, such as LoRaWAN (e.g., the Chirp network), Wi-Fi (e.g., the Wayru network), Bluetooth (e.g., the Nodle network) and hybrid networks that combine multiple wireless technologies (e.g., the Althea network) [25]. However, the cellular market presents DeWi's largest revenue-generating market (large number of cellular network users and collaboration opportunities with legacy cellular networks). Helium Mobile and Pollen Mobile are the two major networks in this sector.

The current state of the art promises a complete rethinking of the MNO paradigm as it currently exists. This new generation of wireless networks provides a revolutionary approach for transforming existing mobile operator networks into a distributed BC-empowered mesh network. This network is made up of antennas built by individuals on their property's rooftops, which give wireless coverage via a backhaul to their home router. When an end user with a DeWi network e-SIM connects to the antenna and uses data units, the antenna owner receives payment from the end user in digital tokens (or dollars) proportional to the amount of data utilized.

The DeWi movement represents an important turning point in the telecoms industry, allowing for the democratization of 5G coverage provision. DeWi removes the need for substantial Capital Expenditures (CapEx) and Operating Expenses (OpEx) as well as real estate acquisition for equipment (towers and antennas) installation, by providing small and cost-effective Radio RAN equipment. This paradigm shift eradicates the historical privilege enjoyed exclusively by large corporation on the scale of MNOs. The broadening of DeWi connectivity results in two-dimensional digital inclusion. First, it provides access to the wireless industry for small-scale and financially constrained businesses who previously faced entrance barriers due to limited financial resources or a lack of prior experience in the mobile telecommunications area. Second, DeWi supports digital inclusion by connecting rural underserved regions and isolated individuals and communities where MNOs have not extended their networks due to a lack of profitability. This will help to accelerate the deployment of 5G/6G technologies and the achievement of their goal of providing extended and ubiquitous coverage.

4 Business and Regulatory Aspects of Using Blockchain and Smart Contracts in Telecommunications

Blockchain technology and smart contracts could significantly disrupt business and regulations in the telecoms industry. Given how quickly this technology is being adopted, businesses and legislation need to adapt their policies in order to capitalize on the potential of such groundbreaking technology.

Blockchain provides effective and dynamic spectrum sharing, which will attract new actors to the market who were previously discouraged by spectrum licensing procedures. The market will see the entry of new players including micro-operators and regional operators. They will use the shared spectrum bands that they lease from MNOs or through spectrum trading markets to conduct their business. They would also operate on the unlicensed bands, thanks to the advanced unlicensed spectrum sharing mechanisms that blockchain and smart contracts introduce, and which help limit interference with other technologies operating on these bands. Regulation should adapt to embrace this transformation by issuing novel policies that would regulate parallel spectrum markets in order to prevent any undervaluation or overvaluation of this asset. When conducting spectrum sharing and trading activities between PUs and SUs, the use of cryptocurrency for spectrum resource payment could potentially be considered. However, the ambiguity and uncertainty that currently surrounds cryptocurrencies may be an obstacle to their use. Regulating cryptocurrency usage and valuation would be a boost factor for actors who are still hesitant to join in blockchain-based systems.

Due to blockchain's ability to eliminate the need for middlemen intervention and to enable high levels of automation through the use of smart contracts, some current telecommunication roles may also disappear, such as the intermediaries that participate in SLA management or in roaming procedures. We may also witness the rise of new actors specialized in the development of telecom services based on blockchain and smart contracts. This would enable the advent of new paradigms such as blockchain-as-a-service, smart-contract-as-a-service, or other integrated solutions with blockchain-based systems packaged as-a-service or on-demand.

While blockchain technology has proven to provide and promote fairness, security, and privacy, regulatory bodies should be more diligent in thoroughly studying its details and regulating its operations. It is important to recognize that potential fraud may occur in several layers of blockchain-based systems rather than the blockchain component itself. As a result, telecom regulatory organizations should take a comprehensive approach when creating rules and policies to safeguard the integrity and reliability of BC-based telecommunication systems.

5 Challenges and Future Research Directions

Benefits like decentralization, trust, and automation are introduced when blockchain technology and smart contracts are put to use in the telecommunications industry. However, there are still several difficulties with this emerging technology.

- *Latency-security trade-off*: Peer validation is a process that must take place before a new block can be added to the blockchain. The larger the blockchain network the more time this process takes. Additionally, the blockchain gets more secure, the more peers participate in the block validation process. And the security of the blockchain is decreased when the number of peers participating in the block validation process is decreased. In order to speed up this procedure, some state-of-the-art solutions propose to delegate a subgroup of participating nodes that would handle block validation; however, this would drastically reduce the system's security. Furthermore, in this particular context of telecommunications, this latency-security trade-off becomes crucial in determining whether to use blockchain or not because the nature of these networks is primarily based on wireless connections, which add to the blockchain's latency. Moreover, these networks are evolving to put even more stress on the strict latency and security requirements.

- *Participation enticement*: The "network effect" happens when individuals join a network in exchange for an alluring income; the more individuals join, the greater the income the network produces, which in turn draws in more participants. Since in a public blockchain the more players join the blockchain the more secure it becomes, blockchain-based systems seek to replicate the "network effect" in order to survive and function better. More systems, however, fail to start the network effect because they lack the first-stage revenue that draws the first users. This is especially true given that blockchain applications in communications are still relatively new, which would not really help with compelling people to participate in these systems. In addition, the rewards in these systems are paid out in a form of cryptocurrency that is less commonly utilized than traditional money and whose functioning is currently uncertain to many.
- *Mining and consensus*: Telecommunications companies must meet strict latency and performance standards. In addition to that, there are limitations on the battery life, processing, storage, and energy consumption of telecommunications devices, particularly user end devices. This necessitates greater study into dedicated consensus algorithms rather than the use of the traditional consensus algorithms (practical Byzantine Fault Tolerance (pBFT), PoW, and PoS), which in telecommunications cases introduce extra latency, calculation complexity, and limit the system's scalability.
- *Scalability*: The blockchain becomes more secure and generates more incentives as more users sign up for it, but this also creates a scalability issue because if the system has a large number of users participating at once, this could cause network congestion and delays, which could harm the overall performance and jeopardize the system's security. Segmenting the blockchain into multiple record could help improve the system's performance, especially in terms of latency and scalability. However, this might open the door to security threats and breaches.

For the telecommunications industry, blockchain and smart contracts alone cannot fully realize the improvement and modernization of the sector. For this reason, it is important to note that in order to take full advantage of blockchain, it must be combined with other 5G/6G standard technologies, such as virtualization, cloud, AI/ML, game theory, and others.

6 Conclusion

Blockchain technology and smart contracts have the potential to completely rethink the current telecommunications architectures by decentralizing processes, opening the ecosystem to multiple actors, and automating numerous processes to do away with the need for middlemen. What is more, blockchain can be considered as an unrivaled competitor and enabler that would enable 6G networks to operate in full distribution and decentralization while also fulfilling their primary aim of self-organizing, self-managing, self-diagnosing, and self-healing automated networks. We surveyed several applications of blockchain and smart contracts in this chapter, including the management of spectrum and network resources, the management of business-level agreements and billing, as well as mobility and signaling. These are the fundamental facets of telecom, and as we have shown, blockchain technology can benefit all of them.

We put forward that telecom standardization bodies should investigate the potential benefits of blockchain technology in optimizing the inherently centralized telecom architectures and processes. Furthermore, this ongoing research, which provides a succinct synthesis of the main contributions that blockchain can offer to consolidate the fundamental components of mobile telecommunications networks, could inspire the development of a blockchain-native 6G network, which would be built on blockchain technology and would facilitate the operation of blockchain-based systems.

References

1. M. Moussaoui, E. Bertin, N. Crespi, 5G shortcomings and beyond-5G/6G requirements, in *International Conference on 6G Networking (6GNet)*, (2022)
2. H. Taherdoost, The role of smart contract blockchain in 6G wireless communication system. Procedia Comput. Sci. **215**, 44–50 (2022)
3. "Digital Regulation Platform." https://digitalregulation.org. Accessed 14 Apr 2023
4. M. Alizadeh, K. Andersson, O. Schelén, Efficient decentralized data storage based on public blockchain and IPFS, in *2020 IEEE Asia-Pacific Conference on Computer Science and Data Engineering (CSDE)*, (IEEE, 2020), pp. 1–8
5. S. Zheng, T. Han, Y. Jiang, X. Ge, Smart contract-based spectrum sharing transactions for multi-operators wireless communication networks. IEEE Access **8**, 88547–88557 (2020)
6. F. Patel, P. Bhattacharya, S. Tanwar, R. Gupta, N. Kumar, M. Guizani, Block6Tel: Blockchain-based spectrum allocation scheme in 6G-envisioned communications, in *2021 International Wireless Communications and Mobile Computing (IWCMC)*, vol. 2021, (IEEE), pp. 1823–1828
7. X. Xie, Z. Hu, M. Chen, Y. Zhao, Y. Bai, An active and passive reputation method for secure wideband spectrum sensing based on blockchain. Electronics **10**(11), 1346 (2021)
8. Y. Xiao et al., Decentralized Spectrum access system: Vision, challenges, and a blockchain solution. IEEE Wirel. Commun. **29**(1), 220–228 (2022)
9. X. Ling, J. Wang, T. Bouchoucha, B.C. Levy, Z. Ding, Blockchain radio access network (B-RAN): Towards decentralized secure radio access paradigm. IEEE Access **7**, 9714–9723 (2019)

10. X. Ling, J. Wang, Y. Le, Z. Ding, X. Gao, Blockchain radio access network beyond 5G. IEEE Wirel. Commun. **27**(6), 160–168 (2020)
11. T. Sachinidis, A.-A.A. Boulogeorgos, P. Sarigiannidis, Dual-hop blockchain radio access networks for advanced coverage expansion, in *2021 10th International Conference on Modern Circuits and Systems Technologies (MOCAST)*, (IEEE, 2021), pp. 1–5
12. E.J. Scheid, B.B. Rodrigues, L.Z. Granville, B. Stiller, Enabling dynamic SLA compensation using blockchain-based smart contracts, in *2019 IFIP/IEEE Symposium on Integrated Network and Service Management (IM)*, (IEEE, 2019), pp. 53–61
13. T. Maksymyuk, M. Volosin, J. Gazda, M. Liyanage, Blockchain-based decentralized service provisioning in local 6G Mobile networks, in *Proceedings of the 19th ACM Conference on Embedded Networked Sensor Systems*, (2021), pp. 516–519
14. R.B. Uriarte, H. Zhou, K. Kritikos, Z. Shi, Z. Zhao, R. De Nicola, Distributed service-level agreement management with smart contracts and blockchain. Concurr. Comput. Pract. Exp. **33**(14), e5800 (2021)
15. R. Ranchal, O. Choudhury, SLAM: A framework for SLA management in multicloud ecosystem using blockchain, in *2020 IEEE Cloud Summit*, (IEEE, 2020), pp. 33–38
16. Z. Shi, V. Ivankovic, S. Farshidi, J. Surbiryala, H. Zhou, Z. Zhao, AWESOME: An auction and witness enhanced SLA model for decentralized cloud marketplaces. J. Cloud Comput. **11**(1), 1–25 (2022)
17. K. Xiao, Z. Geng, Y. He, G. Xu, C. Wang, Y. Tian, A blockchain-based privacy-preserving 5G network slicing service level agreement audit scheme. EURASIP J. Wirel. Commun. Netw. **2021**(1), 1–16 (2021)
18. A. Refaey, K. Hammad, S. Magierowski, E. Hossain, A blockchain policy and charging control framework for roaming in cellular networks. IEEE Netw. **34**(3), 170–177 (2019)
19. K. Xue, X. Luo, H. Tian, J. Hong, D.S. Wei, J. Li, A blockchain based user subscription data management and access control scheme in mobile communication networks. IEEE Trans. Veh. Technol. **71**(3), 3108–3120 (2021)
20. M.R.M. Bailon, L. Materum, International roaming services optimization using private blockchain and smart contracts. Int. J. Adv. Trends Comput. Sci. Eng. **8**(3), 544 (2019)
21. C. Nguyen et al., Blockroam: Blockchain-based roaming management system for future mobile networks. IEEE Trans. Mob. Comput. **21**(11), 3880–3894 (2021)
22. A. Yazdinejad, R.M. Parizi, A. Dehghantanha, K.-K.R. Choo, Blockchain-enabled authentication handover with efficient privacy protection in SDN-based 5G networks. IEEE Trans. Netw. Sci. Eng. **8**(2), 1120–1132 (2019)
23. A. Haleem, A. Allen, A. Thompson, M. Nijdam, R. Garg, A decentralized wireless network, p. 20
24. H. Foundation, Launching the decentralized wireless alliance, Helium Foundation, May 21, 2020. https://medium.com/helium-foundation/launching-the-decentralized-wireless-alliance-8097683ef262. Accessed 7 June 2023
25. The Telecom Cowboys of the Decentralized Wireless Movement., https://messari.io/report/the-telecom-cowboys-of-the-decentralized-wireless-movement. Accessed 9 May 2023

Distributed Ledger Technology for Next-Generation Cellular Networks: A SWOT Analysis

Nischal Aryal, Fariba Ghaffari, Emmanuel Bertin, and Noel Crespi

Abstract Mobile network operators (MNOs) intend to integrate novel technologies into their current and future network solutions to deliver novel, economical, and high-performance services in next-generation networks. However, this will be challenging because these supporting technologies introduce new participants into the ecosystem, increasing network complexity. They also add multiple issues with resource management, security, and privacy, which are existing in the current network. Addressing such difficulties requires a high level of collaboration among cellular network participants, and Distributed Ledger Technologies (DLT) are one potential solution. While DLT-based solutions bring many opportunities, it also brings many inherent weaknesses to the ecosystem. Therefore, it is necessary to identify the strengths and weaknesses of the systems based on DLT, as well as their possible opportunities and threats for future cellular networks. To do this, in this chapter, we first study and categorize the existing challenges and possible DLT-based solutions in different aspects of the cellular network ecosystem. Then, we perform a SWOT analysis to identify the strengths, weaknesses, opportunities, and threats these solutions can bring to the cellular network ecosystem. The result of the SWOT analysis could serve as a guideline for the future use of DLT-based solutions in next-generation cellular networks.

N. Aryal · E. Bertin
Orange Innovation, Caen, France

SAMOVAR, Telecom SudParis, Institut Polytechnique de Paris, Palaiseau, France
e-mail: nischal.aryal@orange.com; emmanuel.bertin@orange.com

F. Ghaffari (✉)
SAMOVAR, Telecom SudParis, Institut Polytechnique de Paris, Palaiseau, France

Institute of Research and Technology b-com, Cesson-Sévigné, France
e-mail: fariba.ghaffari@b-com.com

N. Crespi
SAMOVAR, Telecom SudParis, Institut Polytechnique de Paris, Palaiseau, France
e-mail: noel.crespi@it-sudparis.eu

N. El Madhoun et al. (eds.), *Blockchain and Smart-Contract Technologies for Innovative Applications*, https://doi.org/10.1007/978-3-031-50028-2_10

Keywords DLT · Blockchain · Smart contract · Next-generation cellular network · SWOT analysis

1 Introduction

The Next-Generation Cellular Network (NGN) expects to offer innovative, high-performance services to various vertical domains and businesses, including individual customers. Mobile network operators (MNOs) aim to incorporate many technologies, such as Artificial Intelligence (AI), Blockchain, and cloud/edge computing, to efficiently deliver highly reliable and real-time communication and services.

The current cellular network ecosystem and architecture have several business and technical challenges. For instance, from a business perspective, expensive infrastructure and high cost of market entrance have led to low-paced innovation in cellular networks and delayed the integration of entities in NGNs [1]; the high energy consumption of stand-alone MNOs and the complexity of agreement management are other business challenges. In addition, several technical issues negatively impact user experience and network security. Examples of some of the current issues include suffering from a single point of failure high complexity of IT procedures in MNOs, and security and privacy defects [2].

These problems highlight the need for solutions to decrease operational overhead and provide automation with less complexity, higher security, and more opportunities for innovative services and the integration of technologies. Current solutions, such as Software-Defined Networking (SDN), Network Function Virtualization (NFV), and cloud computing, can help by decoupling the functions from hardware and managing them in the form of software [3]. However, multiple challenges remain in terms of business model flexibility, network operations complexity, security, and privacy [4].

The use of Distributed Ledger Technology (DLT) has gained popularity, with Blockchain being the most well-known implementation, along with its expansion of the concept of smart contracts. Thanks to their immutability, distributed nature, non-repudiation, intrinsic security features, permanence, and traceability, DLTs have been studied by numerous researchers to address the issues in NGNs. These technologies provide new opportunities to validate data transactions while providing traceability in a wide range of scenarios. Examples of its use in telecommunications include offering a multi-vendor environment, access control and authentication, identity management, mobility management, resource/service delivery, new business models, speeding up technology integration, improving system fault tolerance, and increasing automation. The number of research for DLT in the telecommunications area is rapidly growing with the objective of decreasing network cost, maintaining technological leadership, introducing new business models, improved security, and facilitating cooperation.

Despite its unprecedented opportunities, current operational implementations of DLTs suffer from storage complexity, non-real-time transaction validation, maintainability issues regarding smart contract updates, regulation, standardization, etc. However, to meet the distinctive demands of different domains, DLT is rapidly advancing toward less energy-intensive operations, more real-time transactions, accurate standardization, and strong governance. So, analyzing the advantages and disadvantages of using DLT in the telecommunication domain is insightful for businesses and academics that are investigating this field.

We present a complete SWOT analysis of DLT deployment in existing and next-generation networks. SWOT analysis [5] is a process for assessing the Strengths, Weaknesses, Opportunities, and Threats (SWOT) of a new business approach to help in deciding on a successful plan. It is important to note that the SWOT analysis was conducted after a thorough examination of the state of the art in this field. To the best of our knowledge, this is the first research to apply the SWOT analysis method to examine the current and future applications of DLT.

Chapter Organization Section 2 provides the necessary background on the telecommunication domain, its ecosystem, actors, and its architecture. Moreover, this section describes the SWOT method. Section 3 lists the existing challenges to cellular networks, followed by a comprehensive SWOT analysis of DLT-based solutions in NGNs in Sect. 4. Finally, based on the concepts obtained from the SWOT analysis, Sect. 5 discusses the lessons learned and possible future directions.

2 Background

This section provides the required preliminaries regarding the telecommunication ecosystem and architecture, the existing use cases of Blockchain technology in cellular networks and networking applications, and the essentials of SWOT analysis methodology.

2.1 Background on Telecommunication

It is critical to have a general understanding of the ecosystem and current architecture of cellular networks in order to identify the problems with the current cellular network, the drivers for new approaches in next-generation mobile networks, and the potential benefits and drawbacks of new technologies such as DLT in relation to the new requirements. Understanding the cellular network's ecosystem gives us a business perspective on how mobile network operators collaborate with other entities and businesses, while understanding its architecture gives us technical knowledge of the various primary components of cellular networks that work together to provide communication functionalities.

A mobile network operator (MNO) is an independent provider of telecommunications services that owns the entire telecom infrastructure, including Radio Access Networks (RANs) and the core network, which is designed to host and manage mobile communications between users of the same/different wireless/wired telecom networks [6]. From the business viewpoint, the primary actors in the existing MNO ecosystem are the suppliers, MNOs, business-scale clients, and end users. The companies that offer the hardware and software for communication infrastructures are known as MNO suppliers. The vendors offer the base station, antenna, towers, storage, etc., to the MNOs. Other suppliers offer the necessary software, such as network functions. MNOs can deploy the RAN and core network and offer wireless and wired connections to end users and business-scale clients using the infrastructure that vendors have provided. The businesses that use the infrastructure provided by the MNOs to provide various services to their users are considered business-scale customers. These services can be content provisioning, video streaming, remote conferencing, Internet-based calls, smart healthcare, smart vehicles, security services, cloud services, and other Internet-based services, and these companies may have a separate agreement with MNOs to offer these services.

To investigate the technical architecture of the MNOs, a sample entity consists of a RAN and a core network. The RAN provides radio connectivity in a certain area based on its capacity, whereas the core network's primary functions are to create secure connections and transmit user data to and from the data network (DN) to enable communication [7]. Starting from the 14th release of 3GPP,[1] the core network has been divided into user planes and control planes. The user plane carries the network user traffic, while the control plane carries the core network signaling traffic. As an example, in 5G networks, packet controlling, policy management, subscriber management, network resource management, signaling, and location management are handled in the control plane.

2.2 *Blockchain Application in Telecommunication Use Cases*

Examining several uses of Blockchain technology in different telecommunication use cases would give insight into how this technology can affect different markets and technologies. We therefore provide a high-level overview of Blockchain applications in cellular network use cases.

Blockchain applications in telecommunication networks include IoT, smart cities, smart healthcare, smart grids, smart vehicles, content delivery networks (CDNs), cloud computing, information-centric networking (ICN), wireless sensor networks (WSNs), and device-to-device (D2D) communications. Smart cities rely on cellular networks to ensure seamless connectivity between diverse IoT and

[1] 3GPP is an organization that combines seven standard organizations in the field of mobile telecommunications to develop protocols in this environment.

smart devices. Cellular networks support the management of healthcare information and facilitate the secure interchange of health-related information among various healthcare personnel. Cellular network access points are necessary for D2D and smart transportation systems to establish communication between two entities. Electrical companies use cellular networks to manage information about electricity in the context of smart grids.

The hierarchical architecture of Blockchain usage in cellular network applications is represented in Fig. 1. Note that the Blockchain layers in this figure are based on [8]. This architecture consists of five layers: application layer, contract layer, consensus layer, network layer, and data layer.

The lowest layer in the architecture is the *data layer* that encapsulates the underlying block structure. Above that, the *network layer* includes the mechanisms of distributed networking, data propagation and communication, and data verification. This layer distributes, forwards, and verifies Blockchain transactions based on pre-defined structures (e.g., transaction verification via digital signature based on asymmetric cryptography). Next, the *consensus layer* mainly focuses on the consensus protocol of the nodes in the network (e.g., PoW, PBFT, and PoS). These algorithms can have an incentive mechanism to encourage nodes to collaborate in the network. The *contract layer* relates to the solutions based on smart contracts (i.e., not just based on transactions). This layer brings programmability into Blockchain and contains different languages to handle smart contracts. The top layer in the architecture, the *Application layer*, is the representation of the application of Blockchain in the desired context. Authentication, access control, identity management, key management/agreement, resource management, service provisioning, data/resource sharing, mobility management, etc., are examples of different use cases in which Blockchain is used in cellular network applications. Blockchain is used, for instance, in **IoT use cases** for data sharing and controlling access to sensor data. This technology can help **Wireless Sensor Networks (WSNs)** manage the mobility of the sensors within the network. Moreover, Blockchain is used in **cloud computing** to share resources such as processing power, data, and memory. Blockchain is used in several instances for **cellular networks** to manage user mobility, provide self-organized network access, port the user's mobile number between MNOs, enable medium access, share network resources, and create unique user identifiers. Blockchain technology can be used in **smart healthcare** to safely store patient data in the network, share data securely, and protect patient privacy.

2.3 SWOT Analysis

A SWOT analysis is a strategic planning and management framework for the early stages of an organization's decision-making process [9]. It helps in assessing the advantages and disadvantages of implementing an innovative proposal or objective within an established and functioning system [5]. The main idea of SWOT analysis

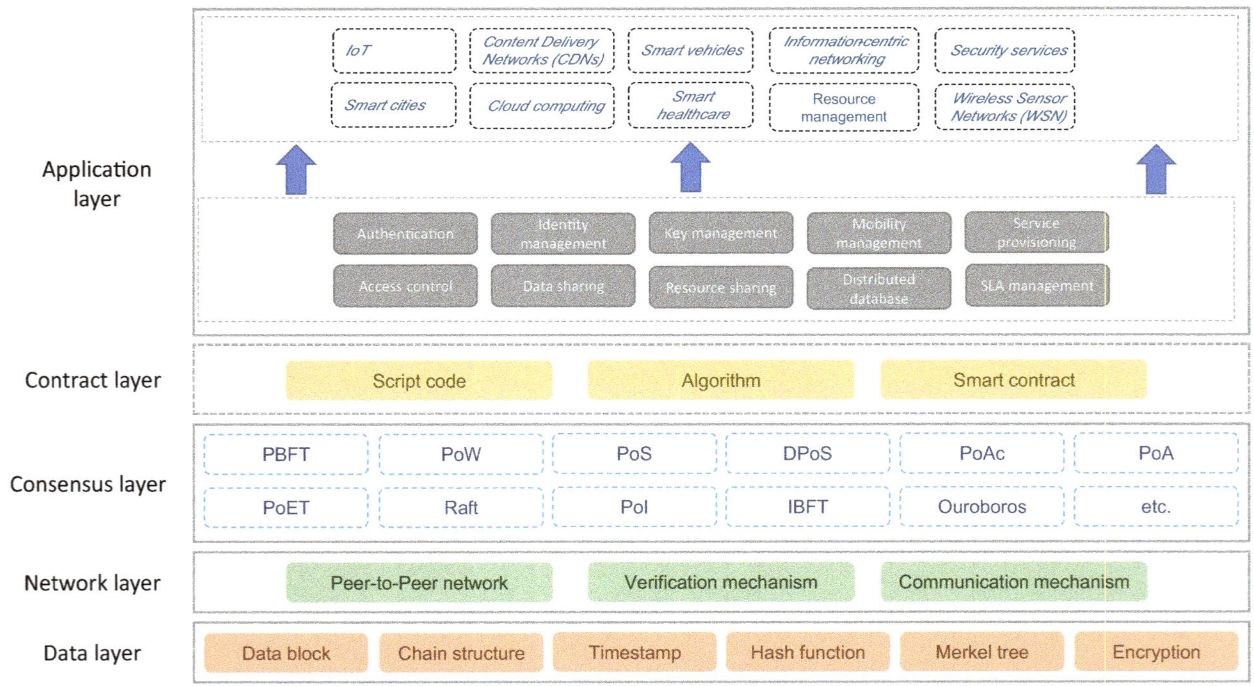

Fig. 1 Blockchain application in telecommunication and cellular networks

is to identify a competitive advantage by integrating the objective into the existing system.

SWOT analysis often keeps track of all the internal and external factors that are beneficial and harmful when aiming to achieve the objective. External factors include all positive and negative scenarios that occur outside the system, whereas internal factors include all positive and negative aspects that could occur within the system. SWOT analysis contains four components that help in the assessment of these internal and external factors: Strengths, Weaknesses, Opportunities, and Threats:

- **Strengths** are internal factors that highlight the benefits of incorporating a new idea into the system.
- **Weaknesses** are internal factors that highlight the disadvantages of incorporating the new idea into the system.
- **Opportunities** are external factors that focus on solutions that could benefit the system.
- **Threats** are external factors that can arise from outside the system and that are designed to cause difficulties to the system.

3 Challenges in Existing Cellular Networks

The cellular network market needs to become more innovative in order to meet the requirements for next-generation networks. It also needs to have a multi-actor ecosystem that enables a large number of entities to work together effectively, efficiently, and with a high level of trust while providing higher levels of privacy, security, availability, reliability, and integrity with less environmental impact. These objectives are challenging given the current cellular network architecture's numerous issues. A list and discussion of the current technology and business concerns follow the presentation of these challenges. A taxonomy of the challenges in the current cellular network is depicted in Fig. 2.

3.1 Business-Related Challenges

As defined earlier, an MNO owns a complete cellular network infrastructure to manage the mobile communication between subscribed users and other users or data networks. Currently, cellular networks are deployed in a stand-alone manner that can lead to the following business challenges:

- **High cost of entrance to the market**: Due to the lack of collaboration among entities (i.e., MNOs, connectivity providers, etc.), every MNO must install the complete infrastructure of the cellular network. Indeed, the deployment of MNO infrastructure is a very expensive investment, both for buying spectrum and for

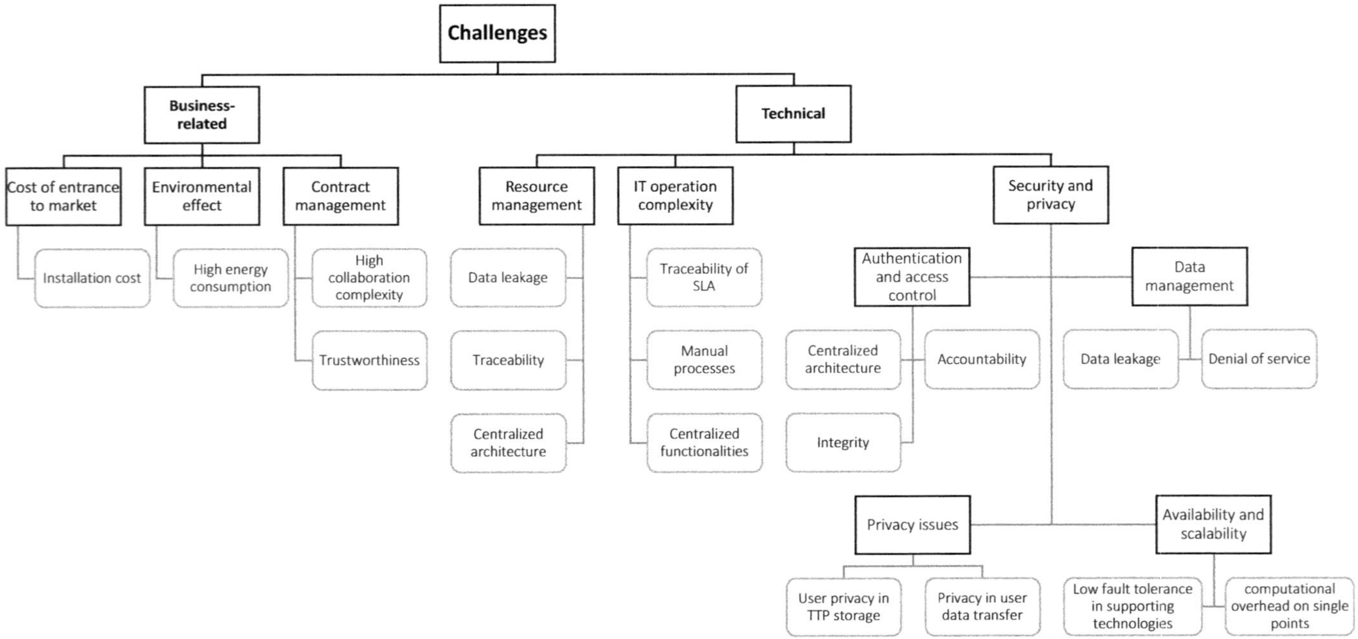

Fig. 2 Taxonomy of existing challenges in telecommunication ecosystem

deploying networking entities. The mutualization of resources such as RAN, storage, etc., could decrease these installation costs. Furthermore, because of the high installation costs, there is no motivation for MNOs to add more value to their products or to upgrade them [1], and, in the case of upgrades from the MNO side, their revenue will be flat, but their investment must increase.

- **Environmental effects and energy consumption of the stand-alone infrastructure** are higher than those of scenarios in which telecommunication resources can be shared between different actors and MNOs.
- **Contract management:** Increasing the number of collaborators in the next-generation cellular networks would result in tremendous growth in the number of contracts and make their management impracticable for MNOs with current stand-alone, non-automatic, and centralized architecture [1]. In other words, managing and tracing the complexity of engagements, agreements, Service-Level Agreements (SLA), and contracts have skyrocketed the operational and financial costs for MNOs. While in today's networks with a limited number of MNOs, managing such agreements is a feasible task in an environment with many providers and new businesses, the number of agreements required to deliver services based on the user's needs from different providers quickly becomes untenable [1]. Moreover, third-party access to user data is inevitable for authentication and billing procedures, which can lead to *privacy concerns* for users. Apart from its complexity, the main goal of next-generation networks and beyond 5G is to provide an open and diverse ecosystem in which different entities can participate and deliver/use services. In this kind of ecosystem, trust establishment is a cornerstone to adopting the technology [10, 11].

3.2 Technical Challenges

Apart from business-related challenges in the existing cellular network ecosystem and architecture, several technical issues negatively impact its security, reliability, scalability, performance, and availability. Some of these challenges are listed in the rest of this subsection.

Generally, these challenges are common to cellular networks, their supporting technologies, and to their use cases (for use cases, see Sect. 2.2). Note that supporting technologies are the group of technologies that have been proposed to improve the performance, functionality, flexibility, and security of the previous generation of cellular networks. Examples of these technologies include cloud computing, Mobile Edge Computing (MEC), Network Function Virtualization (NFV), and Software-Defined Networks (SDN), which are integrated into 5G cellular networks to achieve better performance and scalability. More specifically, these technologies have been introduced to provide greater application responsiveness, better network programmability and management [12] by decoupling hardware and software [13, 14], introducing the micro-operators [15], delivering differentiated services with network slicing [12], providing more flexible and cost-efficient network functions, improving performance and scalability, etc.:

- **Resource Management**: Resource management focuses on managing the allocation, release, maintenance, etc., of different network services and technologies in cellular networks, including network slicing (in 5G) [3], spectrum management [16, 17], data sharing, and resource allocation. Key challenges for resource management in cellular networks include **centralized architecture, data leakage, and insufficient traceability** [3, 18–20]. For instance, in scenarios involving resource and spectrum management and data sharing, data leakage during resource sharing can compromise system integrity [3, 18]. Additionally, exchanging data in untrustworthy settings such as mobile networks can case users to be concerned about their security and privacy (more precisely when their data need to be shared with or retained by third parties) [19]. According to [20], the central authority in the case of resource allocation can be a single point of failure, and a lack of traceability can result in the potential of action denial and non-accountability.
- **Complexity of IT operation:** The centralized architecture of a conventional cellular network results in all the connections being managed by a centralized party, which increases the **processing load and overhead in the central point**, reducing the quality of service (QoS), increasing the complexity of the IT operations [21], and the complexity of SLA management. Moreover, due to the ever-growing complexity of beyond 5G networks, one MNO may not be able to handle its cost and operational overhead [22]. As noted in [23], security management in 5G is more complex due to the various types and the massive number of connected devices that bring higher IT complexity to current networks. Compounded by the complexity of IT operations, resource management also suffers from the high complexity of current MNO architecture. Due to the growing demand for resource provisioning and the increasing variety of services, the management of communication resources (i.e., connection channels, processors, memory, bandwidth, and storage) is faced with an unprecedented level of complexity [23].
- **Security and privacy:** These challenges can be categorized as follows:
  - *Availability and scalability:* MNO architecture relies on a centralized architecture in which the implementation and the hosting of the infrastructure are located in limited sites and managed by one/a limited number of entities. This centralized architecture suffers from single-point failures, which can affect the availability and fault tolerance of a system [23, 24]. Moreover, the overall performance and scalability of the network (to provide communication and security services) are highly dependent on the capacity of the centralized entities [12, 25]. Furthermore, in current MNOs, a centralized unit handles subscribers' profile management and stores their subscription data. As mentioned in [22], centralized storage can be a single point for a data breach or data leakage.

- *Authentication and access control procedure:* Generally, the complexity of authentication, access control, and data integrity in the conventional architecture of MNOs are critical security challenges [26]. Rather than complexity, the centralized authentication and access management process limits the scalability of the AAC procedure in next-generation networks with their high demand on the interconnecting of a huge number of devices (e.g., in IoT environment and smart cities) [12].
- *Data management:* User's personal information in cellular networks is an attractive target for advertisement and intelligent agencies. Privacy is therefore a significant concern for the users [27]. In current cellular networks, a user's privacy can be violated in the storage and applications of third parties, during end-to-end data transmission through several stakeholders, and when user's data are stored in a shred environment [26]. As mentioned by Tahir et al. [22], centralized storage of user's identities is a significant challenge regarding security and privacy.
- *Privacy issues in Trusted Third Parties (TTP)*: Due to the huge amounts of users' PII data (e.g., identity, location, and other sensitive data) stored in TTPs, privacy issues are often raised from the TTP perspective; given that TPPs have the right to access confidential data, malicious entities will attempt to exploit that access to compromise user privacy [10].

4 DLT for Next-Generation Network: SWOT Analysis

In this section, we perform a SWOT analysis to identify the strengths, weaknesses, opportunities, and threats of implementing DLT-based solutions in the next-generation cellular network. It is important to mention that the list of metrics is derived from diverse references that are cited in the following sections (Fig. 3).

4.1 Strength of DLT in NGNs

The first element of the SWOT analysis identifies the strengths associated with the implementation of DLT-based solutions in cellular networks. These strengths can be classified into six categories: Immutability, Decentralization, Reaching consensus, Traceability/Transparency, Non-repudiation, and Data persistence:

- **Immutability:** As mentioned earlier, for DLT (e.g., DAG and Blockchain) to generate a block, all or selected nodes in a system must execute a consensus algorithm. After reaching a consensus, one node will add a block to the ledger. To do this, it computes the hash of each transaction in the current block and inserts a header containing the hash of the preceding block in the chain. This header and the transactions will be combined into one block and added to the

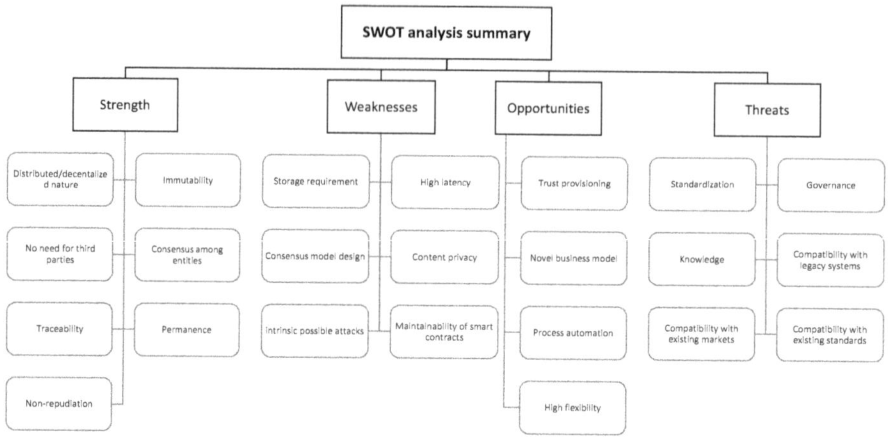

Fig. 3 Identifying the strengths, weaknesses, opportunities, and threats of implementing DLT-based solutions in next-generation networks using the SWOT analysis method

ledger. Every block in the chain links to its previous and subsequent blocks using hash values (this procedure is almost the same in DAGs and TDAGs). The execution of a consensus process for a significant number of preceding and subsequent blocks is required to modify one block. This procedure requires substantial power for the calculation, incurs a financial investment, and takes time. As a result of its complexity, experts acknowledge that DLT is immutable against transaction and block modification. When integrated with NGNs, this feature provides many opportunities to protect privacy [28], offer access to the services [20], provide policy immutability, protect data integrity in edge systems [29], trustful communication in smart transport [30], and data storage in smart grids [31].

- **Decentralization and No Third Parties:** Decentralization in DLT means there is no central authority handling transactions; instead, this control is the responsibility of the distributed network. Because there is no central authority in a DLT system and each node has a copy of the ledger, all network changes require consensus among all or a majority of participating nodes. This feature helps to remove centralized control and decision-making from cloud computing, edge computing, and network functions [32], thereby maintaining or improving the system performance during network slicing [33], and introducing cost-effective spectrum management [34]. DLT also removes the requirement for third parties, resulting in lower costs and an increase in democracy in the system. This feature can increase automation in resource sharing [35], service provisioning [20], etc.
- **Reaching consensus:** In conventional systems, protecting the centralized point and having confidence in it are essential to ensuring security and trustworthiness. However, this common point is removed in DLT to offer distributed management. Therefore, DLT can help to provide security and trustworthiness in a distributed network. In DLT, all nodes can reach a consensus based on algorithms defined

to ensure that all nodes have the latest version of the ledger. Based on these algorithms, this attribute helps to preserve the integrity of data and transactions. Therefore, we can say that the system is secure and that the DLT can create trust even in an environment of distrust. A DLT-based solution in next-generation networks could help with data integrity [36] and safe resource or data sharing thanks to this capability [37], especially in mobile networks such as healthcare systems.

- **Traceability/ Transparency:** All data in a DLT are always available and traceable at any time because the distributed ledger keeps track of all the transactions. This feature can be especially useful in improving trust in data sharing scenarios in D2D systems, increasing security in NFVs, etc. [38].
- **Non-repudiation:** Non-repudiation refers to the inability to deny responsibility for performing a certain action in the system. Every transaction in the DLT system requires the user's signature to verify its authenticity. The user signs the transaction using a private key, which is available only to the user (key owner), eliminating the possibility of action denial. This feature is vital for authentication and access control procedures in NGN [39].
- **Data Persistence:** DLTs utilize distributed ledgers to hold records of every data transaction in the ecosystem, and every node in the system has a copy of the updated information. This feature helps to make data readily available and limits any unauthorized data alterations.

4.2 *Weaknesses of DLT in NGNs*

The second element of the SWOT analysis method addresses the weaknesses associated with using DLT-based solutions in the cellular network. These weaknesses have been categorized into performance and security.

4.2.1 Performance

Weakness describes the challenges that arise in the cellular network when integrating DLT-based solutions and can further be categorized into storage requirements, high latency, and consensus model design:

- **Storage Requirements:** Due to the immutability of the ledger, DLT is an append-only technology in which transactions can only be inserted into the system and cannot be removed. Generally, requests in DLT are in two categories: calls and transactions. The purpose of a call is to retrieve a value from the ledger without changing its state, while transactions update the ledger and must be validated in a consensus procedure. Keeping the updated ledger accessible in all network nodes is critical for security and integrity. Recording all the network transactions and blocks necessitates massive storage capacities in all nodes that hold a copy of the ledger, a requirement that may result in a resource shortage for system contributors.

- **High Latency:** As mentioned before, inserting "transactions" into the ledger and updating its state require agreement from all nodes in a system. In other words, each transaction requires at least one "block time" to be confirmed and inserted into the ledger. Because DLT relies on consensus among nodes, transaction latency is higher than in traditional systems. Latency is affected by the complexity of the consensus puzzle and the size of the block. However, this parameter is adjustable, as there can be a trade-off between security and latency. It is important to note that it is not only "calls" that reduce latency; reading operations also substantially lower latency because data are stored across multiple nodes (compared to centralized SQL) [10]. Thus, DLT can be especially beneficial for "read" operations.
- **Consensus model design:** Due to emerging requirements and the performance of existing consensus models, model design needs to be revolutionized [10]. Moreover, the present consensus methods, such as PoW, have very low throughput, and newly proposed methods, such as Avalanche's snowman [40], or PBFT, are not widely used to verify their applicability and security.

4.2.2 Security

Weaknesses describe the pre-existing challenges of the DLT system and the new challenges that arise due to the integration of DLT-based solutions. This can be categorized into content privacy, DLT attack, and maintainability:

- **Content Privacy:** Indeed, cellular networks are highly sensitive domains for protecting users' privacy, as they contain personally identifiable information about the users (PII). Data leakage due to a bad decision in an AI system of cloud computing [10], curious trusted third parties in the IoT [41], and non-secure communication interfaces [10] are some of the data breach scenarios in conventional systems. One of the inherent characteristics of DLT is its ability to protect the privacy of meta-data (the user's identity). However, from another perspective, the data required for various transactions in cellular networks may necessitate the transmission of the user's private information to the DLT network. Indeed, this technology does not provide privacy for the contents of transactions. So, one topic of debate in the deployment of DLT in cellular networks is the issue of privacy.
- **DLT intrinsic attacks:** One of the major concerns about using DLT in cellular networks is its inherent security flaws. These intrinsic security flaws can be classified as *attacks related to the mathematical techniques* used to create the ledger such as forks, stale blocks, and orphaned blocks and attacks related to the *distributed architecture* of DLT such as selfish mining, 51% attacks, double spending, and Fork After Withholding (FAW) attacks.

- **Maintainability:** Smart contracts are self-contained, self-performed, and self-enforced pieces of code that are immutable after deployment. Such code immutability shortens the lifecycle of smart contracts [42], making maintainability of these systems on top of smart contracts more challenging (e.g., in case of existing security vulnerability, or when updating the system).

4.3 Opportunities of DLT in NGNs

The third element of the SWOT analysis method is the opportunities associated with adopting DLT-based solutions in a cellular network. These opportunities emerge from the strength and weaknesses of DLT adoption in cellular networks (mentioned in previous sections) and from areas outside the cellular network ecosystem. The opportunities have been categorized into process automation, novel business models, trust, and higher flexibility:

- **Process automation:** Smart contracts, which run on top of the Blockchain architecture, can offer several possibilities for process automation in cellular networks. Service providers can define some conditions and network processes in these contracts, and when the conditions are satisfied, the operation gets executed automatically. Two examples of this scenario are RAN access management [43] and mobility management based on pre-defined conditioners and policies in a cellular network.
- **Novel business models:** Due to the consensus procedure in transaction validation and non-repudiation, establishing trust in an untrustworthy environment can lead to new business models for different areas, including billing and service provisioning [20], resource and spectrum sharing [43], new wireless network architecture (e.g., Helium Blockchain), network function outsourcing, mobile number portability (MNP) [44], etc.
- **Trust:** The information stored in a DLT system is difficult to modify, and every action in the system is traceable. This feature helps DLT provide trust in an untrustworthy environment. Some examples of this trusted information include providing WiFi access point connections and Mobile Number Portability (MNP) [44].
- **Higher flexibility:** Smart contracts allow different parties of connections in cellular networks to define their requirements for data sharing policies, payment policies, quality of service requirements, and so on in the form of a contract. These rules and policies would be considered in the network management process when validating transactions generated by smart contracts in DLT [44].

4.4 Threats of DLT in NGNs

The fourth element of the SWOT analysis method is the threats associated with using DLT in the cellular network. These threats typically emerge from areas that fall outside the cellular network ecosystem, and they have been categorized into standardization, compatibility with existing standards, governance, knowledge, and compatibility with legacy systems and existing markets:

- **Standardization:** The lack of standardization in deploying smart contracts, consensus models, DApp implementation, governance, etc., is one of the most challenging issues in utilizing DLT in different use cases in standardized cellular networks (e.g., 5G). However, due to its rapid evolution and the high demand for its utilization, several standardization organizations are investing in standardizing this technology in different aspects. For instance, the following standards and best practices were available at the time of writing this article:

    - ***Standards/Best practices published by ETSI:*** The European Telecommunications Standards Institute (ETSI) is a standardization organization best known for its collaboration with 3GPP for cellular network standardization. ETSI has developed several standards related to Permissioned Distributed Ledgers (PDL)[2] regarding reputation management and trustworthiness, non-repudiation techniques, distributed data management, requirements for smart contracts' architecture and security, interoperability, reference architecture, federated data management, system architecture, and functional specifications, etc.
    - ***Standards/best practices published by ISO:*** The International Organization for Standardization (ISO) is a global federation of national standards bodies that has published several standards in the ISO/TC 307.[3] This category includes, at a minimum, reference architecture standards and best practices for reference architecture, identifiers of subjects and objects in DLT design, privacy, identity management, governance, smart contract design, etc.

    These are not exhaustive lists, as DLT standards and best practices are constantly evolving.
- **Energy consumption and environmental effects:** As stated earlier, reaching consensus in Blockchain is accomplished by executing common consensus algorithms across all/the majority of nodes in a system. Some consensus approaches, the most well-known of which is Proof of Work (PoW), are compute-intensive algorithms that demand a significant amount of computational resources to solve a consensus problem. The two most popular cryptocurrencies that use this consensus process are Bitcoin and Ethereum. According to a White House assessment for 2022, global electricity generation for cryptocurrency mining in

---
[2] https://www.etsi.org/committee/pdl.

[3] https://www.iso.org/committee/6266604/x/catalogue.

the United States emits around 140 30 million metric tons of carbon dioxide per year (Mt CO2/y), or approximately 0.3% of global annual greenhouse gas emissions. This amount is nearly equivalent to one year's worth of emissions from the diesel fuel used for rail transports in the United States [45]. Indeed, these numbers may cause major concerns for company stakeholders looking to migrate to Blockchain-based solutions; however, various efforts have been made to address this issue. For example, there are numerous non-computation-intensive techniques that organizations can utilize to reduce the environmental impact of the consensus algorithm.

Bada et al. [46] compared different consensus mechanisms based on their energy consumption, throughput, the possibility of using clean energy, etc. According to their findings, several consensus models, including Proof of Authority (PoA), which is currently used in Hyperledger Fabric, Quorum, and VeChain, can provide high transaction speed, low latency, high scalability, low energy consumption, and the ability to use clean energy in private Blockchain scenarios. Furthermore, different variations of Byzantine Fault Tolerance, such as BFT, Practical BFT (PBFT), and Delegate BFT (DBFT), have nearly identical features (mostly in terms of clean energy and environmental impact) except for their scalability issues, which makes these models less practical for cellular network applications. PoW consumes the greatest energy on the other end of the energy consumption range. Based on the analyses, we can conclude that the environmental impact of DLT can be managed by carefully selecting a consensus model and reducing the amount of storage required to hold the required data.

- **Compatibility with existing standards:** The current operational cellular networks (i.e., 3G, 4G, and 5G) are standardized by 3GPP, requiring manufacturers to produce mobile/IoT devices according to the standards. As a result, integration of DLT with cellular networks (particularly in core network-related procedures) is highly dependent on current standards [47]. This requirement has yet to be addressed. As a result, integration of DLT with cellular networks at layers lower than service and application layers is not entirely feasible in the current architecture.
- **Governance:** Before implementing DLT-based solutions in NGNs, several concerns related to the DLT system's owner, system policy, the type of DLT to be used, the geographical rules for the system (i.e., whether it is a national or global DLT system), and so on must be addressed. Managing these issues is a complex procedure requiring the agreement of several parties on various political, economic, ecological, and social aspects.
- **Knowledge:** Due to the complexity and immaturity of DLT-based systems, there is not enough information regarding their operation, their genuine benefits, how they can be harmful in a use case, etc. This lack of expertise and information about DLT make it very challenging for stakeholders and business owners to adopt this technology in their organizations.
- **Compatibility with legacy systems:** New regulations to implement DLT-based systems in cellular networks (e.g., 5G systems must be able to work with

Blockchain addresses) will be required, as the DLT infrastructure is frequently incompatible with the current systems.

- **Compatibility with the existing market:** The DLT-based systems proposed for cellular networks will require a transformation in the market structure. While new business models may emerge, the existing business needs will have to adjust to the changes. Convincing the stakeholders of cellular network business and getting approval for such technologies will require the support of all the key providers and market participants.

5 Discussion and Conclusion

This chapter surveys the current business-related and technical challenges in cellular networks and their applications. According to the findings of this study, high complexity in IT operations, trustworthiness in multi-access environments, centralized architecture, security concerns, lack of automated collaboration, and limited business models are some of the key challenges in conventional cellular networks. While several DLT-based solutions have been proposed to address these challenges, more comprehensive and subtle analysis is needed to assure the effectiveness of putting DLT-based solutions into practice. We used the SWOT analysis approach to identify the strengths, weaknesses, opportunities, and threats of DLT when it is integrated with the cellular network ecosystem to assess the viability of DLT-based solutions for NGNs.

As shown in Figs. 4 and 5, DLT, particularly Blockchain, and smart contracts, can provide many unprecedented opportunities in this field due to their immutability, distributed nature, traceability, non-repudiation, consensus convergence, and permanence. However, they can have negative effects as well, mostly due to their high latency, their transparency, the trade-off between scalability and security, and their environmental effects.

In general, DLT could improve the security and privacy of cellular networks by creating a decentralized and distributed system that can authenticate and authorize devices and users. Furthermore, it could enable secure and efficient micropayments and financial transactions within a cellular network. MNOs could also benefit by creating new business models and revenue streams. Blockchain can facilitate resource and spectrum sharing among multiple operators, allowing for more efficient use of network resources. Additionally, some inherent benefits of DLT include enabling collaboration between various entities in the cellular network without requiring trust between them, delivering greater flexibility due to addressing entities' needs in smart contracts, increasing fault tolerance/availability due to its distributed nature, improving process automation due to the execution of smart contracts instead of manual operations, and thereby adding robustness to the cellular system. Despite its unique features and opportunities, enterprises and MNOs would face several challenges when applying this technology in a real-world scenario. For instance, DLT (especially Blockchain) suffers from significant storage requirements,

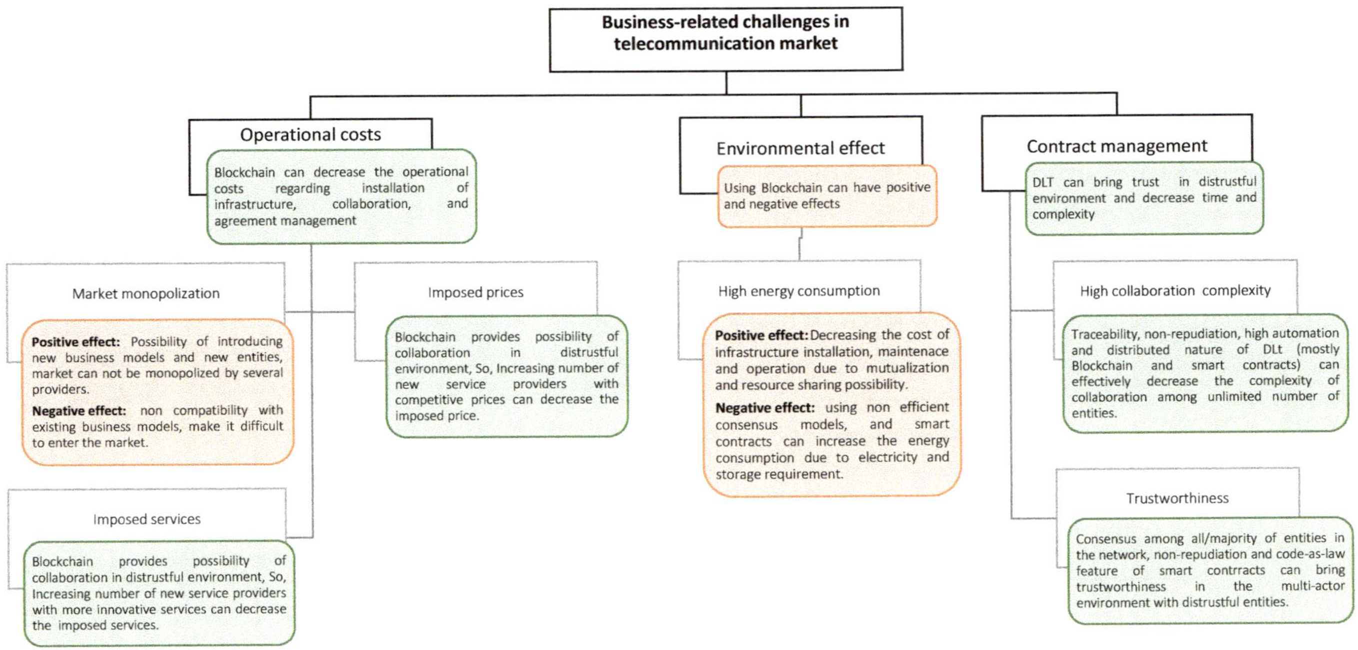

Fig. 4 How Blockchain can have positive/negative effects on business challenges in telecommunication applications. Green boxes indicate that Blockchain can have a beneficial effect on the problem, while orange boxes indicate that Blockchain can have both positive and negative impacts on the problem

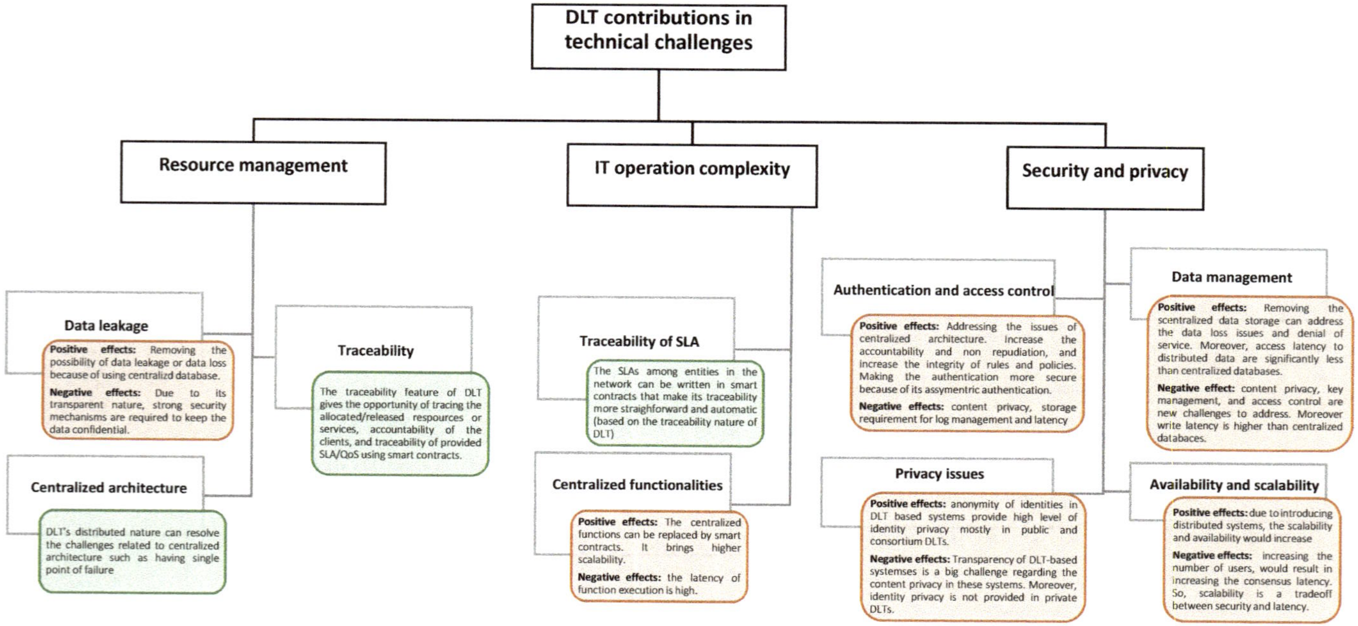

Fig. 5 How Blockchain can have positive/negative effects on technical challenges in telecommunication applications. Green boxes indicate that Blockchain can have a beneficial effect on the problem, while orange boxes indicate that Blockchain can have both positive and negative impacts on the problem

high latency for consensus convergence, smart contract maintainability, and content privacy.

In addition to its inherent challenges (noted as negative effects in Figs. 4 and 5), this technology brings some unforeseen challenges in the telecommunication domains from a regulation perspective. For instance:

- Standardization: As previously stated, the telecommunications domain is mostly standardized by several governance groups such as 3GPP. If it is required to update the design/architecture of cellular networks, migration to DLT-based systems would be not feasible before approval and standardization. Indeed, the standardization process is not agile due to the relevance of its operational verification, security issues, compatibility with existing technologies, and so on. As a result, introducing DLT-based use cases in private non-3GPP cellular networks and DLT-based service-level applications can be other ways to realize practical and quick DLT-based operations.
- Governance: Migration to DLT-based systems, as with standardization, will require further research and clearance from authorization authorities, regulators, and governments. In addition, new obstacles will likely emerge for specific use cases like payments, as that will require the definition of new cryptocurrencies, the exchange of conventional currencies, and the government's acceptance to integrate a new currency into the financial system. Another challenge in governance is determining which companies can participate in the security of DLT-based systems. Indeed, it would be impossible for all entities (i.e., users, stakeholders, vendors, MNOs, service providers, and so on) to participate in consensus procedures, and in some situations, their participation could lead to security difficulties.
- Compatibility: In general, DLT-based systems do not require complex hardware updates in existing systems, but applying this technology will require significant changes not only in software development but also in the design and lifecycle management of the development procedure. Maintainability and updatability of smart contracts, for example, are particularly difficult processes in software lifecycle management; additionally, offering high-performance applications necessitates a reasonable trade-off between scalability, security, and latency. Furthermore, there are few standards and best practices for developing and testing distributed applications in DLT, making operational testing of these services more difficult.

Several solutions have been proposed to address these drawbacks, and there are still ongoing research efforts in this regard. From a business perspective, a thorough investigation of the compatibility of Blockchain technology with the market is required. This investigation should result in a better understanding of the market requirements and the solutions for the progressive integration of Blockchain into the existing market to avoid sudden drifts and changes in the market. From a technical standpoint, proposing lightweight secure encryption solutions, secure communication with off-chain distributed databases, and designing solutions for the trusted processing of user's data are interesting goals with which to address privacy and data leakage challenges. Moreover, since storage complexity and system latency are highly dependent on the implementation scenario, proposing efficient solutions for storage/latency optimization and security improvements would be a rewarding future direction. Finally, a generic proposal for the DLT community could be the design of a telecommunication use-case-specific consensus model and DLT solution. This proposal could be a significant step forward to achieve some of the system requirements, such as low energy consumption, scalability, minimum delay, user privacy, and low resource consumption.

References

1. Z. Luo, S. Fu, M. Theis, S. Hasan, S. Ratnasamy, S. Shenker, Democratizing cellular access with CellBricks, in *Proceedings of the 2021 ACM SIGCOMM 2021 Conference* (2021), pp. 626–640
2. Y. Xiao, N. Zhang, W. Lou, Y.T. Hou, A survey of distributed consensus protocols for blockchain networks. IEEE Commun. Surveys Tutor. **22**(2), 1432–1465 (2020). Conference Name: IEEE Communications Surveys Tutorials
3. S. Zhang, An overview of network slicing for 5G. IEEE Wirel. Commun. **26**(3), 111–117 (2019)
4. P. Ranaweera, A.D. Jurcut, M. Liyanage, Survey on multi-access edge computing security and privacy. IEEE Commun. Surv. Tutor. **23**(2), 1078–1124 (2021)
5. A. Sarsby, *SWOT Analysis*. Spectaris Limited, 2016. [Online]. Available: https://books.google.fr/books?id=Yrp3DQAAQBAJ
6. 3GPP, Technical specification group services and system aspects; network architecture (release 13), European Telecommunications Standards Institute (ETSI), Technical Report, 03 (2016)
7. S. Rommer, P. Hedman, M. Olsson, L. Frid, S. Sultana, C. Mulligan, *5G Core Networks: Powering Digitalization* (Academic, Cambridge, 2019)
8. J. Xie, H. Tang, T. Huang, F.R. Yu, R. Xie, J. Liu, Y. Liu, A survey of blockchain technology applied to smart cities: research issues and challenges. IEEE Commun. Surv. Tutor. **21**(3), 2794–2830 (2019). Publisher: IEEE
9. C. Namugenyi, S.L. Nimmagadda, T. Reiners, Design of a SWOT analysis model and its evaluation in diverse digital business ecosystem contexts. Procedia Comput. Sci. **159**, 1145–1154 (2019)
10. K. Yue, Y. Zhang, Y. Chen, Y. Li, L. Zhao, C. Rong, L. Chen, A survey of decentralizing applications via blockchain: the 5G and beyond perspective. IEEE Commun. Surv. Tutor. **23**, 2191–2217 (2021)
11. C. Benzaïd, T. Taleb, M.Z. Farooqi, Trust in 5G and beyond networks. IEEE Netw. **35**(3), 212–222 (2021)
12. T. Hewa, G. Gür, A. Kalla, M. Ylianttila, A. Bracken, M. Liyanage, The role of blockchain in 6G: Challenges, opportunities and research directions, in *2020 2nd 6G Wireless Summit (6G SUMMIT)* (2020), pp. 1–5
13. J. Costa-Requena, J.L. Santos, V.F. Guasch, K. Ahokas, G. Premsankar, S. Luukkainen, O.L. Pérez, M.U. Itzazelaia, I. Ahmad, M. Liyanage et al., SDN and NFV integration in generalized mobile network architecture, in *2015 European Conference on Networks and Communications (EuCNC)* (IEEE, Piscataway, 2015), pp. 154–158
14. S. Wijethilaka, M. Liyanage, Survey on network slicing for internet of things realization in 5G networks. IEEE Commun. Surv. Tutor. **23**(2), 957–994 (2021)
15. S. Yrjölä et al., Decentralized 6G business models, in *Proceedings of the 6G Wireless Summit, Levi, Finland* (2019), pp. 5–7

16. P. Yang, L. Kong, G. Chen, Spectrum sharing for 5G/6G URLLC: research frontiers and standards. IEEE Commun. Standards Mag. **5**(2), 120–125 (2021)
17. S. Sasipriya, R. Vigneshram, An overview of cognitive radio in 5G wireless communications, in *2016 IEEE International Conference on Computational Intelligence and Computing Research (ICCIC)* (IEEE, Piscataway, 2016), pp. 1–5
18. J. Mitola, J. Guerci, J. Reed, Y.-D. Yao, Y. Chen, T.C. Clancy, J. Dwyer, H. Li, H. Man, R. McGwier et al., Accelerating 5G QoE via public-private spectrum sharing. IEEE Commun. Mag. **52**(5), 77–85 (2014)
19. Y. Kim, J.G. Park, J.-H. Lee, Security threats in 5G edge computing environments, in *2020 International Conference on Information and Communication Technology Convergence (ICTC)* (IEEE, Piscataway, 2020), pp. 905–907
20. F. Ghaffari, E. Bertin, N. Crespi, S. Behrad, J. Hatin, A novel access control method via smart contracts for Internet-based service provisioning. IEEE Access **9**, 81253–81273 (2021)
21. P.K. Agyapong, M. Iwamura, D. Staehle, W. Kiess, A. Benjebbour, Design considerations for a 5G network architecture. IEEE Commun. Mag. **52**(11), 65–75 (2014)
22. M. Tahir, M.H. Habaebi, M. Dabbagh, A. Mughees, A. Ahad, K.I. Ahmed, A review on application of blockchain in 5G and beyond networks: Taxonomy, field-trials, challenges and opportunities. IEEE Access **8**, 115876–115904 (2020)
23. D.C. Nguyen, P.N. Pathirana, M. Ding, A. Seneviratne, Blockchain for 5G and beyond networks: a state of the art survey. J. Netw. Comput. Appl. **166**, 102693 (2020). [Online]. Available: https://www.sciencedirect.com/science/article/pii/S1084804520301673
24. A.H. Khan, N.U. Hassan, C. Yuen, J. Zhao, D. Niyato, Y. Zhang, H.V. Poor, Blockchain and 6G: the future of secure and ubiquitous communication. IEEE Wirel. Commun. **29**(1), 194–201 (2021)
25. Q. Ni, Z. Linfeng, X. Zhu, I. Ali, A novel design method of high throughput blockchain for 6G networks: performance analysis and optimization model. IEEE Int. Things J. **9**, 25643–25659 (2022)
26. R. Khan, P. Kumar, D.N.K. Jayakody, M. Liyanage, A survey on security and privacy of 5G technologies: potential solutions, recent advancements, and future directions. IEEE Commun. Surv. Tutor. **22**(1), 196–248 (2019)
27. H. Khan, K.M. Martin, A survey of subscription privacy on the 5G radio interface-the past, present and future. J. Informat. Secur. Appl. **53**, 102537 (2020)
28. H. Xu, L. Zhang, E. Sun et al., BE-RAN: Blockchain-enabled RAN with decentralized identity management and privacy-preserving communication (2021). Preprint arXiv:2101.10856
29. C. Xu, K. Wang, P. Li, S. Guo, J. Luo, B. Ye, M. Guo, Making big data open in edges: a resource-efficient blockchain-based approach. IEEE Trans. Parall. Distrib. Syst. **30**(4), 870–882 (2018)
30. R. Shrestha, R. Bajracharya, A.P. Shrestha, S.Y. Nam, A new type of blockchain for secure message exchange in VANET. Digital Commun. Netw. **6**(2), 177–186 (2020)
31. Z. Guan, G. Si, X. Zhang, L. Wu, N. Guizani, X. Du, Y. Ma, Privacy-preserving and efficient aggregation based on blockchain for power grid communications in smart communities. IEEE Commun. Mag. **56**(7), 82–88 (2018)
32. P. Sharma, R. Jindal, M.D. Borah, Blockchain-based decentralized architecture for cloud storage system. J. Informat. Secur. Appl. **62**, 102970 (2021)
33. T.M. Hewa, I. Kovacevic, P. Porambage, N. Weerasinghe, M. Liyanage, E. Harjula, M. Ylianttila, Blockchain-based network slice broker to facilitate factory-as-a-service. IEEE Trans. Ind. Informat. **19**, 1–1 (2022)
34. P. Gorla, V. Chamola, V. Hassija, N. Ansari, Blockchain based framework for modeling and evaluating 5G spectrum sharing. IEEE Netw. **35**(2), 229–235 (2020)
35. B. Mafakheri, T. Subramanya, L. Goratti, R. Riggio, Blockchain-based infrastructure sharing in 5G small cell networks, in *2018 14th International Conference on Network and Service Management (CNSM)* (IEEE, 2018), pp. 313–317

36. R. Yang, F.R. Yu, P. Si, Z. Yang, Y. Zhang, Integrated blockchain and edge computing systems: a survey, some research issues and challenges. IEEE Commun. Surv. Tutor. **21**(2), 1508–1532 (2019)
37. D.C. Nguyen, P.N. Pathirana, M. Ding, A. Seneviratne, Blockchain for secure sharing of mobile cloud based e-health systems. IEEE Access **7**, 66792–66806 (2019)
38. N. Aryal, F. Ghaffari, E. Bertin, N. Crespi, Moving towards open radio access networks with blockchain technologies, in *5th Conference on Blockchain Research & Applications for Innovative Networks and Services-BRAINS 2023* (2023)
39. D. Yue, R. Li, Y. Zhang, W. Tian, C. Peng, Blockchain based data integrity verification in P2P cloud storage, in *2018 IEEE 24th International Conference on Parallel and Distributed Systems (ICPADS)* (IEEE, Piscataway, 2018), pp. 561–568
40. T. Rocket, M. Yin, K. Sekniqi, R. van Renesse, E.G. Sirer, Scalable and probabilistic leaderless BFT consensus through metastability (2019). Preprint arXiv:1906.08936
41. L. Babun, K. Denney, Z.B. Celik, P. McDaniel, A.S. Uluagac, A survey on IoT platforms: communication, security, and privacy perspectives. Comput. Netw. **192**, 108040 (2021)
42. J. Chen, X. Xia, D. Lo, J. Grundy, X. Luo, T. Chen, Defining smart contract defects on ethereum. IEEE Trans. Softw. Eng. **48**, 327–345 (2020)
43. X. Ling, J. Wang, T. Bouchoucha, B.C. Levy, Z. Ding, Blockchain radio access network (B-RAN): towards decentralized secure radio access paradigm. IEEE Access **7**, 9714–9723 (2019)
44. F. Ghaffari, E. Bertin, N. Crespi, Blockchain-based user profile and mobile number portability for beyond 5G mobile communication networks, in *2022 4th Conference on Blockchain Research & Applications for Innovative Networks and Services (BRAINS)* (IEEE, Piscataway, 2022), pp. 75–78
45. Fact sheet: Climate and energy implications of crypto-assets in the United States. [Online]. Available: https://www.whitehouse.gov/ostp/news-updates/2022/09/08/fact-sheet-climate-and-energy-implications-of-crypto-assets-in-the-united-states/
46. A.O. Bada, A. Damianou, C.M. Angelopoulos, V. Katos, Towards a green blockchain: A review of consensus mechanisms and their energy consumption, in *2021 17th International Conference on Distributed Computing in Sensor Systems (DCOSS)* (2021), pp. 503–511
47. H. Moudoud, S. Cherkaoui, L. Khoukhi, An overview of blockchain and 5G networks, in *Computational Intelligence in Recent Communication Networks*, ed. by M. Ouaissa, Z. Boulouard, M. Ouaissa, B. Guermah. (Springer International Publishing, Cham, 2022), pp. 1–20. [Online]. Available: https://doi.org/10.1007/978-3-030-77185-0-1

Index

N. El Madhoun et al. (eds.), *Blockchain and Smart-Contract Technologies for Innovative Applications*, https://doi.org/10.1007/978-3-031-50028-2

G

I

M

N

O

P

R

S

T

www.ingramcontent.com/pod-product-compliance
Ingram Content Group UK Ltd.
Pitfield, Milton Keynes, MK11 3LW, UK
UKHW022214051026
15624UKWH00001B/149

* 9 7 8 3 0 3 1 5 0 0 2 7 5 *